Differential Equations & Asymptotic Theory in Mathematical Physics

SERIES IN ANALYSIS

Series Editor: Professor Roderick Wong
City University of Hong Kong, Hong Kong, China

Published

Vol. 1 Wavelet Analysis
edited by Ding-Xuan Zhou

Vol. 2 Differential Equations and Asymptotic Theory in Mathematical Physics
edited by Hua Chen and Roderick S.C. Wong

Series in **Analysis** Vol. 2

Differential Equations & Asymptotic Theory in Mathematical Physics

Wuhan University, Hubei, China 20 - 29 October 2003

Editors

Chen Hua
Wuhan University, China

Roderick Wong
City University of Hong Kong, Hong Kong

World Scientific

NEW JERSEY · LONDON · SINGAPORE · BEIJING · SHANGHAI · HONG KONG

Published by

World Scientific Publishing Co. Pte. Ltd.

5 Toh Tuck Link, Singapore 596224

USA office: 27 Warren Street, Suite 401–402, Hackensack, NJ 07601

UK office: 57 Shelton Street, Covent Garden, London WC2H 9HE

British Library Cataloguing-in-Publication Data
A catalogue record for this book is available from the British Library.

DIFFERENTIAL EQUATIONS AND ASYMPTOTIC THEORY IN MATHEMATICAL PHYSICS

ISBN 981-256-055-6

Printed in Singapore by World Scientific Printers (S) Pte Ltd

PREFACE

This book consists of lecture notes for four mini courses delivered at the Conference on Differential Equations and Asymptotic Theory in Mathematical Physics held in Wuhan University, China in October, 2003. Each course contains five one-hour lectures. These notes give the readers a very good idea about recent developments in the area of asymptotic analysis, singular perturbations, orthogonal polynomials, and application of Gevrey asymptotic expansion to holomorphic dynamical systems. It also includes invited papers presented at the conference. The topics include the asymptotic behaviour for the formal solutions of the singular partial differential equations in complex domains and the applications of the partial differential equations in compound crystals, pricing of real options, hydraulic engineering, diblock copolymer, etc.

The conference has more than one hundred participants. We take this opportunity to thank all of them, and in particular, the authors of the articles included in this book.

The Editors
Hua CHEN
Roderick S.C. WONG

CONTENTS

LECTURES ON ORTHOGONAL POLYNOMIALS

MOURAD E. H. ISMAIL*

Department of Mathematics
University of Central Florida
Orlando, FL USA 32816
E-mail: ismail@math.ucf.edu

These lecture notes contain a brief introduction to orthogonal polynomials and mention some applications which are not readily available in books on special functions and orthogonal polynomials. Since these lectures were prepared for a conference on differential equations and asymptotical theory in mathematical physics, we naturally emphasized differential equations satisfied by orthogonal polynomials and attempted to explain the role asymptotics play in the theory of orthogonal polynomials. Due to space and time limitations we have not treated orthogonal polynomials on the unit circle, nor have we touched the techniques of obtaining asymptotics of large zeros of special functions or largest and smallest zeros of orthogonal polynomials. The article by Rod Wong [65] in these proceedings complements this article and provides details of asymptotic techniques.

1. Construction of Orthogonal Polynomials

Given a probability measure μ supported on a subset of $\mathbb{R}$ and having finite moments, $\int_{\mathbb{R}} x^n d\mu(x)$, we wish to construct a sequence of polynomials $\{p_n(x)\}$ such that

(i) $p_n(x)$ has precise degree n for all n, $p_n(x) = \gamma_n x^n +$ lower order terms, $\gamma_n > 0$,
(ii) $\int_{\mathbb{R}} p_m(x)p_n(x)d\mu(x) = \delta_{m,n}$.

The Gram-Schmidt procedure shows that $\{p_n(x)\}$ exists and is unique.

Let

$$\mu_n = \int_{\mathbb{R}} x^n d\mu(x), \tag{1.1}$$

*Research supported by NSF grant DMS 99-70865.

and define the Hankel determinants

$$D_n = \begin{vmatrix} \mu_0 & \mu_1 & \dots & \mu_n \\ \mu_1 & \mu_2 & \dots & \mu_{n+1} \\ \vdots & \vdots & \vdots & \vdots \\ \mu_n & \mu_{n+1} & \dots & \mu_{2n} \end{vmatrix}. \tag{1.2}$$

It is clear from (1.1) that the quadratic form $\sum_{j,k=0}^{n} \mu_{j+k} x_j \overline{x}_k$ is $\int_{\mathbb{R}} \left|\sum_0^n x_j x^j\right|^2 d\mu(x)$, hence is positive definite. Thus $D_n > 0$ for all $n, n = 0, 1, \dots$.

Theorem 1.1. *The polynomials $p_n(x)$ have the form*

$$p_n(x) = \frac{1}{\sqrt{D_n D_{n-1}}} \begin{vmatrix} \mu_0 & \mu_1 & \dots & \mu_n \\ \mu_1 & \mu_2 & \dots & \mu_{n+1} \\ \vdots & \vdots & & \vdots \\ \mu_{n-1} & \mu_n & \dots & \mu_{2n-1} \\ 1 & x & \dots & x^n \end{vmatrix}. \tag{1.3}$$

Proof. Denote the right-hand side of (1.3) by $q_n(x)$. Thus

$$q_n(x) = \sqrt{\frac{D_{n-1}}{D_n}}\, x^n + \text{lower order terms},$$

whence the leading coefficients in $q_n(x)$ is positive. For $k < n$,

$$\int_{\mathbb{R}} x^k q_n(x) d\mu(x) = \frac{1}{\sqrt{D_n D_{n-1}}} \int_{\mathbb{R}} \begin{vmatrix} \mu_0 & \mu_1 & \dots & \mu_n \\ \vdots & \vdots & \vdots & \vdots \\ \mu_k & \mu_{k+1} & \dots & \mu_{k+n} \\ \vdots & \vdots & \vdots & \vdots \\ \mu_{n-1} & \mu_n & \dots & \mu_{2n-1} \\ x^k & x^{k+1} & \dots & x^{k+n} \end{vmatrix} d\mu(x)$$

which is zero. If $k = n$ then the right-hand side of the above equation is $D_n/\sqrt{D_n D_{n-1}} = \sqrt{D_n/D_{n-1}}$. Therefore

$$\int_{\mathbb{R}} q_n^2(x) d\mu(x) = \sqrt{\frac{D_{n-1}}{D_n}} \int_{\mathbb{R}} (x^n + \dots) q_n(x) d\mu(x) = 1. \qquad \square$$

When μ is absolutely continuous, we write $\mu'(x) = w(x)$ and w is called a weight function.

Exercise. Recall the gamma integral

$$\Gamma(z) = \int_0^\infty t^{z-1} e^{-t} dt, \quad Re\, z > 0.$$

Let μ be absolutely continuous with $\mu'(x) = x^\alpha e^{-x}/\Gamma(\alpha+1)$. Prove that

$$\mu_n = (\alpha+1)_n, \quad D_n = \prod_{j=1}^{n} j!(\alpha+1)_j;$$

where

$$(A)_0 := 1, \quad (A)_n := \prod_{j=0}^{n-1}(A+j) = \Gamma(A+n)/\Gamma(A). \tag{1.4}$$

Moreover

$$p_n(x) = \sqrt{\frac{(\alpha+1)_n}{n!}}\,(-1)^n \sum_{k=0}^{n} \frac{(-n)_k x^k}{k!(\alpha+1)_k}.$$

Note that we proved that

$$\gamma_n = \sqrt{\frac{D_{n-1}}{D_n}}, \quad \text{or} \quad D_n = \frac{1}{\gamma_1^2 \cdots \gamma_n^2}. \tag{1.5}$$

Theorem 1.2. [**Heine**]. *The orthonormal polynomials have the integral representation*

$$p_n(x) = \frac{1/n!}{\sqrt{D_n D_{n-1}}} \int_{\mathbb{R}^n} \prod_{i=1}^{n}(x-\lambda_i) \prod_{1\le j<k\le n} (\lambda_j - \lambda_k)^2 d\mu(\lambda_1)\ldots d\mu(\lambda_n). \tag{1.6}$$

Proof. In the representation (1.3) write the entries $\mu_k, \mu_{k+1}, \ldots, \mu_{k+n}$ in row $k+1$ as $\int_{\mathbb{R}} \lambda_{k+1}^k d\mu(\lambda_{k+1}), \ldots, \int_{\mathbb{R}} \lambda_{k+1}^{k+n} d\mu(\lambda_{k+1})$, for $0 \le k < n$. Thus

$\sqrt{D_n D_{n-1}}\, p_n(x)$ has the representation

$$\int_{\mathbb{R}^n} \begin{vmatrix} \lambda_1^0 & \lambda_1 & \dots & \lambda_1^n \\ \lambda_2 & \lambda_2^2 & \dots & \lambda_2^{n+1} \\ \vdots & & & \\ \lambda_n^{n-1} & \lambda_n^n & \dots & \lambda_n^{2n-1} \\ 1 & x & \dots & x^n \end{vmatrix} d\mu(\lambda_1)\dots d\mu(\lambda_n)$$

$$= \int_{\mathbb{R}^n} \lambda_1^0 \lambda_2 \lambda_3^2 \dots \lambda_n^{n-1} \begin{vmatrix} 1 & \lambda_1 & \dots & \lambda_1^n \\ 1 & \lambda_2 & \dots & \lambda_2^n \\ \vdots & \vdots & \vdots & \vdots \\ 1 & \lambda_n & \dots & \lambda_n^n \\ 1 & x & \dots & x^n \end{vmatrix} d\mu(\lambda_1)\dots d\mu(\lambda_n).$$

Recall the evaluation of the Vandermonde determinant

$$\begin{vmatrix} 1 & x_1 & \dots & x_1^{n-1} \\ 1 & x_2 & \dots & x_2^{n-1} \\ \vdots & & & \\ 1 & x_n & \dots & x_n^{n-1} \end{vmatrix} = \prod_{1 \le i < j \le n} (x_j - x_i).$$

Let σ be a permutation on $1, \dots, n$ and S_n denote the symmetric group. In the above integral we can replace $\lambda_1, \dots, \lambda_n$ by $\lambda_{\sigma(1)}, \dots, \lambda_{\sigma(n)}$ then rearrange the rows in the determinant to have $\lambda_1, \dots, \lambda_n$ in rows $1, 2, \dots, n$. This produces a multiplicative factor sign (σ), sign (σ) being the sign of the permutation σ. By summing over $\sigma \in S_n$ and dividing by the order of the symmetric group $(= n!)$ we see that $\sqrt{D_n D_{n-1}}\, p_n(x)$ is

$$\int_{\mathbb{R}^n} \frac{1}{n!} \sum_{\sigma \in S_n} \operatorname{sign}(\sigma) \lambda_{\sigma(1)}^0 \lambda_{\sigma(n)} \dots \lambda_{\sigma(n)}^{n-1}$$

$$\cdot \left\{ \prod_{j=1}^n (x - \lambda_j) \right\} \prod_{1 \le i < j \le n} (\lambda_j - \lambda_i) d\mu(\lambda_1) \dots d\mu(\lambda_n)$$

$$= \frac{1}{n!} \int_{\mathbb{R}^n} \prod_{j=1}^n (x - \lambda_j) \prod_{1 \le i < j \le n} (\lambda_i - \lambda_j)^2 d\mu(\lambda_1) \dots d\mu(\lambda_n).$$ □

Corollary 1.3. *The Hankel determinants have the representation*

$$D_{n-1} = \frac{1}{n!} \int_{\mathbb{R}^n} \prod_{1 \le i < j \le n} (\lambda_i - \lambda_j)^2 d\mu(\lambda_1) \dots d\mu(\lambda_n). \tag{1.7}$$

Proof. Equate coefficients of x^n in (1.6). □

Remark 1.4. Let $\{\phi_n(x)\}$ be a sequence of polynomials such that ϕ_n has precise degree n. One can express $p_n(x)$ as a linear combination of $\phi_0(x), \phi_1(x), \ldots, \phi_n(x)$ in the following way

$$p_n(x) = \frac{1}{\sqrt{\widetilde{D}_n \widetilde{D}_{n-1}}} \begin{vmatrix} a_{0,0} & a_{0,1} & \cdots & a_{0,n} \\ a_{1,0} & a_{1,1} & \cdots & a_{1,n} \\ \vdots & & & \\ a_{n-1,0} & a_{n-1,1} & \cdots & a_{n-1,n} \\ \phi_0(x) & \phi_1(x) & \ldots & \phi_n(x) \end{vmatrix}, \tag{1.8}$$

with

$$a_{i,j} = \int_{\mathbb{R}} \phi_i(x)\phi_j(x) d\mu(x), \tag{1.9}$$

and

$$\widetilde{D}_n = \det(a_{i,j}), \quad 1 \leq i, j \leq n. \tag{1.10}$$

The determinant representations for orthogonal polynomials seem to be underused by experts in the area. For a clever use of these determinants see Wilson [64].

We next show how Heine's formula appears naturally in random matrices, see Mehta [42]. All matrices considered here are $n \times n$ matrices. Let M be a Hermitian matrix and let $M = U\Lambda U^*$, where Λ is a diagonal matrix (formed by the eigenvalues of M) and U a unitary matrix. Here $U = (u_{jk})$, $U^* = (u^*_{jk})$, where $u^*_{jk} = \overline{u_{jk}}$. The set of all Hermitian matrices is isomorphic to $\mathbb{R}^{n^2}$ because a Hermitian matrix has n real numbers on the diagonal and $n(n-1)/2$ complex numbers on the super diagonal. Thus a Hermitian matrix depends on $n + 2[n(n-1)/2] = n^2$ real entries.

Let V be a polynomial, and let $\mathrm{tr}(M)$ denote the trace of M (= the sum of its diagonal elements). The trace is invariant under similarity, so $\mathrm{tr}(M) = \mathrm{tr}(U\Lambda U^*) = \mathrm{tr}(\Lambda)$. Because $M^j = (U\Lambda U^*)^j = U\Lambda^j U^*$, then $\mathrm{tr}(V(M)) = \mathrm{tr}(V(\Lambda)) = \sum_{j=1}^n V(\lambda_j)$.

We consider the integrals

$$I(F) = \int_{M=\text{Hermitian}} F(M) e^{-\mathrm{tr}(V(M))} dM, \tag{1.11}$$

where dM is the Euclidean measure on the space of Hermitian matrices $(= \mathbb{R}^{n^2})$. If we change the variables in the upper triangular part of M to the eigenvalues $\lambda_1, \ldots, \lambda_n$ of M and the elements of U, the Jacobian will be

$\prod_{1\le i<j\le n}(\lambda_i-\lambda_j)^2$, see Deift [17] for this calculation. We now assume that $F(M)$ depends only on the eigenvalues of M and is a symmetric function of the eigenvalues. Perform integration on the U parameters to obtain

$$I(F) = \text{Constant} \int_{\mathbb{R}^n} F(\lambda_1,\ldots,\lambda_n) \prod_{1\le j<k\le n}(\lambda_j-\lambda_k)^2 \prod_{s=1}^{n} e^{-V(\lambda_s)}d\lambda_s. \tag{1.12}$$

Some details are omitted in the last step regarding the ordering of the λ_j's but the details are in Deift [17]. The representation (1.12) reminds us of Heine's formula (1.6). Indeed (1.6) gives the evaluation of the integral in (1.12) when F is an elementary symmetric function of $\lambda_1,\ldots,\lambda_n$. Let $e_k(\lambda_1,\ldots,\lambda_n)$ denote the kth elementary symmetric function of $\lambda_1,\ldots,\lambda_n$. It is clear that

$$\begin{aligned}
&\binom{n}{k}\int_{\mathbb{R}^n} \lambda_1\ldots\lambda_k \prod_{1\le i<j\le n}(\lambda_i-\lambda_j)^2 \prod_{s=1}^{n} e^{-V(\lambda_s)}d\lambda_s \\
&= \int_{\mathbb{R}^n} e_k(\lambda_1,\ldots\lambda_n) \prod_{1\le i<j\le n}(\lambda_i-\lambda_j)^2 \prod_{s=1}^{n} e^{-V(\lambda_s)}d\lambda_s \\
&= (-1)^k \text{coefficient of } x^{n-k} \text{ in } n!\sqrt{D_nD_{n-1}}\,p_n(x),
\end{aligned}$$

where $\{p_n(x)\}$ are orthonormal with respect to $\exp(-V(x))$ on $\mathbb{R}$.

One important problem in this area is to determine the distribution of the spacing of eigenvalues as $n\to\infty$. Some of the results on this topic are in [42] and [17].

2. Some Properties of Orthogonal Polynomials

One fundamental property of orthogonal polynomials is that they satisfy three term recurrence relations. To see this let

$$xp_n(x) = \sum_{k=i}^{n+1} d_{n,k}p_k(x). \tag{2.1}$$

It is clear that the expansion (2.1) exists since $\{p_n(x)\}$ forms a basis for the space of polynomials. Equating coefficients of x^{n+1} in (2.1) and employing condition (i) of § 1 we see that $\gamma_n = d_{n,n+1}\gamma_{n+1}$, hence $d_{n,n+1}=\gamma_n/\gamma_{n+1} > 0$. If $j<n-1$ then

$$d_{n,j} = \int_{\mathbb{R}} p_j(x)xp_n(x)d\mu(x) = 0,$$

because $p_j(x)x$ has degree $j+1$ which is less than n. It is also clear that

$$d_{n,n+1} = \int_{\mathbb{R}} x p_n(x) p_{n+1}(x) d\mu(x), \quad d_{n,n-1} = \int_{\mathbb{R}} x p_n(x) p_{n-1}(x) d\mu(x).$$

Thus $d_{n,n-1} = d_{n-1,n} = \gamma_{n-1}/\gamma_n$. Moreover $d_{n,n} = \int_{\mathbb{R}} x p_n^2(x) d\mu(x)$, so $d_{n,n} \in \mathbb{R}$. Thus we proved the following result.

Theorem 2.1. *Let $\{p_n(x)\}$ be orthonormal with respect to μ. Then $\{p_n(x)\}$ satisfies the three term recurrence relation*

$$x p_n(x) = a_{n+1} p_{n+1}(x) + b_n p_n(x) + a_n p_{n-1}(x), \tag{2.2}$$

with $b_n \in \mathbb{R}$ and $a_n > 0$. Moreover

$$a_n = \gamma_{n-1}/\gamma_n = \sqrt{D_n D_{n-2}}\,/D_{n-1}. \tag{2.3}$$

Another important normalization is the monic form

$$P_n(x) = x^n + \text{ lower order terms}. \tag{2.4}$$

Thus

$$P_n(x) = p_n(x)/\gamma_n. \tag{2.5}$$

In this normalization the recurrence relation (2.2) becomes

$$x P_n(x) = P_{n+1}(x) + b_n P_n(x) + a_n^2 P_n(x). \tag{2.6}$$

A consequence of (2.2) is the following theorem, see Chapter 2 in Szegő 58.

Theorem 2.2. *The zeros of $p_n(x)$ are real and simple and belong to the smallest interval containing the support of μ. Moreover the zeros of $p_n(x)$ and $p_{n+1}(x)$ interlace.*

One source of information on classical orthogonal polynomials has been that they satisfy a Sturm-Liouville problem. Consider the eigenvalue problem

$$-\frac{1}{w(x)}\frac{d}{dx}\left(p(x)\frac{d}{dx}y(x,\lambda)\right) + Q(x)y(x,\lambda) = \lambda y(x,\lambda), \tag{2.7}$$

for $x \in (a,b)$ and $y \in L^2(a,b;w)$. It is assumed that $w(x) > 0, p(x) > 0$ on $(a,b), w$ and p are continuous on (a,b) with $w(a^+) = w(b^-) = 0$. Further we assume that $\int_a^b |Q_n(x)| w(x)\, dx < \infty$. The eigenvalues of this symmetric eigenvalue problem are all real. If $\lambda_1 \neq \lambda_2$ then

$$\int_a^b w(x) y(x,\lambda_1) y(x,\lambda_2) dx = 0. \tag{2.8}$$

The Jacobi polynomials $\{P_n^{(\alpha,\beta)}(x)\}$ satisfy

$$-\frac{1}{w_{\alpha,\beta}(x)}\frac{d}{dx}\left(w_{\alpha+1,\beta+1}(x)\frac{d}{dx}P_n^{(\alpha,\beta)}(x)\right) \\ = n(n+\alpha+\beta+1)P_n^{(\alpha,\beta)}(x), \tag{2.9}$$

where

$$w_{\alpha,\beta}(x) = (1-x)^\alpha(1+x)^\beta, \quad x \in [-1,1]. \tag{2.10}$$

The ultraspherical polynomials correspond to the case $\alpha = \beta$. The Legendre polynomials occur when $\alpha = \beta = 0$, while the Chebyshev polynomials of the first and second kinds correspond to $\alpha = \beta = \mp\frac{1}{2}$, respectively. The Laguerre polynomials $\{L_n^{(\beta)}(x)\}$ arise when x is scaled as $-1+2x/\alpha$ then we let $\alpha \to \infty$. The Hermite polynomials arise when $\alpha = \beta$ and x is replaced by x/α and we let $\alpha \to \infty$.

As an example of a classical polynomial we derive some of the properties of Laguerre polynomials. We require the use of the Chu-Vandermonde sum, Andrews, Askey, Roy [2]

$$\sum_{k=0}^{k} \frac{(-n)_k(A)_k}{k!(C)_k} = \frac{(C-A)_n}{(C)_n}. \tag{2.11}$$

The weight function is

$$w_\alpha(x) = x^\alpha e^{-x}/\Gamma(\alpha+1), \quad x > 0. \tag{2.12}$$

Let the polynomials be $L_n^{(\alpha)}(x)$,

$$L_n^{(\alpha)}(x) = \sum_{k=0}^{n} \frac{(-n)_k}{k!} c_{n,k} x^k, \tag{2.13}$$

where $\{c_{n,k}\}$ are coefficients to be determined. The factor $(-n)_k$ makes the sum terminate at $k = n$, since

$$(-n)_m = 0, \quad m > n. \tag{2.14}$$

We first evaluate $\int_0^\infty x^m L_n^{(\alpha)}(x)\, e^{-x} x^\alpha dx$ and demand that it vanishes for $m < n$. Clearly

$$\begin{aligned}\int_0^\infty x^m L_n^{(\alpha)}(x) x^\alpha e^{-x} dx &= \sum_{k=0}^{n} \frac{(-n)_k}{k!} c_{n,k} \int_0^\infty x^{k+m+\alpha} e^{-x} dx \\ &= \sum_{k=0}^{n} \frac{(-n)_k}{k!} c_{n,k} \Gamma(\alpha+m+k+1) \\ &= \Gamma(\alpha+m+1) \sum_{k=0}^{n} \frac{(-n)_k(\alpha+m+1)_k}{k!} c_{n,k}.\end{aligned}$$

To evaluate the above sum we take $c_{n,k} = c_n/(A)_k$ in order to apply (2.11). Thus we want, for $m < n$,

$$0 - \sum_{k=0}^{n} \frac{(-n)_k(\alpha+m+1)_k}{k!(A)_k} - \frac{(A-m-\alpha-1)_n}{(A)_n}.$$

From (2.14), it follows that $A = \alpha + 1$ will work. Thus

$$L_n^{(\alpha)}(x) = \frac{(\alpha+1)_n}{n!} \sum_{k=0}^{n} \frac{(-n)_k}{k!} \frac{x^k}{(\alpha+1)_k}, \tag{2.15}$$

where c_n was chosen as $(\alpha+1)_n/n!$. This normalization has been adopted for over a hundred years and $\{L_n^{(\alpha)}(x)\}$ are neither orthonormal nor monic. Observe that the above calculations establish

$$\begin{aligned}\int_0^\infty x^n L_n^{(\alpha)}(x) x^\alpha e^{-x} dx &= \Gamma(\alpha+n+1) \frac{(\alpha+1)_n}{n!} \frac{(-n)_n}{(\alpha+1)_n} \\ &= (-1)^n \Gamma(\alpha+n+1).\end{aligned}$$

To find $\int_0^\infty (L_n^{(\alpha)}(x))^2 x^\alpha e^{-x} dx$, we proceed as follows, where use the above integral evaluation.

$$\begin{aligned}&\int_0^\infty x^\alpha e^{-x} (L_n^{(\alpha)}(x))^2 \, dx \\ &= \frac{(\alpha+1)_n}{n!} \int_0^\infty x^\alpha e^{-x} \left[\frac{(-n)_n x^n}{n!(\alpha+1)_n} + \dots \right] L_n^{(\alpha)}(x) \, dx \\ &= \frac{(\alpha+1)_n}{n!} \int_0^\infty x^\alpha e^{-x} \frac{(-n)_n x^n}{n!(\alpha+1)_n} L_n^{(\alpha)}(x) \, dx \\ &= \frac{\Gamma(\alpha+n+1)}{n!}.\end{aligned}$$

Therefore

$$\int_0^\infty x^\alpha e^{-x} L_m^{(\alpha)} L_n^{(\alpha)}(x)dx = \frac{\Gamma(\alpha+n+1)}{n!}\delta_{m,n}. \tag{2.16}$$

It is important to note that in the above derivation the only integral evaluation used is the gamma function integral. When we expressed $L_n^{(\alpha)}(x)$ as a linear combination of x^k we used the same integral that evaluated the total mass of the measure, namely $\int_0^\infty x^\alpha e^{-x} dx$, with α replaced by $\alpha+n+k$. In other words x^{n+k} attaches nicely to $x^\alpha e^{-x}$. This attachment procedure has been very useful in constructing closed form expressions for orthogonal polynomials, see Andrews and Askey [1] and Berg and Ismail [10].

To find the three term recurrence relation satisfied by $L_n^{(\alpha)}(x)$ we first observe that the coefficient of x^n in $L_n^{(\alpha)}(x)$ is $(-1)^n/n!$. Hence $xL_n^{(\alpha)}(x) + (n+1)L_{n+1}^{(\alpha)}(x)$ is a polynomial of degree at most n. Let

$$-xL_n^{(\alpha)}(x) = (n+1)L_{n+1}^{(\alpha)} + B_n L_n^{(\alpha)}(x) + C_n L_{n-1}^{(\alpha)}(x).$$

By equating coefficients of x^n and x^{n-1} on both sides of the above equation and applying (2.15) we establish the recursion relation

$$-xL_n^{(\alpha)}(x) = (n+1)L_{n+1}^{(\alpha)}(x) - (n+\alpha+1)L_n^{(\alpha)}(x) + (n+\alpha)L_{n-1}^{(\alpha)}(x). \tag{2.17}$$

In § 3 we shall establish the differential equation

$$-x\frac{d}{dx}L_n^{(\alpha)}(x) + (\alpha+1-x)\frac{d}{dx}L_n^{(\alpha)}(x) + nL_n^{(\alpha)}(x) = 0. \tag{2.18}$$

The Christoffel-Darboux formula

$$\sum_{k=0}^n p_k(x)p_k(y) = a_n \frac{p_{n+1}(x)p_n(y) - p_n(x)p_{n+1}(y)}{x-y} \tag{2.19}$$

follows from (2.2). As $y \to x$ we have

$$\sum_{k=0}^n p_n^2(x) = a_n[p_{n+1}'(x)p_n(x) - p_n'(x)p_{n+1}(x)]. \tag{2.20}$$

3. Differential Equations

In this section we derive differential equations and raising and lowering operators for general orthogonal polynomials when

$$d\mu = w(x)dx, \quad \text{and} \quad w(x) = e^{-v(x)}, \quad x \in (a,b). \tag{3.1}$$

Theorem 3.1. *Assume v' is continuous on (a,b) and*

$$\int_a^b \frac{v'(x)-v'(y)}{x-y} y^n dy$$

exists for nonnegative integers n. Define $A_n(x)$, $B_n(x)$ by

$$\frac{A_n(x)}{a_n} = \left.\frac{w(y)p_n^2(y)}{y-x}\right]_{y=a^+}^{y=b^-} + \int_a^b \frac{v'(x)-v'(y)}{x-y} p_n^2(y)w(y)dy, \tag{3.2}$$

$$\frac{B_n(x)}{a_n} = \left.\frac{w(y)p_n(y)p_{n-1}(y)}{y-x}\right]_{y=a^+}^{y=b^-} + \int_a^b \frac{v'(x)-v'(y)}{x-y} p_n(y)p_{n-1}(y)w(y)dy. \tag{3.3}$$

and assume that the boundary terms in (3.2)–(3.3) *exist. Then*

$$\frac{d}{dx} p_n(x) = A_n(x)p_{n-1}(x) - B_n(x)p_n(x). \tag{3.4}$$

Theorem 3.1 follows from the Christoffel-Darboux identity (2.18) and integration by parts. The details are in [15].

Theorem 3.1 is due to Bauldry [9], Bonan and Clark [11], Mhaskar [43], and Chen and Ismail [15].

Define operators $L_{1,n}$ and $L_{2,n}$ by

$$L_{1,n} = \frac{d}{dx} + B_n(x), \quad L_{2,n} = -\frac{d}{dx} + B_n(x) + v'(x). \tag{3.5}$$

Thus we write (3.4) as

$$L_{1,n}p_n(x) = A_n(x)p_{n-1}(x), \tag{3.6}$$

hence $L_{1,n}$ is a lowering or annihilation operator for $\{p_n(x)\}$. By eliminating $p_{n-1}(x)$ between (3.4) and (2.2) we establish

$$L_{2,n}p_{n-1}(x) = \frac{a_n}{a_{n-1}} A_{n-1}(x)p_n(x). \tag{3.7}$$

Thus $L_{2,n}$ is a raising or creation operator. In fact $L_{1,n}$ and $L_{2,n}$ are adjoints with respect to the inner product

$$(f,g) = \int_a^b f(x)\overline{g(x)}w(x)dx,$$

see Chen and Ismail [15]. Combining (3.6) and (3.7) we arrive at

$$L_{2,n}\left(\frac{1}{A_n(x)}L_{1,n}\right)p_n(x) = \frac{a_n}{a_{n-1}} A_{n-1}(x)\,p_n(x), \tag{3.8}$$

which is a second order differential equation satisfied by the polynomials.

One important application of the differential equation (3.8) is that one can apply the Liouville-Green (or WKB) approximation and derive the large

n asymptotic behavior of $p_n(x)$. One can also derive uniform asymptotic expansions. For details see R. Wong 's article in these proceedings [65].

Example 1. From (2.14)–(2.16) we see that the orthonormal Laguerre polynomials are

$$p_n(x) = \sqrt{\frac{n!}{(\alpha+1)_n}}\,(-1)^n L_n^{(\alpha)}(x),$$

and

$$a_n = \sqrt{n(n+\alpha)}\,, \quad w(x) = e^{-x}x^\alpha/\Gamma(\alpha+1), \quad \gamma_n = (n!(\alpha+1)_n)^{-\frac{1}{2}}.$$

Thus we get

$$\begin{aligned}\Gamma(\alpha+1)\frac{A_n(x)}{\sqrt{n(n+\alpha)}} &= \frac{\alpha}{x}\int_0^\infty e^{-y}y^\alpha \frac{1}{y}p_n^2(y)\,dy = \frac{1}{x}\int_0^\infty e^{-y}p_n^2(y)\alpha y^{\alpha-1}\,dy \\ &= \frac{1}{x}\left[0 - \int_0^\infty y^\alpha\left(e^{-y}2p_n(y)p_n'(y) - e^{-y}p_n^2(y)\right)dy\right] \\ &= \Gamma(\alpha+1)/x.\end{aligned}$$

Similarly, integration by parts and the orthogonality relation yield

$$\begin{aligned}\frac{xB_n(x)}{\sqrt{n(n+\alpha)}} &= \int_0^\infty p_n(y)p_{n-1}(y)\frac{\alpha y^{\alpha-1}}{\Gamma(\alpha+1)}e^{-y}dy \\ &= -\int_0^\infty p_{n-1}(y)p_n'(y)w(y)dy \\ &= -\frac{n\gamma_n}{\gamma_{n-1}}\int_0^\infty p_{n-1}(y)p_{n-1}(y)w(y)dy \\ &= -n\,\frac{\gamma_n}{\gamma_{n-1}}.\end{aligned}$$

Therefore

$$A_n(x) = \sqrt{n(n+\alpha)}\,/x, \quad B_n(x) = -n/x.$$

In this case (3.4) leads to

$$-x\frac{d}{dx}L_n^{(\alpha)}(x) = (n+\alpha)L_{n-1}^{(\alpha)}(x) - nL_n^{(\alpha)}(x).$$

The differential equation (3.8) becomes

$$x\frac{d^2}{dx^2}L_n^{(\alpha)}(x) + (1+\alpha-x)\frac{d}{dx}L_n^{(\alpha)}(x) + nL_n^{(\alpha)}(x) = 0, \tag{3.9}$$

which is of Sturm-Liouville type.

Example 2. Consider the weight function Ce^{-x^4} on $\mathbb{R}$, so $v = x^4 - \ln C$. The boundary terms in (3.2)–(3.3) vanish and we have

$$\frac{A_n(x)}{a_n} = \int_{\mathbb{R}} \frac{4x^3 - 4y^3}{x - y} p_n^2(y) w(y) dy = 4 \int_{\mathbb{R}} (x^2 + xy + y^2) p_n^2(y) w(y) dy.$$

Since w is an even function, $p_n(-x) = (-1)^n p_n(x)$ so $p_n^2(x)$ is an even function. Moreover (2.2) implies $b_n = \int_{\mathbb{R}} x p_n^2(x) w(x) dx = 0$. Thus the right-hand side of the above equation is

$$4x^2 + 4 \int_0^\infty [a_{n+1} p_{n+1}(y) + a_n p_n(y)]^2 w(y) dy,$$

and we find

$$A_n(x) = 4a_n(x^2 + a_n^2 + a_{n+1}^2). \tag{3.10}$$

Similarly

$$B_n(x) = 4a_n^2 x. \tag{3.11}$$

Therefore

$$p_n'(x) = 4a_n(x^2 + a_n^2 + a_{n+1}^2) p_{n-1}(x) - 4a_n^2 x p_n(x), \tag{3.12}$$

and

$$\left(4a_n^2 x + 4x^3 - \frac{d}{dx}\right) \left(\frac{1}{x^2 + a_n^2 + a_{n+1}^2} \left(4a_n^2 + \frac{d}{dx}\right)\right) p_n(x) = 16a_n^2(x^2 + a_n^2 + a_{n-1}^2) p_n(x). \tag{3.13}$$

The differential equation (3.13) was first derived by Shohat [54] through a complicated procedure. For applications see Nevai [47]. One can also derive nonlinear relations among the recursion coefficients $\{a_n\}$ from (3.12) by using the recurrence relation

$$x p_n(x) = a_{n+1} p_{n+1}(x) + a_n p_{n-1}(x). \tag{3.14}$$

Clearly (3.14) yields

$$p_n(x) = \frac{x^n}{a_1 a_2 \ldots a_n} + c_n x^{n-2} + \text{lower order terms}. \tag{3.15}$$

Equating the coefficients of x^{n-1} in (3.12) proves

$$\frac{n}{a_1 \ldots a_n} = \frac{4a_n(a_n^2 + a_{n+1}^2)}{a_1 \ldots a_{n-1}} + 4a_n(c_{n-1} - a_n c_n). \tag{3.16}$$

On the other hand, substituting from (3.15) into (3.14) and equating the coefficients of x^{n-1} we find

$$c_n = a_{n+1}c_{n+1} + \frac{a_n}{a_1 \cdots a_{n-1}}.$$

Therefore (3.16) becomes

$$n = 4a_n^2(a_{n-1}^2 + a_n^2 + a_{n+1}^2). \tag{3.17}$$

Many additional nonlinear relations follow from equating coefficients of other powers of x. Freud seems to have been the first to discover such nonlinear equations for recursion coefficients of polynomials orthogonal with respect to exponential weights, $w(x) = \exp(-x^{2m})$. He only studied the case $m = 2$. Formulas like (3.17) are instances of the string equation in Physics. It is clear that one can derive nonlinear relations similar to (3.17) when $v = x^6+$ constant using the same technique. References to the literature on this are in [47]. The treatment presented here is from [15].

Chen and Ismail [15] analyzed the case when v is a polynomial and described the corresponding $A_n(x)$ and $B_n(x)$ functions. Qiu and Wong [49] used the differential equation and the Chen-Ismail analysis to derive large n uniform asymptotics for $p_n(x)$.

The operators $L_{1,n}$ and $L_{2,n}$ generate a Lie algebra where the product of two operators A and B is the Lie bracket $[A, B] = AB - BA$. Finite dimensional Lie algebras are of interest. When $v = x^2+$ constant, $L_{1,n}$ and $L_{2,n}$ generate a three-dimensional Lie algebra called the harmonic oscillator algebra, Miller [45]. Miller [44] characterized all finite dimensional Lie algebras that are generated by first order differential operators. Chen and Ismail [15] proved that when $v(x)$ is a polynomial, $w(x) = e^{-v(x)}$, then $L_{1,n}$ and $L_{2,n}$ generate a Lie algebra of dimension $2m + 1$, $2m$ being the degree of $v(x)$. The converse is not known so we state it as a conjecture.

Conjecture. *If the Lie algebra generated by $L_{1,n}$ and $L_{2,n}$ is finite dimensional and the support of v is $\mathbb{R}$, then the Lie algebra has dimension $2m+1$ and v must be a polynomial of even degree.*

The differential equation (3.8) when expanded out becomes

$$p_n''(x) - [v'(x) + A_n'(x)/A_n(x)]p_n'(x) + S_n(x)p_n(x) = 0, \tag{3.18}$$

where

$$S_n(x) := B_n'(x) - B_n(x)\frac{A_n'(x)}{A_n(x)} - B_n(x)[v'(x) + B_n(x)] + \frac{a_n}{a_{n-1}}A_n(x)A_{n-1}(x). \tag{3.19}$$

4. Electrostatic Equilibrium Problems

We will study Coulomb interactions in the plane where the force is $2e_1e_2/r$, e_1, and e_2 being the charges of two particles and r the distance between them. This arises in $\mathbb{R}^3$ if we have infinite uniformly charged wires perpendicular to the plane under consideration and e_1, and e_2 are the charges per unit length. The potential energy is $-2e_1e_2 \ln r$.

Now consider n movable charged particles each carrying a unit charge. The particles are restricted to a subset E of $\mathbb{R}$ in the presence of an external field with potential energy $V(x)$. Let $x_1, \dots, x_n$ denote the positions of the particles and assume $x_1 > x_2 > \cdots > x_n$. In what follows $\mathbf{x}$ will denote the vector $(x_1, x_2, \dots, x_n)$. The total energy $E(\mathbf{x})$ of this system is

$$E(\mathbf{x}) = \sum_{k=1}^{n} V(x_k) - 2 \sum_{1 \le i < j \le n} \ln |x_i - x_j|. \tag{4.1}$$

The question is to find the equilibrium position of the movable charges, so we must find the stationary points of $E(\mathbf{x})$ and among them choose the points that give a minimum. The equation $\nabla E(\mathbf{x}) = \mathbf{0}$ is equivalent to

$$V'(x_j) = 2 \sum_{\substack{i=1 \\ i \neq j}}^{n} \frac{1}{x_j - x_i}, \quad j = 1, 2, \dots, n. \tag{4.2}$$

For example when $V(x) = x^2+$ constant, (4.2) becomes a nonlinear system of algebraic equations. In general (4.2) is a nonlinear system.

Stieltjes [56] considered the case when

$$V(x) = -\alpha \ln(1 - x) - \beta \ln(1 + x), \quad \text{and} \quad E = (-1, 1). \tag{4.3}$$

In other words we have two fixed charges at ± 1 and the movable particles are restricted to $(-1, 1)$. Stieltjes' idea was to change the nonlinear problem (4.2) to a linear problem. He observed that, for a fixed j,

$$\begin{aligned} \sum_{\substack{i=1 \\ i \neq j}}^{n} \frac{1}{x_j - x_i} &= \lim_{x \to x_j} \left(\sum_{i=1}^{n} \frac{1}{x - x_i} - \frac{1}{x - x_j} \right) \\ &= \lim_{x \to x_j} \left(\frac{f'(x)}{f(x)} - \frac{1}{x - x_j} \right), \end{aligned}$$

where

$$f(x) = \prod_{j=1}^{n} (x - x_j). \tag{4.4}$$

It is a calculus exercise to evaluate the above limit and show that (4.2) is

$$\sum_{\substack{i=1\\ i\neq j}}^{n} \frac{1}{x_j - x_i} = \frac{1}{2}\frac{f''(x_j)}{f'(x_j)}.$$

This transforms the system (4.2) to

$$f''(x) - V'(x)f'(x) = 0, \quad \text{for} \quad x = x_1, \ldots, x_n.$$

The idea now is to choose a suitable function $h(x)$ such that the validity of

$$f''(x) - V'(x)f'(x) + h(x)f(x) = 0, \quad \text{at } x = x_1, \ldots, x_n, \tag{4.5}$$

implies the validity of (4.5) at all x. Of course (4.5) holds for any $h(x)$ as long as $h(x_j)$ is finite, $1 \leq j \leq n$. Observe that (4.5) may have more solutions than the polynomial solutions to

$$y''(x) - V'(x)y'(x) + h(x)y(x) = 0, \quad \text{for all} \quad x \in E, \tag{4.6}$$

E is the set to which the changes are restricted. To connect (4.6) to (3.14), we need to choose v so that $V'(x) = v'(x) + A_n'(x)/A_n$ and choose h as S_n of (3.14). This will work if the electrostatic equilibrium problem has a unique solution and (4.6) has a unique polynomial solution for the chosen h. Stieltjes proved that this holds in the special case (4.3). A more general theorem is the following.

Theorem 4.1. (Ismail [29]). *Assume that $w = e^{-v}$, $v(x)$ is convex on (a, b) and that the n movable charges are restricted to (a, b). If the external field $V(x)$ is $v(x) + \ln(A_n(x)/a_n)$ is convex on (a, b) then the electrostatic equilibrium problem has a unique solution and is given by the zeros of $p_n(x)$.*

To find the value of $E(\mathbf{x})$ at equilibrium we need the concept of discriminants.

Definition 4.1. The discriminant $D(g)$ of a polynomial g,

$$g(x) = \gamma \prod_{j=1}^{n} (x - x_j), \tag{4.7}$$

is

$$D(g) = \gamma^{2n-2} \prod_{1 \leq i < j \leq n} (x_i - x_j)^2. \tag{4.8}$$

Stieltjes [56] and Hilbert [26] proved that the discriminant of a Jacobi polynomial is given by

$$D(P_n^{(\alpha,\beta)}(x)) = 2^{-n(n-1)} \prod_{j=1}^{n} j^{j-2n+2}(j+\alpha)^{j-1}(j+\beta)^{j-1}(n+j+\alpha+\beta)^{n-j}. \tag{4.9}$$

In [28] we generalized (4.9) to polynomials orthogonal with respect to a weight function $w = e^{-v}$. The key was to combine our (3.4) with an idea of Schur, see § 6.71 in [58]. The result is the following theorem.

Theorem 4.2. *The discriminant of p_n is given by*

$$D(p_n(x)) = \left\{ \prod_{j=1}^{n} \frac{A_n(x_j)}{a_n} \right\} \prod_{k=1}^{n} a_k^{2k-2n+2}, \tag{4.10}$$

where $x_1, x_2, \ldots, x_n$ are the zeros of p_n.

It is clear from (4.1) that at equilibrium the term $-2\sum_{1\le i<j\le n} \ln|x_i - x_j|$ is $-\ln D(f)$ with f as in (4.4).

Theorem 4.3. (Ismail [29]). *Let $\{p_n\}$ be orthonormal with respect to w and define an external field with potential $V(x)$,*

$$V(x) = v(x) + \ln A_n(x)/a_n. \tag{4.11}$$

Under the assumptions of Theorem 4.1, the total energy at equilibrium is given by

$$E(x_1, \ldots, x_n) = \sum_{j=1}^{n} v(x_j) - 2\sum_{j=1}^{n} j \ln a_j. \tag{4.12}$$

In other words

$$\exp(-E(x_1, \ldots, x_n)) = \prod_{j=1}^{n} w(x_j) \prod_{j=1}^{n} a_j^{2n}. \tag{4.13}$$

The standard electrostatic equilibrium problem uses v as the potential of the external field and the equilibrium occurs at the zeros of a Fekete polynomial, see Saff and Totik [51]. In general the Fekete polynomials are not orthogonal, so we cannot generate them through a convenient recursion relation. Our thesis is that the introduction of the term $\ln(A_n(x)/a_n)$ does not change the energy at equilibrium in any significant way but has the added advantage of having a closed form expression (4.12) for the energy at equilibrium. In fact when we take $v(x) = x^4 +$ constant, we proved in [29] that the energy at equilibrium is

$$c_1 n^2 \ln n + c_2 n^2 + c_3 n \ln n + \dots, \quad \text{as} \quad n \to \infty, \tag{4.14}$$

where c_1 and c_2 are the same whether the external potential is $v(x)$ or $V(x)$. Moreover our formulation leads to closed form expressions for the energy and to the identification of the positions of the particles. Computationally the zeros of orthogonal polynomials are stable under slight variations of the recursion coefficients.

It is our opinion that Theorems 4.1 and 4.3 are in the same spirit as the work of Stieltjes [56], [57]. To explain this latter point consider the Jacobi polynomials $\{P_n^{(\alpha,\beta)}(x)\}$. In this case the weight function is

$$\begin{aligned} w_{\alpha,\beta}(x) &= C(1-x)^\alpha(1+x)^\beta, \\ v = v_{\alpha,\beta}(x) &= -\alpha\ln(1-x) - \beta\ln(1+x) - \ln C_1. \end{aligned} \tag{4.15}$$

Here C_1 is a constant. In this case

$$\frac{A_n(x)}{a_n} = \frac{(\alpha+\beta+2n+1)}{1-x^2}, \tag{4.16}$$

$$a_n = \frac{2}{\alpha+\beta+2n}\sqrt{\frac{n(n+\alpha)(n+\beta)(n+\alpha_\beta)}{(\alpha+\beta-1+2n)(\alpha+\beta+1+2n)}}. \tag{4.17}$$

Therefore $V(x) = V_{\alpha,\beta}(x)$ is

$$V_{\alpha,\beta}(x) = v_{\alpha+1,\beta+1}(x) + \text{constant},$$

and Stieltjes proved Theorem 4.3 in this case. Moreover it is clear that computing the second product on the right-hand side of (4.10) is straightforward. To compute the first product in (4.10), we only need to compute $\prod_{j=1}^n (1-x_j^2)$. On the other hand

$$P_n^{(\alpha,\beta)}(x) = \frac{(\alpha+\beta+n+1)_n}{2^n n!}\, x^n + \dots = c_n x^n + \dots$$

hence

$$\prod_{j=1}^n (1-x_j)^2 = \frac{1}{c_n^2} P_n^{(\alpha,\beta)}(1)(-1)^n P_n^{(\alpha,\beta)}(-1).$$

It is known that $P_n^{(\alpha,\beta)}(1) = (-1)^n P_n^{(\beta,\alpha)}(-1) = (\alpha+1)_n/n!$. Therefore the discriminant of the orthonormal Jacobi polynomials can be found in closed form. The interested reader is encouraged to carry out the details and verify (4.9).

To state a recent result of Ercoloni and McLaughlin [20] we need to remind the reader of the Laplace asymptotic method. Let $[a,b] \subset \mathbb{R}$ where a function f has unique local minimum at a with $f''(a) > 0$ and assume that f is real analytic in a neighborhood of a and continuous on S. Thus $f(x) = f(a) + (x-a)^2 f''(a)/2! + \dots$. We wish to determine the large n behavior of the integral

$$I_n = \int_a^b e^{-nf(x)}\, g(x)\, dx.$$

The idea is that as $n \to \infty$, $e^{-nf(x)}$ can be well-approximated by $e^{-nf(a)} \exp(-n(x-a)^2 f''(a)/2)$. After the change of variable $x = a + \sqrt{2/n}\, t/\sqrt{f''}(a)$, we see that the integral over $[a,b]$ can be replaced by $\int_{\mathbb{R}}$ and the result is that

$$I_n = e^{-nf(a)} \sqrt{\frac{2\pi}{nf''(a)}}\, g(a)\, [1 + o(1)], \quad \text{as} \quad n \to \infty,$$

provided that $g(a) \neq 0$.

A multidimensional analogue of I_n arose in the work of Lax and Livermore [40] who considered the integral

$$J_{n,N} = \int_{\mathbb{R}^n} \exp\left(-N \sum_{j=1}^{n} v(\lambda j)\right) \prod_{1 \le j < k \le n} (\lambda_j - \lambda_k)^2 d\lambda_1 \dots d\lambda_n. \quad (4.18)$$

Assuming that $N = cn, c \in (0,1)$, the problem is to determine the asymptotics of $J_{n,N}$ as $n \to \infty$ and fixed c. Over the years this problem has attracted the attention of many mathematicians who contributed to its solution. The final step was taken by Ercolani and McLaughlin [20]. They wrote the integrand as $\exp(-E)$, and minimize E, then apply the nonlinear steepest descent technique of Deift and Zhou [18]. It will be interesting to consider the integral

$$I_{n,N} = \int_{\mathbb{R}^n} \frac{\exp(-N \sum_{j=1}^{n} v(\lambda j))}{\prod_{i=1}^{n} A_n(\lambda_i; N))} \prod_{1 \le j < k \le n} (\lambda_j - \lambda_k)^2 d\lambda_1 \dots d\lambda_n. \quad (4.19)$$

In (4.19), $\{p_n(x;N)\}$ is a sequence of polynomials orthonormal with respect to $\exp(-Nv(x))$, with $\lambda_1, \dots, \lambda_n$ denotes the zeros of $p_n(x;N)$. Moreover $A_n(x;N)$ is the A_n function in Theorem 3.1. As in (4.18), we take $N = cn$, with $c \in (0,1)$.

5. Generating Functions and Asymptotics

In this section we discuss techniques of finding the large n behavior of orthogonal polynomials. A generating function of a sequence of polynomials $\{p_n(x)\}$ is a function $G(x,t)$ which has the expansion

$$G(x,t) = \sum_{0}^{\infty} c_n p_n(x) t^n, \tag{5.1}$$

where $\{c_n\}$ are suitably chosen constants to make $G(x,t)$ have a closed form. The Poisson kernel is the bilinear generating function

$$P_t(x,y) = \sum_{n=0}^{\infty} p_n(x) p_n(y) t^n. \tag{5.2}$$

It is not clear where the Poisson kernel converge, but its convergence or divergence clearly depends on the large n behavior of $p_n(x)p_n(y)$.

The first asymptotic technique we treat is Darboux's asymptotic method which depends on finding a reasonable generating function. As an example we first derive a generating function for the ultraspherical (Gegenbauer) polynomials $\{C_n^\nu(x)\}$. The polynomials $\{C_n^\nu(x)\}$ are generated by

$$C_0^\nu(x) = 1, \quad C_1^\nu(x) = 2\nu x, \tag{5.3}$$

$$2(x+\nu)C_n^\nu(x) = (n+1)C_{n+1}^\nu(x) + (2\nu+n-1)C_{n-1}^\nu(x), \tag{5.4}$$

for $n > 0$. The Legendre polynomials correspond to $\nu = 1/2$. Let

$$G(x,t) = \sum_{n=0}^{\infty} C_n^\nu(x) t^n. \tag{5.5}$$

Multiply (5.4) by t^n and add for $n = 1, 2, \ldots$. In view of (5.3) and after replacing nt^n by $t\partial_t t^n$, we establish the differential equation

$$2\nu x G(x,t) + 2xt\partial_t G(x,t) = \partial_t G(x,t) + 2\nu t G(x,t) + t^2 \partial_t G(x,t).$$

Therefore

$$\frac{\partial_t G(x,t)}{G(x,t)} = \frac{2\nu(x-t)}{1-2xt+t^2},$$

hence $G(x,t) = f(x)(1-2xt+t^2)^{-\nu}$. Now $f(x) = 1$ because $1 = C_0^\nu(x) = G(x,0) = f(x)$. This proves

$$\sum_{n=0}^{\infty} C_n^\nu(x) t^n = (1-2xt+t^2)^{-\nu}. \tag{5.6}$$

To justify the above formal steps we start with the answer and reverse the steps until we reach (5.3)–(5.4).

Theorem 5.1. (Darboux's Method). *Let* $f(t) = \sum_0^\infty f_n t^n$, $g(z) = \sum_0^\infty g_n t^n$ *be analytic in* $|t| < r$ *and* $f - g$ *be continuous on* $|t| \le r$. *Then*

$$f_n - g_n = o(r^{-n}). \tag{5.7}$$

This theorem follows from the Riemann-Lebesgue lemma, see Olver [48]. Given f, the function g is called a comparison function and it normally consists of the singular part of f.

Example. Consider the Legendre Polynomials. Their generating function is

$$\sum_{n=0}^\infty P_n(x)t^n = (1 - 2xt + t^2)^{-1/2}, \tag{5.8}$$

since the Legendre polynomials correspond to $\nu = 1/2$ in $\{C_n^\nu(x)\}$. The notation $P_n(x)$ is standard in the literature but P_n is not monic. The t-singularities of the generating function (5.8) are at $t = \alpha$, $t = \beta$, where

$$\alpha, \beta = x \pm \sqrt{x^2 - 1}\,, \quad \text{and} \quad |\beta| \le |\alpha|.$$

Here $x \in \mathbb{C}$. Note that $\alpha = \beta$ if and only if $x = \pm 1$. Moreover $|\alpha| = |\beta|$ if and only if $x \in [-1, 1]$.

Case I: $x = \pm 1$. In this case the right-hand side of (5.8) is $(1 \mp t)^{-1}$, hence $P_n(\pm 1) = (\pm 1)^n$ and no asymptotic analysis is needed.

Case II: $x \in \mathbb{C} \setminus [-1, 1]$. In this case $|\beta| < |\alpha|$. Let

$$g(t) = \frac{(1 - t/\beta)^{-1/2}}{(1 - \beta/\alpha)^{1/2}} = \sum_0^\infty \frac{\Gamma(n + 1/2)}{\Gamma(1/2)\, n!} \frac{\beta^{-n} t^n}{(1 - \beta/\alpha)^{1/2}},$$

where we used the binomial theorem in the last equation. It is easy to see that $\sum_0^\infty P_n(x)t^n - g(t)$ is continuous in $|t| \le |\beta|$ and, $g(t)$ and $\sum_0^\infty P_n(x)t^n$ are analytic in $|t| < |\beta|$. Therefore Darboux's method yield

$$P_n(x) \approx \frac{(1 - \beta/\alpha)^{-1/2}}{\beta^n} \frac{\Gamma(n + 1/2)}{\Gamma(1/2) n!}.$$

Using $n^{b-a}\Gamma(a + n)/\Gamma(b + n) \to 1$ as $n \to \infty$, we see from the above asymptotic formula that

$$P_n(x) \approx (1 - \beta/\alpha)^{-1/2} \beta^{-n} / \sqrt{\pi n}, \tag{5.9}$$

since $\Gamma(1/2) = \sqrt{\pi}$.

Case III. $x \in (-1,1)$, so $|\alpha| = |\beta|$, $\alpha \neq \beta$. In this case a comparison function is

$$g(t) = \frac{(1-t/\beta)^{-1/2}}{(1-\beta/\alpha)^{1/2}} + \frac{(1-t/\alpha)^{-1/2}}{(1-\alpha/\beta)^{1/2}}.$$

With $\beta = e^{-i\theta}$, $\alpha = e^{i\theta}$, so $x = \cos\theta$, and we obtain

$$\begin{aligned} g(t) =& (1-e^{-2i\theta})^{-1/2} \sum_0^\infty t^n e^{in\theta} \frac{\Gamma(n+1/2)}{n!\,\Gamma(1/2)} \\ &+ (1-e^{2i\theta})^{-1/2} \sum_0^\infty t^n e^{-in\theta} \frac{\Gamma(n+1/2)}{n!\,\Gamma(1/2)}. \end{aligned}$$

Therefore

$$P_n(\cos\theta) \approx \frac{e^{in\theta} e^{i\theta/2}}{\sqrt{n\pi}\sqrt{2i\sin\theta}} + \frac{e^{-in\theta} e^{-i\theta/2}}{\sqrt{n\pi}\sqrt{-2i\sin\theta}},$$

or equivalently

$$P_n(\cos\theta) \approx \sqrt{\frac{2}{n\pi}} \frac{\cos((n+1/2)\theta - \pi/4)}{\sqrt{\sin\theta}}, \quad 0 < \theta < \pi. \tag{5.10}$$

In fact (5.10) is uniform in θ for $\epsilon \leq \theta \leq \pi - \epsilon$ and $\epsilon \in (0,\pi)$.

The above example is typical. Darboux's asymptotic method is useful when we know the singular part of a generating function.

Next we discuss Poisson kernels. Let $f \in L^2(E,\mu)$. Then f will have the orthogonal expansion

$$f(x) \sim \sum_{n=0}^\infty f_n p_n(x), \quad f_n = \int_E f(x) p_n(x) d\mu(x), \tag{5.11}$$

and $\{p_n\}$ are orthonormal with respect to μ and complete in $L^2(E,\mu)$. The expansion (5.11) is an expansion in $L^2(E,\mu)$. We try to write an integral representation for the orthogonal series. Let us explore a formal approach, and write

$$\begin{aligned} \sum_{n=0}^\infty f_n p_n(x) &= \sum_0^\infty \int_E f(y) p_n(y) p_n(x) d\mu(y) \\ &= \int_E f(y) \sum_{n=0}^\infty p_n(x) p_n(y) d\mu(y). \end{aligned}$$

It is not clear how to produce f from the above right-hand side unless the integral is $\int_E \delta(y-x) f(y) dy$, δ being a Dirac delta function. To remedy this we introduce a damping factor into the series by replacing $p_n(x)p_n(y)$ by $p_n(x)p_n(y)t^n$, $0 < t < 1$ and hope for

$$f(x) = \lim_{t\to 1^-} \int_E f(y)P_t(x,y)d\mu(y), \tag{5.12}$$

with $P_t(x,y)$ defined in (5.2). The result is that (5.12) holds under some general conditions and the analysis will be greatly simplified if $P_t(x,y)$ is nonnegative for $x, y \in E$, and $t \in (0,1)$. For Hermite polynomials the Poisson kernel is

$$\sum_{n=0}^{\infty} \frac{H_n(x)H_n(y)}{2^n n! \sqrt{\pi}} t^n = \frac{1}{\sqrt{\pi}}(1-t^2)^{-1/2} \exp\left(\frac{2xyt - (x^2+y^2)t^2}{1-t^2}\right). \tag{5.13}$$

Formula (5.13) is called the Mehler formula. Clearly $P_t(x;y) \geq 0$ for $x, y \in \mathbb{R}$, $0 < t < 1$. Indeed (5.12) is the real inversion formula for the Gauss-Weierstrass transform [63]. Poisson kernels are bilinear generating functions.

The Poisson kernel for Laguerre polynomials is a multiple of

$$\begin{aligned} &\sum_{n=0}^{\infty} \frac{n!\, t^n}{(\alpha+1)_n} L_n^\alpha(x) L_n^\alpha(y) \\ &= (1-t)^{-1-\alpha} \exp\left(\frac{-(x+y)t}{1-t}\right) F\left(\frac{xyt}{(1-t)^2}\right), \end{aligned} \tag{5.14}$$

where F is

$$F(z) = \sum_{n=0}^{\infty} \frac{z^n}{n!\,(\alpha+1)_n} \tag{5.15}$$

and is related to the modified Bessel function I_α, see [58], [50]. It is clear that the closed form (5.14), known in the literature as the Hille-Hardy formula, exhibits the positivity of the Poisson kernel for Laguerre polynomials when $t \in [0,1)$. Motivated by the forms (5.13) Sarmanov and Bratoeva [53] considered bilinear series such that

$$\sum_{n=0}^{\infty} \frac{c_n}{2^n\, n!} H_n(x)H_n(y) \geq 0, \tag{5.16}$$

for $c_n \in \mathbb{R}$.

Theorem 5.2. (Sarmanov and Bratoeva) *If $\sum_{n=0}^{\infty} c_n^2$ converges then the condition* (5.16) *holds for all $(x,y) \in \mathbb{R}\times\mathbb{R}$ if and only if there is a positive measure μ such that*

$$c_n = \int_{-1}^{1} t^n d\mu(t), \quad n = 0, 1, \dots. \tag{5.17}$$

It is clear that the sequences $\{c_n\}$ characterized by Theorem 5.2 form a convex set whose extreme points are the sequences $\{t^n\}$, for $t \in (-1, 1)$; which correspond to measures having a single atom at t. Sarmanov [52] proved the corresponding theorem for the Laguerre series by requiring

$$\sum_{n=0}^{\infty} \frac{c_n n!}{\Gamma(\alpha+n+1)} L_n^{\alpha}(x) L_n^{\alpha}(y) \geq 0, \tag{5.18}$$

for all sequences $\{c_n\}$ such that $\sum_{n=0}^{\infty} c_n^2 < \infty$, and $\alpha > -1$. Sarmanov's theorem states that (5.18) holds for all $x \geq 0, y \geq 0$ if and only if $c_n = \int_0^1 t^n \, d\mu(t)$ with μ a positive measure. Askey [3] gave an informal argument which helps explain these results. Here again the extreme points of the set of sequences $\{c_n\}$ is $\{r^n : 0 \leq r < 1\}$ and establishe the positivity of the series side in the Hille-Hardy formula (5.14).

The series for the Poisson kernel for Jacobi polynomial is stated on page 102 of Bailey [7] as

$$\begin{aligned} &\sum_{n=0}^{\infty} \frac{n!\,(\alpha+\beta+1)_n}{(\alpha+1)_n(\beta+1)_n} (2n+\alpha+\beta+1)\, t^n P_n^{(\alpha,\beta)}(x) P_n^{(\alpha,\beta)}(y) \\ &= \frac{(\alpha+\beta+2)(1-t)}{(1+t)^{\alpha+\beta+1}} F_4 \left(\left. \begin{matrix} (\alpha+\beta+2)/2, (\alpha+\beta+3)/2 \\ \alpha+1, \beta+1 \end{matrix} \right| A, B \right), \end{aligned} \tag{5.19}$$

where

$$A = \frac{t(1-x)(1-y)}{(1+t)^2}, \qquad B = \frac{t(1+x)(1+y)}{(1+t)^2} \tag{5.20}$$

The representation (5.19) shows the positivity of the Poisson kernel for $x, y \in [-1, 1], t \in [0, 1)$ in the full range of the parameters $\alpha > -1, \beta > -1$. Bailey [7] also states the kernel

$$\begin{aligned} &\sum_{n=0}^{\infty} \frac{n!\,(\alpha+\beta+1)_n}{(\alpha+1)_n(\beta+1)_n} t^n P_n^{(\alpha,\beta)}(x) P_n^{(\alpha,\beta)}(y) \\ &= (1+t)^{-\alpha-\beta-1} F_4 \left(\left. \begin{matrix} (\alpha+\beta+1)/2, (\alpha+\beta+2)/2 \\ \alpha+1, \beta+1 \end{matrix} \right| A, B \right), \end{aligned} \tag{5.21}$$

which is positive for $x, y \in [-1, 1], t \in [0, 1)$ with $\alpha > -1, \beta > -1$ but with the additional constraint $\alpha + \beta > -1$.

The question of proving the positivity of $P_t(x, y)$ when the polynomials depend on parameters have been the subject of many research papers, see the monograph by Askey [4], the treatise by Gasper and Rahman [22] and the references in them.

Observe that the derivation of (5.9) and (5.10) did not use the orthogonality of $P_n(x)$. In fact we started with the recursive definition (5.3)–(5.4) and applied Darboux's method. Indeed (5.9) and (5.10) show that $P_n(x)$ is oscillatory when $x \in (-1,1)$ and has exponential growth in $\mathbb{C} \setminus [-1,1]$. The oscillatory nature of P_n indicates that the zeros are dense in $[-1,1]$.

Several mathematicians worked on the problem of determining the asymptotic behavior of $p_n(x)$ for n large from the qualitative of the recurrence coefficients $\{a_n\}$ and $\{b_n\}$ in (2.2). The model polynomials for this analysis are the Chebyshev polynomials of the first kind $\{T_n(x)\}$,

$$T_n(\cos\theta) = \cos n\theta, \tag{5.22}$$

$$xT_n(x) = \frac{1}{2}T_{n+1}(x) + \frac{1}{2}T_{n-1}(x). \tag{5.23}$$

The normalized weight function is $(1-x^2)^{1/2}/\pi$. A sample of results in this direction is the following theorem of Nevai from [46].

Theorem 5.3. *Assume that* $b_n \to 0$, *and* $a_n \to 0$ *in such a way that* $\sum_{n=1}^{\infty} n(|a_n - 1/2| + |b_n|)$ *converges. Then* $\{p_n(x)\}$ *are orthonormal with respect to a probability measure* μ, *and* μ' *is supported on* $[-1,1]$. *The discrete part of* μ *is finite (may be empty) and lies outside* $(-1,1)$. *Moreover*

$$\sqrt{\mu'(\cos\theta)}\, p_n(\cos\theta) \approx \sqrt{\frac{2}{\pi}}\sqrt{\frac{1}{\sqrt{1-x^2}}}\, \cos(n\theta + \varphi(\theta) + o(1)), \tag{5.24}$$

holds for $x = \cos\theta$ *for* $0 < \theta < \pi$ *and* φ *depend on* θ *but not on* n.

Note that the normalized weight function for $\{T_n\}$ is $(1-x^2)^{-1/2}/\pi$, and for $n > 0$, $\{\sqrt{2}\, T_n(x)\}$ are the orthonormal polynomial. Therefore (5.24) reduces the asymptotics of general $\{p_n(x)\}$ to the asymptotics of $T_n(\cos\theta)$ with θ shifted by $\varphi(\theta)/n + o(1/n)$. In other words $\sqrt{\mu'(x)}\, p_n(x)$ behaves as if p_n is $\sqrt{2}\, T_n$. Another theorem of Nevai[46] is the following.

Theorem 5.4. *Assume that* $\sum_1^{\infty}(|a_n - 1/2| + |b_n|) < \infty$. *Then* $\mu'(x)$ *is supported on* $[-1,1]$ *and* μ *may have a discrete part outside* $(-1,1)$. *Moreover with* $x = \cos\theta, 0 < \theta < \pi$, *the limiting relation*

$$\limsup_n \left[\sqrt{\mu'(x)}\, p_n(x) - \sqrt{\frac{2}{\pi}}\, \frac{\cos(n\theta + \varphi(\theta) + o(1))}{(1-x^2)^{1/4}}\right] = 0, \tag{5.25}$$

holds.

We next derive an integral representation for a general Jacobi polynomial. This representation has the form $\int_C e^{nf(z)}g(z)dz$, so one can apply the method of steepest descent to this integral and determine the large n asymptotics of the polynomials. The details of the steepest descent method are in Wong's article in these proceedings, [65].

The Jacobi polynomials $\{P_n^{(\alpha,\beta)}(x)\}$ have the series representation

$$P_n^{(\alpha,\beta)}(x) = \frac{(\alpha+1)_n}{n!}\sum_{k=1}^{n}\frac{(-n)_k(\alpha+\beta+n+1)_k}{k!(\alpha+1)_k}\left(\frac{1-x}{2}\right)^n. \tag{5.26}$$

The Pfaff-Kummer transformation [2] is

$$\sum_{k=0}^{\infty}\frac{(A)_k(B)_k}{k!(C)_k}z^k = (1-z)^{-A}\sum_{k=0}^{\infty}\frac{(A)_k(C-B)_k}{k!(C)_k}z^k, \tag{5.27}$$

provided that the series on both sides are convergent. Now apply (5.27) to (5.26) to get

$$P_n^{(\alpha,\beta)}(x) = \frac{(\alpha+1)_n}{n!}\left(\frac{1+x}{2}\right)^n\sum_{k=0}^{n}\frac{(-n)_k(-\beta-n)_k}{k!(\alpha+1)_k}\left(\frac{x-1}{x+1}\right)^k. \tag{5.28}$$

Applying

$$\frac{(\alpha+1)_n}{(\alpha+1)_k} = (-1)^{n-k}(-\alpha-n)_{n-k}, \qquad (-n)_k = \frac{(-1)^k\, n!}{(n-k)!},$$

we establish

$$P_n^{(\alpha,\beta)}(x) = (-1)^n\sum_{k=0}^{n}\frac{(-\beta-n)_k}{k!}\left(\frac{x-1}{2}\right)^k\frac{(-\alpha-n)_{n-k}}{(n-k)!}\left(\frac{x+1}{2}\right)^{n-k}.$$

The use of the binomial theorem and Cauchy's theorem leads to

$$P_n^{(\alpha,\beta)}(x) = \frac{1}{2\pi i}\int_C (1+(x+1)z/2)^{n+\alpha}(1+(x-1)z/2)^{n+\beta}\frac{dz}{z^{n+1}}, \tag{5.29}$$

where C is a closed contour so that $-2(x\pm1)^{-1}$ be in the exterior of C. The integral representation (5.29) has a form suitable for the application of the steepest descent method. In fact the integral representation (5.29) can be used to derive the generating function

$$\sum_{n=0}^{\infty}P_n^{(\alpha,\beta)}(x)t^n = \frac{2^{\alpha+\beta}R^{-1}}{(1-t+R)^{\alpha}(1+t+R)^{\beta}}, \tag{5.30}$$

where $R=\sqrt{1-2xt+t^2}$, [58]. A different proof is in [50]. One can apply Darboux's method to the generating function (5.30) and determine the asymptotic behavior of $P_n^{(\alpha,\beta)}(x)$ for large n and fixed x in different parts of the complex x-plane. The interested reader is strongly encouraged to apply Darboux's method to (5.30), or apply the steepest descent method to (5.29) and establish the following theorem.

Theorem 5.5. *Let α and β be real. Then*

$$\begin{aligned} P_n^{(\alpha,\beta)}(x) &\approx (x-1)^{-\alpha/2}(x+1)^{-\beta/2}\left[(x+1)^{1/2}+(x-1)^{1/2}\right]^{\alpha+\beta} \\ &\quad\times \frac{1}{\sqrt{2n\pi}}\frac{1}{(x^2-1)^{1/4}}\left[x+(x^2-1)^{1/2}\right]^{n+1/2}, \end{aligned} \tag{5.31}$$

for $x \in \mathbb{C}\setminus[-1,1]$. On the other hand if $0<\theta<\pi$, then

$$P_n^{(\alpha,\beta)}(\cos\theta) = \frac{k(\theta)}{\sqrt{n\pi}}\cos(N\theta+\gamma) + O\left(n^{-3/2}\right), \tag{5.32}$$

where

$$\begin{aligned} k(\theta) &= (\sin(\theta/2))^{-\alpha-1/2}(\cos(\theta/2))^{-\beta-1/2}, \\ N &= n+(\alpha+\beta+1)/2, \quad \gamma = -(\alpha+1/2)\frac{\pi}{2}. \end{aligned} \tag{5.33}$$

When $\alpha=\beta$, the generating function (5.30) reduces to a generating function for the ultraspherical polynomials since

$$C_n^{\nu}(x) = \frac{(2\nu)_n}{(\nu+1/2)_n}P_n^{(\nu-1/2,\nu-1/2)}(x). \tag{5.34}$$

We made no attempt to cover the application of Riemann-Hilbert techniques to deriving large degree asymptotics of orthogonal polynomials. This is a powerful technique and is covered in Deift's monograph [17]. The state of the art of this approach is covered in the recent survey article of Arno Kuijlaars [39].

We conclude this section by describing a technique to recover the orthogonality measure(s) μ when the recurrence relation (2.2) is given.

Theorem 5.6. *Let $p_0=1$, $p_1=(x-b_0)/a_1$, and $b_n\in\mathbb{R}, n\geq 0, a_n>0, n>0$, and assume that $\{p_n(x)\}$ is generated by*

$$xp_n(x) = a_{n+1}p_{n+1}(x)+b_np_n(x)+a_np_{n-1}(x), \quad n>0.$$

Then there exists a probability measure μ such that $\int_{\mathbb{R}} p_m(x)p_n(x)\,d\mu(x) = \delta_{m,n}$.

Theorem 5.6 gives a converse to the fact that orthogonal polynomials satisfy three term recurrence relations. We shall refer to Theorem 5.6 as the spectral theorem for orthogonal polynomials. Some authors call it Favard's theorem but it was in the literature before Favard's paper was written. The probability measure μ whose existence is guaranteed by Theorem 5.6 may not be unique.

We now introduce a second solution $\{p_n^*(x)\}$ to (2.2) by the initial conditions

$$p_0^*(x) = 0, \quad p_1^*(x) = 1/a_1, \tag{5.35}$$

and requiring p_n^* satisfies (2.2).

Theorem 5.7. (Markov). *Assume that the recursion coefficients $\{a_n\}$ and $\{b_n\}$ are bounded. Then the measure in Theorem 5.4 is unique, supported on a compact set $E \subset \mathbb{R}$ and satisfies*

$$\lim_{n\to\infty} \frac{p_n^*(z)}{p_n(z)} = \int_{\mathbb{R}} \frac{d\mu(t)}{z-t}, \quad z \notin E. \tag{5.36}$$

Moreover, the convergence in (5.36) is uniform on compact subsets of $\mathbb{C}\backslash E$.

Note that if μ is unique then (5.36) continues to holds but $E = \text{supp}(\mu)$ may not be bounded. The convergence in (5.36) is still uniform on compact subsets of $\mathbb{C} \setminus E$.

The Perron-Stieltjes inversion formula is

$$F(z) = \int_{\mathbb{R}} \frac{d\mu(t)}{z-t} \tag{5.37}$$

if and only if

$$\mu(t) - \mu(s) = \lim_{\epsilon \to 0+} \int_s^t \frac{F(u-i\epsilon) - F(u+i\epsilon)}{2\pi i} du. \tag{5.38}$$

Theorem 5.7 is an instance where the asymptotics of polynomials give crucial information about their orthogonality measure.

6. Applications

This section is devoted to some problems which naturally lead to three term recurrence relations and determining the orthogonality measure μ or its support sheds some light on the original problem.

The first example is birth and death processes. Consider a stationary Markov chain whose state space is the nonnegative integers. Let $p_{m,n}(t)$ be the transition probability to go from state m to state n in time t. Here

$p_{m,n}(t)$ does not depend on the initial time. We assume that

$$p_{m,n}(t) = \begin{cases} \lambda_m t + o(t), & n = m+1, \\ \mu_n t + o(t), & n = m-1, \\ 1 - (\lambda_m + \mu_m)t + o(t) & n = m, \end{cases} \tag{6.1}$$

and $p_{m,n}(t) = o(t)$, if $|m-n| > 1$. The birth rates λ_m and death rates μ_m satisfy $\mu_0 \geq 0$, $\mu_m > 0, m > 0$, and $\lambda_m > 0$ for $m \geq 0$.

The Chapman-Kolmogorov equations which describe this process are

$$\dot{p}_{m,n}(t) = \lambda_{n-1} p_{m,n-1}(t) + \mu_{n+1} p_{m,n+1}(t) - (\lambda_n + \mu_n) p_{m,n}(t), \tag{6.2}$$

$$\dot{p}_{m,n}(t) = \lambda_m p_{m+1,n}(t) + \mu_m p_{m-1,n}(t) - (\lambda_m + \mu_m) p_{m,n}(t), \tag{6.3}$$

where $\dot{y}$ means $\frac{dy}{dt}$. Let us attempt to solve (6.2)–(6.3) by the separation of variables

$$p_{m,n}(t) = f(t) Q_m F_n.$$

Therefore $f'(t)/f(t)$ is independent of t, so set it equal to $-x$. Now (6.2) indicates that F_n depends on x, so we denote it by $F_n(x)$. ¿From (6.2) we get

$$-x F_n(x) = \lambda_{n-1} F_{n-1}(x) - (\lambda_n + \mu_n) F_n(x) + \mu_{n+1} F_{n+1}(x), \tag{6.4}$$

with $F_0(x)$ arbitrary, so we take

$$F_{-1}(x) := 0 \quad \text{and} \quad F_0(x) = 1. \tag{6.5}$$

Similarly $Q_{-1}(x) = 0$ and $Q_0(x) = 1$, and

$$-x Q_n(x) = \lambda_n Q_{n+1}(x) - (\lambda_n + \mu_n) Q_n(x) + \mu_n Q_{n-1}(x). \tag{6.6}$$

Indeed (6.4), (6.6) and the initial conditions imply

$$F_n(x) = \frac{\lambda_0 \lambda_1 \ldots \lambda_{n-1}}{\mu_1 \mu_2 \ldots \mu_n} Q_n(x). \tag{6.7}$$

Thus we have determined a solution and we multiply by separation constants and sum or integrate over all choices of x. Therefore

$$p_{m,n}(t) = \frac{1}{\xi_m} \int e^{-xt} F_m(x) F_n(x) d\mu(x),$$

where

$$\xi_n = \frac{\lambda_0 \lambda_1 \ldots \lambda_{n-1}}{\mu_1 \mu_2 \ldots \mu_n}. \tag{6.8}$$

As $t \to 0^+, \mu_1\mu_2\ldots\mu_n p_{m,n}(t) \to \delta_{m,n}$. Thus

$$\int F_m(x)F_n(x)d\mu(x) = \xi_m\delta_{m,n}, \tag{6.9}$$

so it seems that $\{F_n\}$ are orthogonal with respect to μ. At $t \to \infty$ our $p_{m,n}(t)$ cannot have exponential growth, hence $x \in (0, \infty)$, and we obtain

$$p_{m,n}(t) = \frac{1}{\xi_m}\int_0^\infty e^{-xt}F_m(x)F_n(x)d\mu(x). \tag{6.10}$$

Observe that (6.10) gives the solution in a factored form and it is not difficult to analyze the large t behavior of $p_{m,n}(t)$ from (6.10).

It turns out the μ must be a positive, hence $\{F_n\}$ are orthogonal with respect to μ. Furthermore the application of the chain sequence techniques of §7 to (6.5) and (6.6) show that all zeros of F_n and Q_n lie in $(0, \infty)$.

The integral representation (6.10) has been established by Karlin and McGregor in [35], [36]. Later Karlin and McGregor [38] studied random walks on the state space of the nonnegative integers and defined another sequence of orthogonal polynomials. The random walk polynomials are generated by

$$R_0(x) = 1, \quad R_1(x) = (1 + \mu_0/\lambda_0)x, \tag{6.11}$$

$$xR_n(x) = \frac{\lambda_n}{\lambda_n + \mu_n}R_{n+1}(x) + \frac{\mu_n}{\lambda_n + \mu_n}R_{n-1}(x). \tag{6.12}$$

Since the recurrence coefficients in (6.12) are bounded. Theorem 5.7 shows that $\{R_n(x)\}$ are orthogonal with respect to a measure supported on a compact set. Theorem 7.5 proves that all the zeros of R_n belong to $(-1, 1)$, for all n. From this fact, one can prove that $\{R_n(x)\}$ are orthogonal with respect to a measure supported on a subset of $[-1, 1]$.

The Laguerre polynomials are $\{F_n\}$ polynomials when $\mu_n = n, \lambda_n = n + \alpha + 1$. The ultraspherical polynomials are multiples of random walk polynomials with $\mu_n = n, \lambda_n = n + 2\nu$. In fact the Jacobi, Laguerre, Hahn, Meixner, the Charlier polynomials, or there various special cases, are birth and death process polynomials or random walk polynomials.

The random walk polynomials corresponding to

$$\lambda_n = cn + \beta, \quad \mu_n = n, \quad \beta > 0, c > 0 \tag{6.13}$$

are very interesting. The case $c = 1$ is the ultraspherical polynomials. In the case $c = 0$, the polynomials $\{R_n(x)\}$ are orthogonal with respect to a discrete measure. This measure together with some explicit representations were found in 1958, independently and using completely different techniques, by Carlitz [14] and, Karlin and McGregor [37]. In 1984 Askey and Ismail [5] analyzed the full model (6.13). They proved that the orthogonality measure is absolutely continuous on $[-2\sqrt{c}/(c+1), 2\sqrt{c}/(c+1)]$. When $c \neq 1$ it has an infinite discrete part supported in $[-1, -2\sqrt{c}/(c+1)], [2\sqrt{c}/(c+1), 1]$. The points $\pm 2\sqrt{c}/(c+1)$, do not support positive masses but are the only limit points of the points supporting positive point masses. Moreover [5] contains explicit and asymptotic formulas for the polynomials and their generating functions. The techniques used are Markov's theorem, Darboux's method, as well as standard special function techniques.

Another problem which leads to orthogonal polynomials is the so called J-Matrix Method. The idea is to start with a Schrödinger operator on $\mathbb{R}$,

$$T := -\frac{d^2}{dx^2} + V(x). \tag{6.14}$$

The operator T is densely defined on $L^2(\mathbb{R})$ and is symmetric. The idea is to find a complete orthonormal basis in the domain of T such that T the matrix representation of T in $\{\varphi_n(x)\}$ is tridiagonal, that is

$$\int_{\mathbb{R}} \overline{\varphi}_m T \varphi_n dx = 0, \quad \text{if} \quad |m-n| > 1.$$

We now diagonalize T, that is let $T\psi_E = E\psi_E$ and assume

$$\psi_E(x) \sim \sum_{n=0}^{\infty} \varphi_n(x)\psi_n(E). \tag{6.15}$$

Observe that

$$\begin{aligned} E\psi_n(E) &= (E\psi_E, \varphi_n) = (T\psi_E, \varphi_n) \\ &= (T\varphi_{n-1}\psi_{n-1} + T\varphi_n\psi_n + T\varphi_{n+1}\psi_{n+1}\varphi_n, \varphi_n). \end{aligned}$$

Therefore

$$\begin{aligned} E\psi_n(E) =& \psi_{n+1}(E)(T\varphi_{n+1}, \varphi_n) + \psi_n(E)(T\varphi_n, \varphi_n) \\ &+ \psi_{n-1}(E)(T\varphi_{n-1}, \varphi_n). \end{aligned} \tag{6.16}$$

If $(T\varphi_n, \varphi_{n\pm 1}) \neq 0$, then (6.16) is a recurrence relation for a sequence of polynomials $\{\psi_n(E)\}$. It is easy to see that (6.16) can be reduced to (2.2) if and only if

$$(T\varphi, \varphi_{n-1})(T\varphi_{n-1}, \varphi_n) > 0.$$

This is indeed the case since $(T\varphi_n, \varphi_{n-1}) = (\varphi_n, T\varphi_{n-1}) = \overline{(T\varphi_{n-1}, \varphi_n)}$. Heller, Reinhardt and Yamani [24] introduced this method and applied it to physical problems. In particular they identified the orthogonal polynomials $\{\psi_n(E)\}$ which arise in the harmonic oscillator, the Morse oscillator, and the hydrogen like atom. The polynomials $\{\psi_n(E)\}$ are Meixner polynomials in the harmonic oscillator model. The hydrogen atom with a Coulomb potential was more challenging. In the case when the Coulomb potential is repulsive, i.e. the electron and nucleus have charges of the same sign, $\psi_n(E)$ are the polynomials of Szegő and Pollaczek, see [58]. In the physical case when the Coulomb potential is attractive, the $\psi_n(E)$ are the Askey-Ismail polynomials [5] generated by (6.12)–(6.13).

7. Zeros of Orthogonal Polynomials and Eigenvalues

Let A_∞ be the infinite Jacobi matrix

$$A_\infty = \begin{pmatrix} b_0 & a_1 & 0 & 0 & \dots \\ a_1 & b_1 & a_2 & 0 & \dots \\ 0 & a_2 & b_2 & a_3 & \dots \\ \vdots & \vdots & \vdots & \vdots & \dots \end{pmatrix},$$

and let A_N be the $N \times N$ truncation, that is delete from A_∞ the columns $N+1, N+2, \dots$, and rows $N+1, N+2, \dots$, and A_N is what is left.

Theorem 7.1. *We have*

$$P_N(\lambda) = \det(\lambda I - A_N), \tag{7.1}$$

where "det" stands for "determinant". In other words $P_N(\lambda)$ is the characteristic polynomial of A_N.

Proof. Let $Q_N(\lambda) = \det(\lambda I - A_N)$. Verify that $P_1(\lambda) = Q_1(\lambda), Q_2(\lambda) = P_2(\lambda)$. For $N > 1$, expand $Q_{N+1}(\lambda)$ about the last row then expand the cofactor of $a_N\lambda - b_N$ about its last column. The result is

$$Q_{N+1}(\lambda) = (\lambda - b_N)Q_N(\lambda) - a_N^2 Q_{N-1}(\lambda),$$

which is the same recurrence relation satisfied by $P_n(\lambda)$, see (2.6). □

Theorem 7.1 relates zeros of orthogonal polynomials to eigenvalues of tridiagonal or Jacobi matrices, where linear algebra techniques are available to study the locations of eigenvalues, Horn and Johnson [27].

Some of the questions in this area of research are:

1. Find approximations to the zeros of $P_n(x)$.

2. Let

$$x_{n,1} > x_{n,2} > \cdots > x_{n,n} \tag{7.2}$$

be the zeros of $P_n(x)$. Find the large n asymptotics for $x_{n,k}$ and $x_{n,n-k}$ for fixed k.
3. If $P_n(x)$ depends on a parameter, what happens to $x_{n,k}$ as the parameter increases (decreases).
4. Find inequalities and bounds for the zeros of $P_n(x)$.
5. Ask the same questions for zeros of special functions.

Let us address first the question of monotonicity of zeros of a transcendental function by turning the problem to monotonicity of eigenvalues of a symmetric eigenvalue problem. Consider the Bessel function

$$J_\nu(x) = \sum_{n=0}^{\infty} \frac{(-1)^n (z/2)^{\nu+2n}}{n!\,\Gamma(n+\nu+1)}. \tag{7.3}$$

It satisfies the differential equation

$$x^2 y'' + xy' + (x^2 - \nu^2)y = 0. \tag{7.4}$$

Replace x by $\sqrt{\lambda}\, x$ and write (7.4) as

$$-\frac{1}{x}\frac{d}{dx}\left(x\frac{dy}{dx}\right) + \frac{\nu^2}{x^2} y = \lambda y. \tag{7.5}$$

For $\nu > 0$ consider the eigenvalue problem consisting of (7.5) and the boundary conditions

$$y(0) = 0, \quad y(1) = 0. \tag{7.6}$$

For $\nu > 0$, the Bessel equation has one unbounded solution ($= Y_\nu(x)$) and $J_\nu(x)$ is the bounded solution. Hence the solution of (7.5)–(7.6) must be $y(x) = cJ_\nu(\sqrt{\lambda}\, x)$ and $\lambda > 0$ is such that $J_\nu(\sqrt{\lambda}) = 0$. The function $J_\nu(x)$ may have a trivial zero at $x = 0$ but for $\nu > -1$, it has only real and simple zeros which are symmetric about $x = 0$. Let

$$0 < j_{\nu,1} < j_{\nu,2} < \ldots,$$

be the positive zeros of $J_\nu(x)$. The question we wish to address is how does $j_{\nu,k}$ change with ν when $\nu > 0$. To address this question let us consider a symmetric operator T_ν depending on a parameter ν and is densely defined on $L^2(wdx, E)$, where

$$(f, g) = \int_E w(x) f(x) \overline{g(x)} dx, \quad w \geq 0 \text{ on } E. \tag{7.7}$$

The symmetry of T_ν means $(T_\nu x, y) = (x, T_\nu y)$ whenever x and y belong to the domain of T_ν. Let $T_\nu y = \lambda(v)y$. We now give a heuristic argument to prove the Hellmann-Feynman theorem:

$$\frac{\partial \lambda(\nu)}{\partial \nu} = \left(\frac{\partial T_\nu}{\partial \nu} y, y\right), \quad \text{when} \quad (y,y) = \|y\|^2 = 1. \tag{7.8}$$

To prove (7.8) differentiate the relationship $\lambda(\nu) = \int_E w(x)[T_\nu y(x)]\overline{y(x)}dx$, to get

$$\begin{aligned}
\frac{\partial \lambda(\nu)}{\partial \nu} &= \int_E w(x)\left(\frac{\partial T_\nu}{\partial \nu} y(x)\right)\overline{y(x)}dx + \int_E w(x)\left(T_\nu \frac{\partial y(x)}{\partial \nu}\right)\overline{y(x)}dx \\
&\quad + \int_E w(x)\,(T_\nu y)\,\frac{\partial}{\partial \nu}\overline{y(x)}dx \\
&= \left(\frac{\partial T_\nu}{\partial \nu} y, y\right) + \left(T_\nu \frac{\partial y}{\partial \nu}, y\right) + \left(T_\nu y, \frac{\partial y}{\partial \nu}\right) \\
&= \left(\frac{\partial T_\nu}{\partial \nu} y, y\right) + \left(\frac{\partial y}{\partial \nu}, T_\nu y\right) + \left(T_\nu y, \frac{\partial y}{\partial \nu}\right) \\
&= \left(\frac{\partial T_\nu}{\partial \nu} y, y\right) + \lambda\left[\frac{\partial}{\partial \nu}(y,y)\right],
\end{aligned}$$

from which (7.8) follows.

A similar argument shows that (7.8) will continue to hold if the inner product (7.7) is replaced by

$$(f,g) = \sum_{n=0}^{\infty} w_n f_n g_n, \tag{7.9}$$

provided that $w_n \geq 0, f = \{f_n\}, g = \{g_n\}$.

Exercise: Find a rigorous justification of (7.8) by defining $\frac{\partial T_\nu}{\partial \nu}$ in a suitable way. The difficulty is in justifying the differentiability of $\lambda(\nu)$ and interchanging differentiating and integration.

The Hellmann-Feynman theorem was introduced in [25], [21]. For recent works see [34], [32]. In particular [32] contains applications to q-Bessel functions and certain classes of orthogonal polynomials which do not satisfy differential equations of Sturm-Liouville type, or Schrödinger type (6.14).

It is important to note that the Hellmann-Feynman theorem uses only the eigenvectors (eigenfunctions) and not their derivatives. The derivatives of the eigenfunctions are usually complicated.

Example 1. Consider the Bessel operator

$$T_\nu y := -\frac{1}{x}\frac{d}{dx}\left(x\frac{dy}{dx}\right) + \frac{\nu^2}{x^2}y, \quad \nu > 0. \tag{7.10}$$

The eigenvalue problem consisting of $T_\nu y = \lambda y$ together with the boundary conditions (7.6) is symmetric with respect to the inner product (7.7) with $E = [0,1], w(x) = x$. Let $\lambda = j^2_{\nu,k}$. Thus (7.8) gives

$$j_{\nu,k}\frac{\partial}{\partial \nu} j_{\nu,k} = \nu \int_0^1 J_\nu^2(j_{\nu,k}x)\frac{dx}{x}. \tag{7.11}$$

The right-hand side of (7.11) is clearly positive, hence $j_{\nu,k}$ increases with ν for $\nu > 0$.

Example 2. The Bessel functions satisfy the three term recurrence relation

$$\frac{2\nu}{z} J_\nu(z) = J_{\nu+1}(z) + J_{\nu-1}(z). \tag{7.12}$$

Let

$$f_0(x,\nu) = 0, \quad f_n(x,\nu) = \sqrt{\nu+n}\, J_{\nu+n}(x), \quad n = 1,2,\ldots, \nu > -1. \tag{7.13}$$

With $\lambda = 1/j_{\nu,k}$, (7.12)–(7.13) give

$$\lambda f_n(j_{\nu,k},\nu) = \frac{1}{2}\frac{f_{n+1}(j_{\nu,k},\nu)}{\sqrt{(\nu+2)_2}} + \frac{1}{2}\frac{f_{n-1}(j_{\nu,k},\nu)}{\sqrt{(\nu+n-1)_2}}, \tag{7.14}$$

$n = 1,2,\ldots,\nu > -1$. Clearly (7.3) and (7.13) imply $\sum_1^\infty f_n^2(j_{\nu,k},\nu) < \infty$, thus $c(f_1, f_2, \ldots)^T$ is an eigenvector of A_∞ with $a_n = 2^{-1}((\nu+n)(\nu+n-1))^{-1/2}$, and $b_n = 0$. Here c is a constant to make the eigenvector have norm $= 1$. Thus (7.8) implies

$$\begin{aligned}\frac{\partial}{\partial\nu}\left(\frac{1}{j_{\nu,k}}\right) = -\frac{c^2}{4}\sum_{n=1}^\infty f_n(j_{\nu,k},\nu)\Bigg[&\frac{(2\nu+2n+1)}{((\nu+n)_2)^{3/2}} f_{n+1}(j_{\nu,k},\nu)\\ &+\frac{(2\nu+2n-1)}{((\nu+n-1)_2)^{3/2}} f_{n-1}(j_{\nu,k},\nu)\Bigg].\end{aligned}$$

The right-hand side is

$$\begin{aligned}&-\frac{c^2}{2}\sum_{n=1}^\infty \frac{f_n(j_{\nu,k},\nu)}{(\nu+n)}\left\{\frac{f_{n+1}(j_{\nu,k},\nu)}{\sqrt{(n+\nu)_2}} + \frac{f_{n-1}(j_{\nu,k},\nu)}{\sqrt{(n+\nu-1)_2}}\right\}\\ &+\frac{c^2}{4}\sum_{n=1}^\infty f_n(j_{\nu,k},\nu)\left\{\frac{f_{n+1}(j_{\nu,k},\nu)}{((n+\nu)_2)^{3/2}} - \frac{f_{n-1}(j_{\nu,k},\nu)}{((n+\nu-1)_2)^{3/2}}\right\}.\end{aligned}$$

The second sum is a telescoping sum whose value is zero. In the first sum apply (7.14) and evaluate c to obtain

$$\frac{\partial j_{\nu,k}}{\partial\nu} = j_{\nu,k}\left[\sum_{n=1}^\infty J^2_{\nu+n}(j_{\nu,k})\right] \Bigg/ \sum_{n=1}^\infty (n+\nu) J^2_{\nu+n}(j_{\nu,k}). \tag{7.15}$$

Theorem 7.2. *We have*

$$\frac{1}{j_{\nu,k}}\frac{\partial j_{\nu,k}}{\partial \nu} \le \frac{1}{(\nu+1)}, \tag{7.16}$$

$$\begin{aligned} j_{\nu,k} &\le \frac{2k}{3}\pi(\nu+1), \quad \nu \ge \frac{1}{2}, \\ j_{\nu,k} &\le (2k-1)\pi(\nu+1), \quad \nu \ge -\frac{1}{2}. \end{aligned} \tag{7.17}$$

Proof. Since $\nu + n \ge \nu + 1$, we see that the right-hand side of (7.15) is majorized by $j_{\nu,k}/(\nu+1)$ and (7.16) follows. Integrate (7.16) between 1/2 and ν and use $j_{1/2,k} = k\pi$ to derive the first formula in (7.17). The second formula in (7.17) follows by integration and using $j_{-1/2,k} = (k-1/2)\pi$. □

Theorem 7.3. *If $\frac{\partial T_\nu}{\partial \nu}$ is positive semi-definite (definite) then the eigenvalues of T_ν are increasing (strictly increasing) functions of ν.*

Proof. Apply the Hellmann-Feynman theorem. □

We conclude the material on the Hellmann-Feynman theorem by giving a proof of a theorem of Loewner [41] using Theorem 7.3. Loewner's theorem asserts that if A and B are $n \times n$ Hermitian matrices and $B - A$ is positive definite, then $\lambda_k(B) > \lambda_k(A)$, where $\lambda_1(A) \ge \lambda_2(A) \ge \cdots \ge \lambda_n(A)$ are the eigenvalues of A, and similarly for B. To prove this introduce a ν-dependent matrices $C(\nu), C(\nu) = \nu B + (1-\nu)A$. Clearly $\frac{\partial C(\nu)}{\partial \nu} = B - A$, and Theorem 7.3 establishes $\frac{\partial \lambda(\nu)}{\partial \nu} > 0$, since $B - A$ is positive definite. The theorem follows because $C(1) = B, C(0) = A$. If the positive definiteness of $B - A$ is replaced by positive semidefiniteness then we conclude that $\lambda_k(B) \ge \lambda_k(A)$. Another proof is in [27].

In the rest of this section we introduce and apply chain sequences to locate intervals containing all zeros of orthogonal polynomials. Chain sequences were first introduced by Marion Wetzel in her doctoral dissertation and she, jointly with Wall, applied them to continued fraction, [61], [62]. It was Ted Chihara, however, who popularized chain sequences among researchers in orthogonal polynomials through his papers, lectures and his influential monograph [16].

Definition 7.1. A sequence $\{c_n\}_1^\infty$ is a chain sequence if there is a parameter sequence $\{g_n\}_0^\infty$ such that $0 \le g_0 < 1, 0 < g_n < 1$ for $n > 0$, and $c_n = g_n(1 - g_{n-1}), n > 0$. If $\{c_n\}_1^N$ is a finite sequence we call it a finite chain sequence. The sequence $\{g_n\}$ is called a parameter sequence for $\{c_n\}$.

It is clear that $g_n = 1/2$ is a parameter sequence for $\{\frac{1}{4}\}$. The parameter sequence however, may not be unique. Indeed $\{\frac{n}{2(n+1)}\}$ is also a parameter sequence for $\{\frac{1}{4}\}$.

Theorem 7.4. (Wall-Wetzel). *The Jacobi matrix A_N is positive definite if and only if* (i) *and* (ii),

(i) $b_n > 0, 0 \le n < N$,
(ii) $\{a_n^2/b_n b_{n-1} : 1 \le n < N\}$ *is a chain sequence,*

hold.

The proof consists of performing row operations to reduce A_N to an upper triangular matrix. The positive definiteness is equivalent to the positivity of the diagonal elements (pivots) in the reduced upper triangular matrix. If A_N is positive definite then (i) holds and g_n can be defined recursively by $g_0 = 0$, then prove that $a_k^2/[b_k b_{k-1}(1 - g_k)] \in (0, 1)$, so we set $g_{k+1} = a_k^2/[b_k b_{k-1}(1 - g_k)]$. The converse can be easily verified.

Exercise. Write down the details in the above proof, or read it in [33]. Another proof is in [16]

Theorem 7.5. *A sequence $\{c_n\}$ is a chain sequence if $0 < c_n \le d_n, n > 0$, and $\{d_n\}$ is a chain sequence.*

Theorem 7.6. *All the eigenvalues of A_N belong to (a, b) if and only if the following two conditions hold:*

(i) $a < b_n < b$
(ii) $a_n^2/[(x - b_{n-1})(x - b_n)], n = 1, \ldots,$ *forms a chain sequence at $x = a$ and at $x = b$.*

Proof. Clearly all the eigenvalues of A_N belong to (a, b) if and only if $A_N - aI$ and $bI - A$ are positive definite. The theorem then follows from Theorem 7.3. □

Example. The Hermite polynomials $\{H_n(x)\}$ satisfy

$$2xH_n(x) = H_{n+1}(x) + 2nH_{n-1}(x).$$

The orthonormal polynomials are $2^{-n/2}H_n(x)/\sqrt{n!}$ and $w(x) = e^{-x^2}/\sqrt{\pi}, x \in \mathbb{R}$. The zeros are symmetric around $x = 0$. Moreover

$b_n = 0, a_n = \sqrt{n/2}$. Let

$$\frac{(N-1)/2}{\Lambda^2} \leq \frac{1}{4}, \quad \Lambda > 0. \tag{7.18}$$

Therefore $\{a_n^2/\Lambda^2\}, n = 1, \ldots, N-1$ is a finite chain sequence, since $\{1/4\}$ is a chain sequence, and Theorem 7.3 implies that all the zeros of $H_N(x)$ lie in $(-\Lambda, \Lambda)$, that is $(-\sqrt{2(N-1)}, \sqrt{2(N-1)})$.

Recall that the Chebyshev polynomials of the second kind $\{U_n(x)\}$,

$$U_n(x) = \sin((n+1)\theta)/\sin\theta, \quad x = \cos\theta. \tag{7.19}$$

They satisfy

$$xU_n(x) = \frac{1}{2}U_{n+1}(x) + \frac{1}{2}U_{n-1}(x). \tag{7.20}$$

The largest and smallest zeros of $U_N(x)$ are $\pm\cos[\pi/(N+1)]$. Therefore

$$\frac{1}{4\cos^2[\pi/(N+1)]}, \quad n = 1, \ldots, N-1,$$

is a chain sequence for every $\epsilon > 0$. After replacing $1/4$ in (7.18) by the above finite chain sequence we see that at the zeros of $H_N(x)$ lie in $[-\Lambda_N, \Lambda_N]$,

$$\Lambda_N = \cos\left(\frac{\pi}{N+1}\right)\sqrt{2(N-1)}\,.$$

Ismail and Li [30] used Theorems 7.3 and 7.4 to prove the following.

Theorem 7.7. *Let $\{C_n\}_1^{N-1}$ be a finite chain sequence and let x_n, y_n be the roots of $(x-b_n)(x-b_{n-1})C_n = a_n^2$, and $x_n \geq y_n$. Set*

$$B = \max\{x_n : 0 < n < N\}, \quad A = \min\{y_n : 0 < n < N\}.$$

Then the zeros of $P_N(x)$ lie in (A, B).

For the use of chain sequences to derive bounds on the zeros of special orthogonal polynomials the reader is advised to consult [33].

Chapter 6 in Szegő contains a treatment of the Sturm comparison theorem and its applications to zeros of orthogonal polynomials, where monotonicity and convexity of zeros of Jacobi polynomials as functions of their parameters can be proved. Moreover bounds and asypmtotics of zeros of Jacobi, Hermite and Laguerre polynomials are derived.

We conclude this section by stating several open problems related to the material presented here. In some of the problems we included comments explaining the source from which the problem originated.

Problem 1. The chain sequences and Theorem 7.3 give information about the largest and smallest zero (eigenvalue) of $P_N(x)$ (the matrix A_N). Find an effective technique to give bounds on the second largest and second smallest zeros (eigenvalues).

Problem 2. The Hellmann-Feynman theorem

$$\frac{\partial \lambda}{\partial \nu} = \left(\frac{\partial T_\nu}{\partial \nu} y, y \right), \quad T_\nu y = \lambda y,$$

implies that λ increases (decreases) with ν if $\frac{\partial T_\nu}{\partial \nu}$ is positive (negative) semidefinite, see Theorem 7.3. The assumptions of semi-definiteness or definiteness of $\frac{\partial T_\nu}{\partial \nu}$ are too stringent because one wants to determine the sign of $\left(\frac{\partial T_\nu}{\partial \nu} y, y\right)$ on the eigenspaces and not necessarily on the whole domain of $\frac{\partial T_\nu}{\partial \nu}$. Formulate precise weaker conditions which imply the monontonicity of the eigenvalues.

Problem 3. Try to determine the large n asymptotics of the largest and smallest zeros of $P_n(x)$ from the knowledge of the three term recurrence relation. It is even more desirable to determine uniform asymptotic expansions for $P_n(x)$. This requires a development of a turning point theory for difference equations. Wang and Wong [60] have already started developing this theory by treating the cases when a_n and b_n have polynomial growth. The cases of exponential growth are interesting, important, and challenging. Two interesting concrete models when the coefficients grow exponentially are

$$\begin{aligned} \text{Model I:} \quad b_n = 0, \quad a_n^2 &= \frac{1}{4}(q^{-n} - 1), \\ \text{Model II:} \quad b_n = 0, \quad a_n^2 &= \frac{1}{4}(q^{-n} - q^n). \end{aligned}$$

Model I has been studied by Ismail and Masson [31] while model II is in Borzov, Damashinski, Kulish [12], and Stanton [55]. The pointwise large n asymptotics are known for model I but not for model II. Since the largest zero of both models is $q^{-n/2}(1 + o(1))$, it is highly desirable to scale x by $xq^{-n/2}$ and develop a large n asymptotic series for $p_n(xq^{-n/2})$ in both models. More importantly a turning point theory for recurrence relations with exponentially growing coefficients will be most interesting.

In Model I, explicit formulas for the orthogonal polynomials and their measures of orthogonality as well as generating functions and Poisson kernels are available in [31]. Very little is known about Model II. Stanton [55] found closed form expressions for the linearization coefficients in

$$p_m(x)p_n(x) = \sum_k c_{m,n,k} p_k(x) \tag{7.21}$$

where p_n are the orthogonal polynomials in Model II.

Acknowledgments. I wish to thank professors Chen and Liu for the hospitality during my visit to Wuhan University and for all their help and for being very accommodating for my special needs. Thanks to Erik Koelink for his detailed comments on the paper. I am very grateful to Richard Askey who suggested many improvments and supplied me with the information on Poisson kernels contained in the important references [53] and [52].

References

1. G. E. Andrews and R. A. Askey, Classical orthogonal polynomials, in "Polynômes Orthogonaux et Applications," C. Brezinski et al. eds., Lecture Notes in Mathematics, **1171**, Springer-Verlag, Berlin, 1985, pp. 36–62.
2. G. E. Andrews, R. A. Askey and R. Roy, *Special Functions,* Cambridge University Press, Cambridge, 1999.
3. R. A. Askey, Orthogonal plynomials and positivity, in "Special Functions and Wave Probagation", edited by D. Ludwig and F. W. J. Olver, Studies in Applied Mathematics **6**,Society for Industrial and Applied Mathematics, Philadelphia, 1970, pp. 64–85.
4. R. A. Askey, *Orthogonal polynomials and Special Functions*, Society for Industrial and Applied Mathematics, Philadelphia, 1975.
5. R. A. Askey and M. E. H. Ismail, Recurrence relations, continued fractions, and orthogonal polynomials, *Mem. Amer. Math. Soc.* Number 300 (1984), 108 pages.
6. F. V. Atkinson, *Discrete and Continuous Boundary Problems*, Academic Press, New York, 1964.
7. W. N. Bailey, Generalized Hypergeometric Series, Cambridge University Press, Cambridge, 1935.
8. E. Bank and M. E. H. Ismail, The attractive Coulomb potential polynomials, *Constructive Approx.* **1** (1985), 103–119.
9. W. Bauldry, Estimates of asymptotic Freud polynomials on the real line, *J. Approx. Theory* **63** (1990) 225–237.
10. C. Berg and M. E. H. Ismail, q-Hermite polynomials and classical orthogonal polynomials, *Canadian J. Math.* **48** (1996), 43–63.
11. S. S. Bonan and D. S. Clark, Estimates of the Hermite and Freud polynomials, *J. Approx. Theory* **63** (1990), 210–224.
12. V. V. Borzov, E. V. Damashinki and P. P. Kulish, Construction of the spectral measures for a deformed oscillator position operator in the case of undetermined moment problem, *Reviews in Math. Phys.* **12**, (2000), 691–710.

13. J. T. Broad, Gauss quadrature generalized by diagonalization of H in finite L^2 bases, *Phys. Rev. A* (3), **18** (1978), 1012–1027.
14. L. Carlitz, On some polynomials of Tricomi, *Boll. Un. Mat. Ital.* (3) **13** (1958), 56–64.
15. Y. Chen and M. E. H. Ismail, Ladder operators and differential equations for orthogonal polynomials, *J. Phys. A* **30** (1997), 7818–7829.
16. T. S. Chihara, *An Introduction to Orthogonal Polynomials*, Gordon and Breach, New York, 1978.
17. P. Deift, *Orthogonal polynomials and random matrices: A Riemann-Hilbert approach, Amer. Math. Soc.*, Providence, 2000.
18. P. Deift and X. Zhou, A steepest descent method for oscillatory Riemann-Hilbert problems. Asymptotics for the MKdV equation, *Ann. of Math.* (2), **137** (1993), 295–368.
19. D. J. Diestler, The discretization of continuous infinite sets of coupled ordinary differential equations: Applications to the collision-induced dissociation of a diatomic molecule by an atom, in "Numerical Integration of Differential Equations and Large Linear Systems," J. Hinze (ed.), *Lecture Notes in Math.* **968**, Springer-Verlag, Berlin, 1982, pp. 40–52.
20. N. M. Ercolani and K. D. T-R McLaughlin, Asymptotics of the partition function for random matrices via Riemann-Hilbert techniques and applications to graphical enumeration, *Int. Math. Res. Not.* (2003) Number 14, 755–820.
21. R. Feynman, Forces in molecules, *Phys. Rev.* **56** (1939), 340–343.
22. G. Gasper and M. Rahman, *Basic Hypergeometric Series*, Cambridge University Press, Cambridge, 1990.
23. E. J. Heller, Theory of J-matrix Green's functions with applications to atomic polarability and phase-shift error bounds, *Phys. Rev.* **12** (1975), 1222–1231.
24. E. J. Heller, W. P. Reinhardt and H. A. Yamani, On an equivalent quadrature calculation of matrix elements of $(z - p^2/2m)$ using an L^2-expansion technique, *J. Comp. Phys.* **13** (1973), 536–549.
25. E. Hellmann, *Einführung in die Quantenchemie*, Deuticke, Vienna 1937.
26. D. Hilbert, Über die discriminante der in endlichen abbrechenden hypergeometrischen reihe, *J. Für die reine and angewandte Matematik* **103** (1885), 337–345.
27. R. A. Horn and C. R. Johnson, *Matrix Analysis*, corrected reprint of the 1985 original, Cambridge University Press, Cambridge, 1991.
28. M. E. H. Ismail, Discriminants and functions of the second kind of orthogonal polynomials, *Results in Math.* **34** (1998), 132–149.
29. M. E. H. Ismail, An electrostatic model for zeros of orthogonal polynomials, *Pac. J. Math.* **193** (2000), 355–369.
30. M. E. H. Ismail and X. Li, Bounds for extreme zeros of orthogonal polynomials, *Proc. Amer. Math. Soc* **115** (1992), 131–140.
31. M. E. H. Ismail and D. R. Masson, q-Hermite polynomials, biorthogonal rational functions, and q-beta integrals, *Trans. Amer. Math. Soc.* **346** (1994), 63–116.
32. M. E. H. Ismail and M. E. Muldoon, On the variation with respect to a parameter of zeros of Bessel functions and q-Bessel functions, *J. Math. Anal.*

Appl. **135** (1988), 187–207.

33. M. E. H. Ismail and M. E. Muldoon, A discrete approach to monotonicity of zeros of orthogonal polynomials, *Trans. Amer. Math. Soc.* **323** (1991), 65–78.
34. M. E. H. Ismail and R. Zhang, On the Hellmann-Feynman theorem and zeros of special functions, *Adv. Appl. Math.* **9** (1988), 439–436.
35. S. Karlin and J. McGregor, The classification of birth and death processes, *Trans. Amer. Math. Soc.* **86** (1957), 366–400.
36. S. Karlin and J. McGregor, The differential equations of birth and death processes and the Stieltjes moment problem, *Trans. Amer. Math. Soc.* **85** (1957), 489–546.
37. S. Karlin and J. McGregor, Many server queueing processes with Poisson input and exponential service times, Pac. J. Math. **8** (1958), 87–118.
38. S. Karlin and J. McGregor, Random walks, *Illinois J. Math.* **3** (1959), 66–81.
39. A. B. J Kuijlaars, Riemann-Hilbert analysis for orthogonal polynomials, in "*Orthogonal Polynomials and Special Functions*", E. Koelink and W. Van Assche, eds., Springer-Verlag, Berlin, 2003, pp. 167–210.
40. P. D. Lax and C. D. Livermore, The zero dispersion limit for the Korteweg-de Vries KdV equation, *Proc. Nat. Acad. Sci. U.S.A.* **76** (1979), 3602–3606.
41. C. Loewner, Über monotone Matrixfunktionen, Math Zeit **38** (1934), 177-216, reprinted in Charles Loewner Collected Papers, L. Bers, ed., Birkhaüser Boston, 1988, pp 65-104.
42. M. L. Mehta, *Random Matrices*, second edition, Academic Press, Boston, 1999.
43. H. L. Mhaskar, Bounds for certain Freud polynomials, *J. Approx. Theory* **63** (1990), 238–254.
44. W. Miller, *Lie Theory and Hypergeometric Functions*, Academic Press, New York, 1968.
45. W. Miller, *Symmetry Groups and Their Applications*, Academic Press, New York, 1974.
46. P. G. Nevai, Orthogonal polynomials, *Mem. Amer. Math. Soc.* Number 213 (1979), 185 pages.
47. P. G. Nevai, Géza Freud, Orthogonal polynomials and Christoffel functions, A case study, *J. Approximation Theory* **48** (1986), 3–167.
48. F. W. J. Olver, *Asymptotics and Special Functions*, Academic Press, New York, 1974.
49. Q. Y. Qiu and R. Wong , Uniform asymptotic formula for orthogonal polynomials with exponential weight, *SIAM J. Math. Anal.* **31** (2000), 992–1029.
50. E. D. Rainville, *Special Functions*, Chelsea, the Bronx, 1971.
51. E. B. Saff and V. Totik, *Logarithmic Potentials with External Field*, Springer-Verlag, New York, 1997.
52. I. O. Saramanov, generalized symmetric gamma correlation, Soviet Math. Dokl. **9** (1968), 547–550.
53. I. O. Saramanov and Z. N. Bratoeva, Probabilistic properties of bilinear expansions of Hermite polynomials, Theor. Probability and Appl. **12** (1967), 470–481.
54. J. Shohat, A differential equation for orthogonal polynomials, *Duke Math. J.*

5 (1939), 401–417.
55. D. W. Stanton, private communication, December 2000.
56. T. J. Stieltjes, Sur les polynômes de Jacobi, *Comptes Rendus* **100** (1885), 442–444.
57. T. J. Stieltjes, Sur les racines de l'équation $X_n = 0$, *Acta Mathematica* **9** (1886), 385–400.
58. G. Szegő, *Orthogonal Polynomials*, 4th edition, Amer. Math. Soc., Providence, 1975.
59. F. G. Tricomi, *Integral Equations*, Interscience, New York, 1957.
60. Z. Wang and R. Wong, Asymptotic expansions for second-order linear difference equations with a turning point, *Numer. Math.* **94** (2003), 147–194.
61. H. S. Wall and M. Wetzel, Quadratic forms and convergence regions for continued fractions, *Duke Math. J.* **11** (1944), 89–102.
62. H. S. Wall and M. Wetzel, Contributions to the analytic theory of J-fractions, *Tran. Amer. Math. Soc.* **55** (1944), 373–392.
63. D. V. Widder, *Transform Theory*, Academic Press, New York, 1970.
64. J. A. Wilson, Orthogonal functions from Gram determinants, *SIAM J. Math. Anal.* **22** (1991), 1147–1155.
65. R. Wong, Five lectures on asymptotics, these proceedings.

GEVREY ASYMPTOTICS AND APPLICATIONS TO HOLOMORPHIC ORDINARY DIFFERENTIAL EQUATIONS

JEAN-PIERRE RAMIS

Lab. E. PICARD (UMR - CNRS 5580) and Inst. Univ. de France
UFR MIG, Université Paul Sabatier de Toulouse
118, route de Narbonne
31062 Toulouse cedex 4, France
E-mail: ramis@picard.ups-tlse.fr

0. Introduction

Our main problem is the quantitative and qualitative study of solutions of *functional equations*. We will mainly consider the case of *one* independent variable, the unknown being a vector variable.

Historically it begins with *algebraic* functional equations. Here are some examples:

$$y'' + y = 0, \quad y' = \frac{1}{x}, \quad xf(x) = f(x+1), \quad f(2x) = f(x)^2, \; \ldots$$

(homogeneous linear ordinary differential equations, non-homogeneous linear ordinary differential equations, homogeneous linear difference equations, q-difference equations, ...). Later mathematicians studied *analytic* functional equations like

$$x'' + (\sin t)x = 0$$

(pendulum equation), and later C^∞, C^m, $\cdots$ equations.

In these notes, we will limit ourselves to the case of *analytic* equations. Our technics are efficient *only* in this case, which is already very difficult and very interesting for the *applications*.

Our main concern will be a **local study**, the study of solutions of analytic functional equations in a neighborhood of a point. This point will be a *regular* point or a *singular* point, the second case being the more difficult but the more interesting. It is extremely important to distinguish between *formal* (or *symbolic*) *solutions* and *actual solutions.*

Formal Solutions. We search (by hand or using a computer algebra system like *Maple*) symbolic solutions, like *formal power series* solutions

$$y = \sum_{n=0}^{+\infty} a_n x^n$$

or more general solutions like

$$y = e^{-\frac{1}{x}} + (\log x)^2 \sum_{n=0}^{+\infty} a_n x^n.$$

Typically a pair (Functional Equation + Initial Condition) gives rise to a *program* generating a sequence of coefficients $(a_0,\ a_1, \cdots, a_n,\ \cdots)$.

Actual Solutions. There are different ways to handle actual solutions. We can use a *theoretical approach* (using in general *Functional Analysis*) or a numerical and graphical approach (with a computer), typically a Runge-Kutta method for Ordinary Differential Equations.

Our aim is to build locally (in a neighborhood of a regular or singular point) a *good dictionary* between *Formal* and *Actual* Solutions:

$$\text{Formal solutions} \longleftrightarrow \text{Actual solutions}$$

The elementary situation. In each case, we will first consider "regular points". The *formal* power series solutions at a regular point will *converge automatically*, and, if f is the *classical sum* of the formal power series solution $\hat{f}$, then f will be an *actual solution.*

Example 0.1. Cauchy theorem for Ordinary Differential Equation (equations or systems)

$$\frac{dY}{dx} = F(x, Y) \quad (F \quad \text{analytic})$$

A nearly elementary situation. Some singular parts of analytic functional equations are "not too bad". In some sense they are similar to regular points: more complicated symbolically but not analytically. At these points the power series expansions appearing in symbolic solutions will automatically *converge.* A very important example of this situation and historically the first one is the *Fuchs theory* for linear ordinary differential equation discovered by the German mathematician L. Fuchs in XIX-th centaury. A linear analytic equation *fuchsian* (or *regular singular*) at the origin of the *complex plane* can be written:

$$(x\frac{d}{dx})^n y + a_{n-1}(x)(x\frac{d}{dx})^{n-1} y + \cdots + a_0(x)y = 0,$$

the $a_0, a_1, \cdots, a_{n-1}$ being holomorphic near zero.

An extremely important example is the hypergeometric Ordinary Differential Equation of Euler and Gauss:

$$E(a,b,c,z) \qquad z(1-z)y'' + [c-(a+b+1)z]y' - aby = 0.$$

It admits singularities on the Riemann sphere $\mathbb{P}^1(\mathbb{C}) = \mathbb{C}\bigcup\{\infty\}$ only at the points 0, 1, ∞, and these singularities are *fuchsian.*

The difficult situation. When we have eliminated regular points (the "generic" case) and regular-singular points, it remains (perhaps) the so-called *irregular singular* points. At such points the power series expansions appearing in the symbolic solutions are "in general" *divergent* (they will converge "exceptionally").

Example 0.2. For the following equation

$$x^2y' + y = 0$$

we have a symbolic (and actual $\cdots$) solution $e^{1/x}$, its coefficient is the power series $\hat{f} \equiv 1$ which of course converges !

Example 0.3. The equation

$$x^2y' + y = x$$

will be called *Euler Equation.* It was introduced by L. Euler during the XVIII-th century when he tried to give a sense to the expansion $\sum\limits_{n=0}^{+\infty}(-1)^n n!$ (and reached some success in this attempt $\cdots$).

Euler differential equation admits a *unique* power series expansion solution (at zero): the Euler series

$$\hat{f}(x) = \sum_{n=0}^{+\infty}(-1)^n n! x^{n+1}.$$

This series is evidently *divergent.* The general symbolic solution of Euler Equation is of course

$$\hat{f}(x) + ce^{1/x} \quad (c \in \mathbb{C}).$$

There are many such examples in special functions theory: cf. the symbolic solutions of Borel, Kummer, Whittaker ordinary differential equations at infinity [13], [34].

This divergence phenomena is puzzling. It is of course a serious problem, but it is very important because it appears in applications, in particular in physics (caustics, energy levels of hydrogen atom $\cdots$). Finally as we will see it is possible to overcome completely the difficulty. Moreover there is a very interesting phenomena: in some case, when one performs *actual computations* with *divergent* series, one gets a lot more precise results (numerically) than with *convergent* series. This was discovered by Stokes[1] for the computation of fringes for optical caustics [24]. The reader will find an extremely striking example in [8].

1. Asymptotic expansions

1.1. *Classical asymptotics (or Poincaré asymptotics)*

Asymptotics were introduced by the French mathematician H. Poincaré. It is worth to go back to his original paper [22] (1882). Here we will start from the introduction of *Mécanique Céleste*, part 2 [23]. It is important to notice that the notion of *classical asymptotics* or *Poincaré asymptotics* in some sense misses the original problems risen by Poincaré.

We start from a power series expansion *formal* $\sum_{n=0}^{+\infty} a_n x^n \in \mathbb{C}[[x]]$, purely (the a_n's are *arbitrary complex* numbers). The problem is to give a reasonable sense to the sum of the series:

$$\sum_{n=0}^{+\infty} a_n x^n = \textit{Some} \text{ actual function}$$

This problem rises from the fact that, in some cases, formal power series solutions are solutions of ordinary differential equation or partial differential equation, but are divergent (cf. Section 0). This happens in particular in applications, like Celestial mechanics (astronomy). In such cases, we can hope that the summation process will give an actual solution of the ordinary differential equation or the partial differential equation. We will see that it is possible to get such processes, but only after a hard work.

In [23], H. Poincaré describes *two* summation processes:

[1] cf. below 1.2

1. Summation "in the geometers sense": it is the *usual sum* of *convergent* power series expansions $\sum_{n=0}^{+\infty} a_n x^n \in \mathbb{C}\{x\}$, when $|a_n| < CA^n$ for some positive constants $C,\ A > 0$, independently of $n \in \mathbb{N}$ (Geometric growth).

2. Summation "in the astronomers sense": it is an *old method*, already used by Euler, Lacroix ..., called the *summation at the smallest term.*

The principle is the following. The following situation appears in many applications. The power series expansion $\sum_{n=0}^{+\infty} a_n x^n$ is *divergent*, but $|a_n x^n|$ decays from $n = 0$ to $n = N$, and growth with n for $n \geq N$.

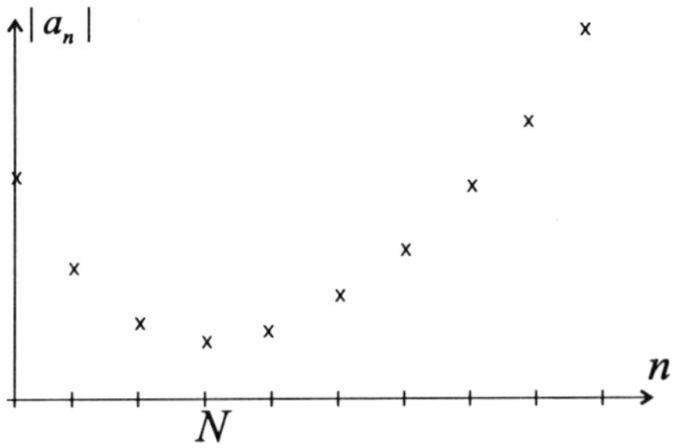

Fig 1.1.1

Be careful here: $N = N(x)$ *depends* on x. Then we can *decide* that the *finite* sum $\sum_{n=0}^{N} a_n x^n \in \mathbb{C}$ *is the sum* of the power series expansion $\sum_{n=0}^{+\infty} a_n x^n$.

A typical example discussed by Poincaré [23] (and related to Euler series, cf 0. Introduction) is the power series $\sum_{n=0}^{+\infty} n! x^n$ with $x = \frac{1}{1000}$. Here $a_n = n!$. We have $I_n = \frac{a_{n+1} x^{n+1}}{a_n x^n}$; $0 < I_n \leq 1$ for $n + 1 \leq 1000$, $I_n \geq 1$ for $n + 1 \geq 1000$. The smallest term corresponds to $n = N = 1000 = \frac{1}{x}$ (For general values, if $x > 0$, we will choose the entire part $[\frac{1}{x}]$ of $\frac{1}{x}$). Using Stokes description of similar power series, we can say that the series *first converges* and *after diverges*.

If now we consider the power series $\sum_{n=0}^{+\infty} \frac{x^n}{n!}$ with $x = 1000$, we see that $|a_n x^n| = \frac{x^n}{n!}$ first increases with $n = N = 1000$ and after decreases. This series is convergent in the usual (geometers) sense. In Stokes style we can say that it is *firstly divergent* and *after convergent.*

H. Poincaré stresses upon the fact that the summation process **2** (astronomers) is in many *practical* situations *numerically very* accurate: the power series expansion $\sum_{n=0}^{+\infty} a_n x^n$ *represents symbolically* an *actual function* f and the *sum* $\sum_{n=0}^{+\infty} a_n x^n$ (at the smallest term) gives a *very precise* numerical value of $f(x)$. It is quite surprising that this method based upon *divergent expansions* and unfounded theoretically (up to now, cf. below) is more efficient for numerical computations than the use of *convergent* power series expansion. (cf. the striking example of computation of fringes near a caustic by Stokes [24].) We will explain below why in the classical example of the integral (cf. [21])

$$f(x) = \int_0^{+\infty} \frac{e^{-t/x}}{1+t} dt.$$

We return now to Poincaré ideas. He was convinced that a formal power series expansion which is a (formal) solution of a problem (from mathematics, physics, astronomy, ... stated in *analytical* terms) *must represent* an *actual function solution* of the problem (*holomorphic* on some convenient domain). Apparently this idea has been shared by many scientists before Poincaré: Euler, Cauchy, Briot and Bouquet, Stokes ..., but some important mathematicians in Poincaré times were thinking the contrary (like E. Picard). This conviction of Poincaré led him to his fundamental definition of asymptotics[2].

Let $\sum_{n=0}^{+\infty} a_n x^n \in \mathbb{C}[[x]]$ be an arbitrary power series expansion. Let V be an *open sector* in $\mathbb{C}^* = \mathbb{C} - \{0\}$, with its vertex at the origin $0 \in \mathbb{C}$:

$$V = \{x \in \mathbb{C}^* / |x| < R, \arg x \in (\alpha, \beta)\}.$$

(More generally we will consider sector on the Riemann surface of the logarithm, that is why we will allow $|\beta - \alpha| > 2\pi$.) We will denote by $O(V)$ the $\mathbb{C}$−algebra of *holomorphic* scalar functions on V.

[2]We insist on the fact that with this definition Poincaré *missed completely* the subtle phenomenon of the summation at the smallest term.

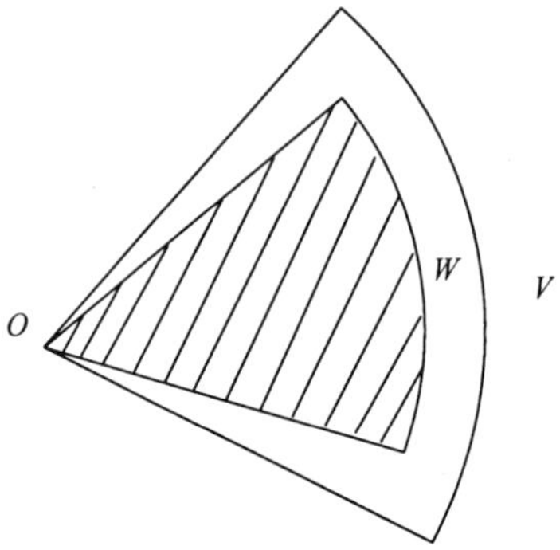

Fig 1.1.2

Definition 1.1. Let $f \in O(V)$ (holomorphic on the open sector V). Let $\hat{f} \in \sum_{n=0}^{+\infty} a_n x^n \in \mathbb{C}[[x]]$. We will say that f is asymptotic to $\hat{f}$ on V (in the classical or Poincaré's sense) if, for every strict[3] subsector $W \prec V$, there exists some positive constants $M_{W,n} (n \in \mathbb{N})$ such that, for every $x \in W$, the following estimates hold:

$$|x|^{-n} \left| f(x) - \sum_{p=0}^{n-1} a_p x^p \right| \leq M_{W,n}$$

We will denote $f \in \mathcal{A}(V)$ and $f \sim \hat{f}$ on V. The set $\mathcal{A}(V)$ of functions in $O(V)$ admitting an asymptotic expansion is clearly a sub$-\mathbb{C}-$algebra of $O(V)$. Moreover, if $f \in \mathcal{A}(V)$ admits $\hat{f} \in \mathbb{C}[[x]]$ as an asymptotic expansion, then its derivative f' admits $\hat{f}'$ as an asymptotic expansion. For elementary properties of asymptotic expansions, we refer the reader to W.Wasow's book [33].

The asymptotic expansion, when it exists is clearly *unique*. The Taylor map $J : \mathcal{A}(v) \to \mathbb{C}[[x]]$, $f \mapsto \hat{f}$ is an homomorphism of differential $\mathbb{C}$-algebras. The basic idea of H. Poincaré was that if $f \in \mathcal{A}(v)$ admits $\hat{f} \in \mathbb{C}[[x]]$ as an asymptotic expansion, then in some sense, we can consider f as a sum of $\hat{f}$. Unfortunately such a sum is not unique. The non-uniqueness corresponds to the "Error Space" $\ker J = \mathcal{A}^{<0}(V)$, which is the differential ideal of $\mathcal{A}(V)$ of the infinitely flat holomorphic functions on V.

Example 1.1. $V = \{\mathrm{Re}\, x > 0\}$, $f(x) = \exp^{-\frac{1}{x}}$, $f \sim 0$.

[3] A subsector $W \subset V$ is *strict* if $\overline{W} - \{0\} \subset V$ ($\overline{W}$ being the closure of W in $\mathbb{C}$).

We insist on the derivation property. It works because we deal with holomorphic functions and because in the definition of asymptotics we asked uniqueness only on strict subsectors.

The proof follows immediately from the following lemma.

Lemma 1.1. *Let $g \in \mathcal{O}(V)$. If, for every strict subsectors $W \prec V$, we have estimates $|g(x)| < C_W |x|^{n+1}$, then we have similar estimates on the derivative: $|g'(x)| < C'_W |x|^n$.*

Idea of the proof. Let $W' \prec W'' \prec V$. We denote by $\overline{D}(x, d)$ the closed disc (in $\mathbb{C}$) with center x and radius $d > 0$. There exists $\delta > 0$ such that, for every $x \in W'$, we have $\overline{D}(x, \delta|x|) \subset W''$. Then we can write a Cauchy integral

$$g'(x) = \frac{1}{2\pi i} \int_{\partial \overline{D}} \frac{g(\xi)}{\{x - \xi\}^2} d\xi$$

and our lemma follows.

It is easy to prove that $f \in \mathcal{A}(V)$ if and only if, for any $W \prec V$, we have

$$\lim_{\substack{x \to 0 \\ x \in W}} f^{(n)}(x)/n! = a_n.$$

Then $f \sim \hat{f} = \sum_{n=0}^{+\infty} a_n x^n$.

We have seen that $\ker J = \mathcal{A}^{<0}(V)$ is an obstruction for the uniqueness of our "summation". What about the "existence", that is the surjectivity of

$$J : \mathcal{A}(V) \to \mathbb{C}[[x]].$$

Is it possible to "sum" any power series expansion $\hat{f} \in \mathbb{C}[[x]]$ using asymptotics? The answer is YES.

Theorem 1.1. *(Borel - Ritt Theorem) The Taylor map $J : \mathcal{A}(V) \to \mathbb{C}[[x]]$ is surjective. (It extends to any "ramified sector" V.)*

Proof. cf. [33] (Theorem 9.3, p. 43).

If $\hat{f} \in \mathbb{R}[[x]]$, we can choose $f \in \mathcal{A}(V)$ real on the real axis.

We can summarize the situation by the short exact sequence:

$$0 \to \mathcal{A}^{<0}(V) \to \mathcal{A}(V) \xrightarrow{J} \mathbb{C}[[x]] \to 0$$

We can dream of a "good summation" theory which will allow us to invert the Taylor map J:

$$SUM : \mathbb{C}[[x]] \to \mathcal{A}(V)$$

Unfortunately the "error space" $\mathcal{A}^{<0}(V)$ is an "obstruction" to the existence of such a map. Therefore we are naturally led to the following problems, which must be related to domains of application.

Question 1.1. Reduce the "error space" to a space as small as possible of "unavoidable small corrections".

Question 1.2. Reduce the "error space" to the ZERO SPACE and get a theory of EXACT ASYMPTOTICS.

We will essentially solve these two problems. The first step will be from Poincaré Asymptotics to Gevrey Asymptotics (Question 1.1)). The second step will be from Gevrey Asymptotics to Exact Asymptotics (Question 1.2).

We will see that Gevrey Asymptotics is quite universal in the applications[4]. The domain of application of exact asymptotics is extremely large (Singular analytic ordinary differential equations and partial differential equations, normal forms, algebraic quantum oscillators ...), however such asymptotics are not the rule for interesting domains like singular perturbations or averaging problems. In such domains these exist in general non trivial unavoidable exponentially flat small corrections.

In these notes we will present mainly our personal viewpoint about matters like Gevrey Asymptotics and Exact Asymptotics, but it is important to notice that many theories have been developed during the last 25 years by various schools:

- Resurgence Theory (J. Ecalle and after him F. Pham and coworkers like E. Delabaere, D. Sauzin ...)
- Exponential asymptotics (M. Kruskal, O. Costin, ...)
- Hyperasymptotics (M. Berry, Howles,...)

All these approaches are strongly interrelated and are in fact different aspects of a *same theory* which is strongly emerging, with a lot of beautiful applications. Here we will limit ourselves to an *elementary* presentation.

---

[4]There are some exceptions like q-difference equations [24] but in such cases, we can use adapted generalizations [35].

When Exact Asymptotics apply, a given formal power series expansion can have *several* exact sums. (It is in fact the rule if the expansion is *divergent*: it is in fact the source of the divergence phenomena.) These sums (as it was already noticed by Stokes for Airy function) are like the "branches" of an algebraic function like $\sqrt{z}$. This is related to functional Galois theories generalizing classical Galois theory [24]: Galois groups correspond to some "allowed" permutations of branches.

Our next step will be Gevrey Asymptotics.

1.2. *Gevrey asymptotics*

For this paragraph the reader can use the following references: [24], [14]. We begin with some **observations** ([1], [2]).

1. Starting from a "genuine problem" (coming from Maths, Physics, ...), we want to compute numerically a value $f(x)$ of a function f related to this problem, knowing the function f formally:

$$f \sim \hat{f} = \sum a_n x^n$$

(This knowledge comes in general from some theoretical considerations.) After this computation, we will compare the result with some measurements of some "exact value". Our first observation is that if we compute $f(x)$ using a summation at the smallest term, then the result agree with measurements (or exact value) up to a "very small error", more precisely an exponentially small error[5].

A typical example is Quantum Electrodynamics [8] (Feymann describes a physical example[6] where such a error is of the order of width of a hair compared to the distance from New-York to Los Angeles)

2. For almost all the cases, power series issued from "practical problems" satisfy Gevrey estimates: $\hat{f}(x) = \sum_{n=0}^{+\infty} a_n x^n$, with $|a_n| < CA^n(n!)^s$ (for some positive constants C, A, s independent of n).

It is easy to find examples. We ask the reader to open any "bible" of special functions, like [13], haphazardly. Then if an explicit power series expansion appears, it will satisfy such an estimate.

---

[5] $\leq e^{-\frac{a}{|x|^k}}$; $a > 0, k > 0$ being independent of x; k is the order.

[6] In this example, Feymann do not use a summation at the smallest term (nobody knows it ...) but a summation truncated at some terms. In general we can choose imprecisely the "small term" with good results.

3. In "practical problems", if knowing a power series expansion $\hat{f}$, there are some (theoretical or numerical) ambiguities on the corresponding actual function, then these ambiguities have an exponential decay (of some order $k > 0$), when $x \to 0$, $|\varphi(x)| \leq K e^{-\frac{a}{|x|^k}}$ (K, $a > 0$ independent of x).

Considering these three observations, it is natural to try to replace Poincaré Asymptotics by a new asymptotic theory explaining 1, 2, 3. The good news is that such a theory exists: it is Gevrey Asymptotics. In fact "more or less":

$$1 \Longleftrightarrow 2 \Longleftrightarrow 3 \Longleftrightarrow \text{ we are in the case of Gevrey Asymptotics.}$$

Gevrey Asymptotics were discovered by G. Watson at the beginning of the XXth century. But unfortunately it meet more or less no success and was forgotten. I rediscovered it (and gave it its name ..., in relation with M. Gevrey work on partial differential equation) at the end of the 70's, and developed it systematically in relation with the applications [24].

G. Watson's work was rejected because mathematicians was thinking that its field of applications was extremely narrow. (G. Watson applied his theory only to some special functions: Γ-function, Bessel functions ...). In fact, as I will explain later, its field of application is today extremely large, containing whole families of analytic functional equations (ordinary differential equations: without restrictions, singular perturbations of ordinary differential equations, some problems of partial differential equations ...). If G. Watson's work was forgotten for a long time, it is worth to notice that there is however a "red thread" going from G. Watson to S. Mandelbrojt (though some works of R. Nevanlinna, Carleman and Denjoy).

Gevrey asymptotics is an essential step towards exact asymptotics. Moreover it is exactly the good asymptotics for singular perturbations and it allows us to understand phenomena like delay in bifurcations, ducks phenomena, Ackerberg-O'Malley resonance, ... or perturbations of Hamiltonian systems (adiabatic invariants, Nekhorosev estimates, ...)

I will begin with my favorite example (Euler series):

$$\hat{f}(x) = \sum_{n=0}^{+\infty} (-1)^n n! x^{n+1} = ?$$

It was introduced by L. Euler in his paper "De seriebus divergentibus". His aim was to "compute" numerically the infinite sum

$$\hat{f}(1) = \sum_{n=0}^{+\infty} (-1)^n n!$$

In relation with this problem, Euler considered the linear differential equation $x^2 y' + y = x$. Solving it by the "variation of constants", we get:

$$y = e^{\frac{1}{x}} \int e^{\frac{1}{\zeta}} \frac{d\zeta}{\zeta}$$

and (setting $\frac{1}{\zeta} = \frac{1+t}{x}$)

$$y = \int_0^{+\infty} \frac{e^{-\frac{t}{x}}}{1+t} dt + Ce^{\frac{1}{x}}.$$

We remark that the integral

$$f(x) = \int_0^{+\infty} \frac{e^{-\frac{t}{x}}}{1+t} dt$$

is the unique solution of our differential equation which is bounded on $\mathbb{R}^+$. Euler concluded from this remark that we can consider $f(x)$ as the sum of the power series expansion $\hat{f}(x)$ (which is a formal solution of the equation). Finally he got $\sum_{n \geq 0} (-1)^n n! = \hat{f}(1)$[7].

We will explain why the idea of Euler is reasonable ($x > 0$). We set

$$f_n(x) = x - 1!x^2 + \ldots + (-1)^{n-1}(n-1)!x^n \in \mathbb{C}[x]$$

and

$$R_n(x) = (-1)^n \int_0^{+\infty} \frac{t^n e^{-\frac{t}{x}}}{1+t} dt.$$

We have $f(x) = f_n(x) + R_n(x)$ for every $n \in \mathbb{N}$, and

$$|R_n(x)| < \int_0^{+\infty} t^n e^{-t/x} dt = n!x^{n+1}.$$

The sign of R_n is the same than the sign of $(-1)^n n! x^{n+1}$ which is the first term we omit when we stop the summation after n terms. Therefore we have

$$f_{2p}(x) < f(x) < f_{2p+1}(x).$$

[7]Moreover he described four other methods of summation numerically coherent with this one.

Let $x > 0$ be fixed. Then $f_{2p+1}(x) - f_{2p}(x) = (2p)!x^{2p+1}$ first decrease, then growth when p growth. The smallest difference gives the best approximation for $f(x)$. This corresponds to the summation at the smallest term. We have the following estimates:

$$\frac{n!x^{n+1}}{(n-1)!x^n} = nx$$

and the smallest term is reached at $N \approx \frac{1}{x}$, $N = [\frac{1}{x}]$. From Stirling formula:

$$N! \sim \sqrt{2\pi N} e^{-N} N^N, \qquad N!x^N \sim N!N^{-N} \sim \sqrt{\frac{2\pi}{x}} e^{-\frac{1}{x}}.$$

Therefore, when $x \xrightarrow{>} 0$, the equality of our approximation is exponentially good. We can understand with this example why divergent series are better than convergent series for numerical applications.

The case of $\log \Gamma$ is similar (A. Cauchy). We will discuss a striking example due to Stokes.

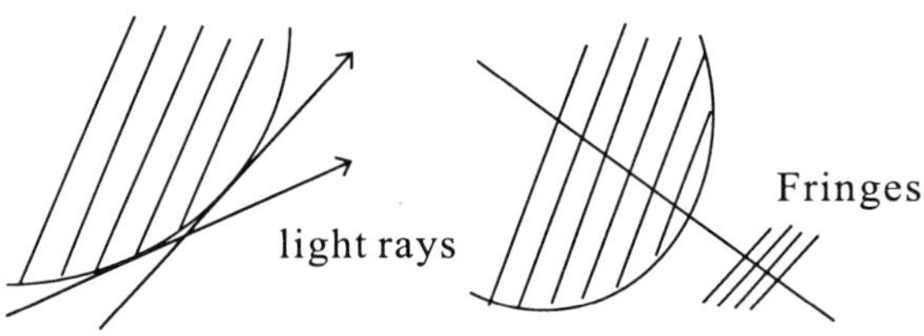

Fig 1.2.1

We start from a caustic in classical optics. The caustic is wrapping the light rays and separating lighted zone and darkness zone. If we consider caustics in wave optics, then we can observe interference fringes. The problem considered by Stokes was to compute the distance between the fringes using theoretical ways and to compare with experiments. (Physicists measured 30 fringes with an accuracy of 4 digits.)

The intensity of the light along a small piece of line transversal to the caustic has been computed by Airy. With a good choice of unities it is $(Ai(z))^2$, where $Ai(z)^2$ is Airy function (z being a parameter transversally to the caustic). We have:

$$Ai(z) = \frac{1}{\pi} \int_0^{+\infty} \cos(\frac{1}{3}t^3 + 3t)dt$$

and this function is a solution of the Airy's equation

$$y'' - zy = 0.$$

Airy function admits a convergent asymptotic expansion at the origin $z = 0$:

$$Ai(z) = 3^{-\frac{2}{3}} \sum_{n=0}^{+\infty} \frac{z^{3n}}{9^n n! \Gamma(n+\frac{2}{3})} - 3^{-\frac{4}{3}} \sum_{n=0}^{+\infty} \frac{z^{3n+1}}{3^n n! \Gamma(n+\frac{4}{3})}$$

Airy tried to compute the fringes using this convergent expansion. He could only get one fringe (with 4 digits).

Conversely to Airy, Stokes used expansions at infinity ($z = \infty$). Using stationary phases method, we get the expansion

$$Ai(z) \sim \frac{1}{\pi^{\frac{3}{2}}} z^{-\frac{1}{4}} e^{-\frac{2}{3} z^{\frac{3}{2}}} 3^{-\frac{2}{3}} \sum_{n=0}^{+\infty} \Gamma(n+\frac{5}{6}) \Gamma(n+\frac{1}{6}) (\frac{3}{4})^n (-z)^{-\frac{3n}{2}}.$$

The power series appearing in this expansion is clearly divergent. However, using this expansion and summation at the smallest term (in fact an improved version), Stokes was able to compute all the 30 fringes with an accuracy of 4 digits, compared with physical experiments. (He failed only for the first fringe: he got only 3 digits...). We can see on this example the apparition of Gevrey estimates and exponentially small functions.

Returning back to this example several years later, Stokes discovered what is called today Stokes Phenomenon (March 17, 1857, three a.m.; cf. [24]).

We will give now some precise definitions.

Definition 1.2. Let V be an open sector of $\mathbb{C}$ (with its vertex at the origin). Let $f \in \mathcal{O}(V)$. Let $\hat{f} \in \mathbb{C}[[x]]$ and $s \in \mathbb{R}$. We will say that f is asymptotic to $\hat{f}$ in Gevrey order s sense, if, for every strict subsector $W \prec V$, there exists positive constants $C_W > 0$ and $A_W > 0$ such that, for every $n \in \mathbb{N}$, $x \in W$:

$$|x|^{-n} |f(x) - \sum_{p=0}^{n-1} a_p x^p| \leq C_W A_W^n (n!)^{\varsigma}.$$

We will denote $\hat{f} \in \mathbb{C}[[x]]_s$ and $f \in \mathcal{A}_s(V)$.

Remark 1.1. • If $s = 0$, then $\hat{f} \in \mathbb{C}x$: $\hat{f}$ is convergent and conversely.

• If $s < 0$, we set $s = -\frac{1}{k} (k > 0)$. Then $\hat{f} \in \mathbb{C}[[x]]_s$ if and only if f is an entire function with an exponential growth order $\leq k$ at ∞.

By definition, we will set $\mathbb{C}[[x]]_\infty = \mathbb{C}[[x]]$.

Definition 1.3. If $\hat{f} = \sum\limits_{n\geq 0} a_n x^n$ satisfies

$$|a_n| < CA^n(n!)^s$$

for some positive constants C, $A > 0$, then we will say that $\hat{f}$ is Gevrey of order s and we will denote $\hat{f} \in \mathbb{C}[[x]]_s$.

It is ideal that the image of $\mathcal{A}_s(V)$ by J is contained in $\mathbb{C}[[x]]_s$. We will consider the map

$$\mathcal{A}_s(V) \xrightarrow{J} \mathbb{C}[[x]]_s.$$

We have the following result [24], [1], [2], [31].

Proposition 1.1. *(i)* $\mathbb{C}[[x]]_s$ *is a sub-differential algebra of* $\mathbb{C}[[x]]$.
(ii) $\mathcal{A}_s(V)$ *is a sub-differential algebra of* $\mathcal{A}(V)$.

It is natural to study the subjectivity of the map $J : \mathcal{A}_s(V) \to \mathbb{C}[[x]]_s$. The answer depends on the opening of the sector V. It is yes for narrow sector (*i.e.* if opening of $V < \frac{\pi}{k}$; $k = \frac{1}{s}$) and no for large sectors (opening of $V > \frac{\pi}{k}$).

Theorem 1.2. *(Borel - Ritt Theorem, due to Ramis) If* V *is a narrow sector, that is the opening of* V *is strictly smallest than* $\frac{\pi}{k} = \pi s$*, then the map* $J : \mathcal{A}_s(V) \to \mathbb{C}[[x]]_s$ *is surjective.*

We will give an idea of a proof (following an idea of B. Malgrange). We will limit ourselves to the case $k = s = 1$. We will use an incomplete Laplace transform.

Let $\hat{f} = \sum\limits_{n=1}^{+\infty} a_n x^n$. By definition its formal Borel transform is the power series

$$\hat{B}\hat{f} = \sum_{n=0}^{+\infty} \frac{a_{n+1}}{(n+1)!} t^n.$$

Then $\hat{f} \in \mathbb{C}[[x]]_1 \longleftrightarrow \hat{B}\hat{f} \in \mathbb{C}\{t\}$ (*i.e.* convergent).

Formally, the Laplace transform of a function $t \mapsto \varphi(t)$ is the function

$$x \mapsto \int_0^{+\infty} \varphi(t) e^{-\frac{t}{x}} dt.$$

Here, we denote by $\varphi(t)$ the sum of $\hat{B}\hat{f} = \hat{\varphi}(t)$. Unfortunately, in general we cannot extend φ analytically along $\mathbb{R}^+$. Moreover, even if this extension exists, in general the Laplace integral does not converge ... But, if $R > 0$ is the radius of convergence of $\hat{\varphi}$ (in the t-plane), then the integral

$$\mathcal{L}_r\varphi(x) = \int_0^r \varphi(t)e^{-\frac{t}{x}}\,dt$$

will exist for $0 < r < R$, $x \in \mathbb{C}$, and will define a function of x asymptotic to $\hat{f}$ on V in Gevrey 1 sense.

For other values of $s = \frac{1}{k} > 0$, we can adapt the proof. We replace the incomplete Laplace transform by an incomplete k-Laplace transform. Formally we set

$$\rho_k f(x) = f(x^{\frac{1}{k}}), \qquad \mathcal{L}_k = \rho_k^{-1}\mathcal{L}\rho_k, \qquad \hat{B}_k = \rho_k^{-1}\hat{B}\rho_k.$$

We denote $\mathcal{A}^{\leq -k}(V)$ the kernel of the map $J : \mathcal{A}_s(V) \to \mathbb{C}[[x]]_s$.

Heuristics. With narrow sectors, Gevrey asymptotics is very similar to Poincare's asymptotics. It is in some sense a smooth theory (typically, we will have existence theorem, but non unicity). With large sectors, Gevrey asymptotics appears as a completely new asymptotic theory. We get exact asymptotics. It is a rigid theory (typically existence theorem are exceptional, but we have unicity like for elementary Cauchy theory of holomorphic functions).

For large sectors, the following lemma is crucial.

Lemma 1.2. *(Watson Lemma) Let $k > 0$. If the opening of the sector $V > \frac{\pi}{k}$ and if $f \in \mathcal{A}^{\leq -\frac{1}{k}}(V)$, then f is identically zero on V.*

This lemma is a simple consequence of Phragmen-Lindelöf maximum principle. It follows from Watson lemma that, for large sectors, the map $J : \mathcal{A}_s(V) \to \mathbb{C}[[x]]_s$ is injective. As we will see later it is no longer surjective:

$$0 \to \mathcal{A}_s(V) \xrightarrow{J} \mathbb{C}[[x]]_s.$$

The power series belonging to the image of J are very special (this is related to the notion of k-summability, for $k = 1/s$, cf. next paragraph).

For narrow sectors, the situation is in some sense the opposite. We have a short exact sequence ($s = 1/k$):

$$0 \to \mathcal{A}^{\leq -k}(V) \to \mathcal{A}_s(V) \xrightarrow{J} \mathbb{C}[[x]]_s.$$

Definition 1.4. Let $k > 0$. We will say that $f \in \mathcal{O}(V)$ admits an exponential decay of order k on V (when $x \to 0$), if, for every strict subsector $W \prec V$, there exist positive constants $K_W > 0$ and $a_W > 0$ such that

$$|f(x)| \leq K_W e^{-a_W/|x|^k}$$

for $x \in W$.

Proposition 1.2. *The following conditions are equivalent:*
(i)$f \in \mathcal{A}^{\leq -k}(V)$
(ii)f *admits an exponential decay of order* k *on* V.

1.3. k-*summability*

Let's begin by the following definition.

Definition 1.5. (Ramis) Let $k > 0$. Let $\hat{f} \in \mathbb{C}[[x]]$ such that there exists an open sector V whose opening $> \frac{\pi}{k}$ and a holomorphic function $f \in \mathcal{A}_{\frac{1}{k}}(V)$ such that f is asymptotic to $\hat{f}$ on V in Gevrey $\frac{1}{k}$ sense. We will say that $\hat{f}$ is k-summable in the direction d (d being the bisecting line of V).

Then we will say that f is the k-sum of $\hat{f}$ in the direction d (d being the bisecting line of the sector V).

In this case we remark that f is unique (up to opening and the radius of the sector V, d remaining fixed). Moreover, we must have $\hat{f} \in \mathbb{C}[[x]]_{\frac{1}{k}}$.

Notation. $\hat{f} \in \mathbb{C}\{x\}_{\frac{1}{k},d}$.

We will see that the inclusion $\mathbb{C}\{x\}_{\frac{1}{k},d} \subset \mathbb{C}[[x]]_{\frac{1}{k}}$ is strict. Of course, if $\hat{f}$ is convergent, $\hat{f} \in \mathbb{C}\{x\}$, then $\hat{f}$ is k-summable in the direction d (any $k > 0$ and any d) and the classical sum and k-sum coincide on their respective domain of definition : $\mathbb{C}\{x\} \subset \mathbb{C}\{x\}_{\frac{1}{k},d}$.

Remark 1.2. This summability definition sounds quite abstract. In fact it is extremely useful:

• With it, it is easy to prove the elementary properties of summability (sum, products, differentiation ...).

• It is easy to prove in many cases issued from dynamical systems problems that some formal power series solution $\hat{f}$ is k-summable, looking at the underlying "geometry" of the system. (Ramis-Sibuya theorem is an efficient tool ; cf. below).

• What is apparently bad with our definition is that it is seems a priori impossible to compute the sum. With other definitions of summability (by

E. Borel and his followers, like Leroy), conversely we have explicit formulae for the sums but summability seems extremely difficult to check[8]

The good news is that the two definitions are in fact equivalent. This is very convenient because we can choose one or the other following the applications we have in mind.

Theorem 1.3. *(Ramis) For $\hat{f} \in \mathbb{C}[[x]]$, k-summability in the direction d is equivalent to Borel-Leroy-Nevanlinna summability in the direction d.*

Proof. cf. [14], [1].

We recall the principle of Borel-Leroy-Nevanlinna summation. We will limit ourselves to the case of $k = 1$ (Borel[9]) It is easy to derive the general case (Leroy, R. Nevanlinna, using an operator ρ_k). We choose also $d = \mathbb{R}^+$ for simplicity.

The principle is the following. We start from a power series expansion: $\hat{f} = \sum_{n=1}^{+\infty} a_n x^n$, we associate to $\hat{f}$ a new power series expansion (in the t variable):

$$\hat{\varphi} = \sum_{n=0}^{+\infty} \frac{a_{n+1}}{(n+1)!} t^n.$$

We suppose that:

a) $\hat{\varphi}$ is convergent ($\longleftrightarrow \hat{f} \in \mathbb{C}[[x]]_1$);

b) the sum φ of $\hat{\varphi}$ in an open disc centered at $t = 0$ can be analytically extended in an open sectorial neighborhood of $\mathbb{R}^+$;

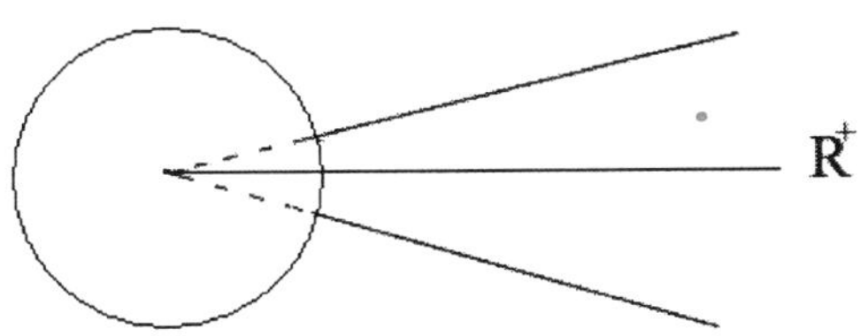

Fig 1.3.1

[8]There was a strong prejudice against Borel-summation due to this fact (Mittag-Leffler, 1900)

[9]In fact Borel summation is a little more general than 1-summation.

c) the integral

$$f(x) = \mathcal{L}\varphi(x) = \int_0^{+\infty} \varphi(t)e^{-\frac{t}{x}}\, dt$$

converges for $x \in V$ (the sector V being bisected by $\mathbb{R}^+$ with opening $> \pi = \frac{\pi}{k}$). Then, by definition $f(x) = \mathcal{L}\varphi(x)$ is the Borel-sum of $\hat{f}$ in the direction $\mathbb{R}^+$.

In the next paragraph, we will recall basic facts about Laplace transform.

Using the first (abstract) definition of k-summability, it is evident to check the following.

Proposition 1.3. *Let $k > 0$. Let d a fixed direction. The set $\mathbb{C}\{x\}_{\frac{1}{k},d}$ is a $\mathbb{C}$-sub-differential algebra of $\mathbb{C}[[x]]_{\frac{1}{k}}$.*

Definition 1.6. If $\hat{f} \in \mathbb{C}[[x]]$ is k-summable in all the directions but perhaps a finite set, we will say that it is k-summable and will denote $\hat{f} \in \mathbb{C}\{x\}_{\frac{1}{k}}$.

Proposition 1.4. *(i) If $\hat{f}$ is k-summable in all the directions, then $\hat{f}$ is convergent (and conversely).*

(ii) $\mathbb{C}\{x\}_{\frac{1}{k}}$ is a sub-differential algebra of $\mathbb{C}[[x]]_{\frac{1}{k}}$.

Remark 1.3. The inclusion $\mathbb{C}\{x\}_{1;d} \subset \mathbb{C}[[x]]_1$ is strict.

Indeed, consider $\hat{f} \in \mathbb{C}[[x]]_1$ such that $\hat{B}\hat{f} = \hat{\varphi}$ admits a radius of convergence 1 with the circle of center O and radius one as a natural cut. Then it is not possible to extend φ analytically along d and $\hat{f} \notin \mathbb{C}\{x\}_{1,d}$. A similar argument can be used to prove that the inclusions $\mathbb{C}\{x\}_{\frac{1}{k},d} \subset \mathbb{C}[[x]]_{\frac{1}{k}}$ are all strict.

1.4. *Laplace transform*

Fourier transform. We will use *real* conjugated variables t and ξ. Let $\varphi \in L^1(\mathbb{R})$ (t-line). We set

$$\mathcal{F}\varphi(\xi) = \hat{\varphi}(\xi) = \int_{-\infty}^{+\infty} \varphi(t)e^{-it\xi}\, dt.$$

An important result is the *inversion formula*:

$$\varphi(t) = \frac{1}{2\pi}\int_0^{+\infty} \hat{\varphi}(\xi)e^{it\xi}\, d\xi.$$

A *sufficient* hypothesis is that φ *and* $\hat{\varphi} \in L^1(\mathbb{R})$. It is sufficient to suppose that $\varphi \in \mathcal{C}^2(\mathbb{R})$ and that $\varphi,\ \varphi',\ \varphi'' \in L^1(\mathbb{R})$ (then $\hat{\varphi} \in O(1/\xi^2)$).

Laplace transform. Let $\varphi \in \mathcal{C}^2([0,+\infty))$. We will suppose that

$$|\varphi(t)| \leq Ae^{Bt} \qquad \text{on} \quad \mathbb{R}^+ \tag{1.4.1}$$

for some positive $A,\ B > 0$. Then the integral

$$\mathcal{L}\varphi(z) = \int_0^{+\infty} \varphi(t)e^{-tz}dt (= \mathcal{F}[\varphi(t)e^{-at}I_{\mathbb{R}^+}](z))$$

is convergent for $z \in \mathbb{C}$, $\operatorname{Re} z > B$. Here, we set $z = a + i\xi$ and we denote by $I_{\mathbb{R}^+}$ the characteristic function of $\mathbb{R}^+ \subset \mathbb{R}$.

We suppose that we have for $\varphi,\ \varphi',\ \varphi''$ estimates of type (1.4.1), and moreover: $\varphi(0) = \varphi'(0) = \varphi''(0) = 0$. Then using *Fourier inversion formula*, we get

$$\varphi(t) = \frac{1}{2\pi i}\int_{a-i\infty}^{a+i\infty} \mathcal{L}\varphi(z)e^{zt}dz.$$

By direct computations:

$$\mathcal{L}(1) = \frac{1}{z}, \qquad \mathcal{L}(t) = \frac{1}{z^2}$$

and

$$\frac{1}{2\pi i}\int_{a-i\infty}^{a+i\infty} \frac{e^{tz}}{z^2}dz = \operatorname{Res}(0, \frac{e^{tz}}{z^2}) = t \quad (a > 0);$$

$$\mathcal{L}(t^2) = \frac{2}{z^3}, \qquad \frac{1}{2\pi i}\int_{a-i\infty}^{a+i\infty} 2\frac{e^{tz}}{z^3}dz = t^2.$$

The Laplace transform in the complex plane

Proposition 1.5. *The Laplace transform $\mathcal{L} : \varphi \to g$ is an isomorphism between two spaces:*

(i) Holomorphic functions φ on an unprecise (narrow) sector $\sum = \{|Arg t| < \varepsilon\}$ $(\varepsilon > 0)$ with an exponential growth of order one and finite type at infinity in this sector:

$$|\varphi(t)| \leq Ae^{B|t|}$$

and admitting an asymptotic expansion at 0:

$$|\varphi(t) - \sum_{p=0}^{m} b_p t^p| \leq c_p|t|^{n+1} \qquad (n \in \mathbb{N}).$$

(ii) Holomorphic functions g in an unprecise sector

$$V' = \{|Arg(z - A')| < \frac{\pi}{2} + \varepsilon'\} \qquad (\varepsilon' > 0)$$

and admitting an asymptotic expansion at ∞ *in this sector:*

$$g(z) \sim \sum_{n \geq 1} \frac{a_n}{z^n}.$$

Moreover we have $a_{n+1} = n!\, b_n$.

We have (cf. Fig. 1.4.1):

$$\varphi(t) = \frac{1}{2\pi i} \int_{a-i\infty}^{a+i\infty} \mathcal{L}\varphi(z) e^{tz} dz; \qquad g = \mathcal{L}\varphi.$$

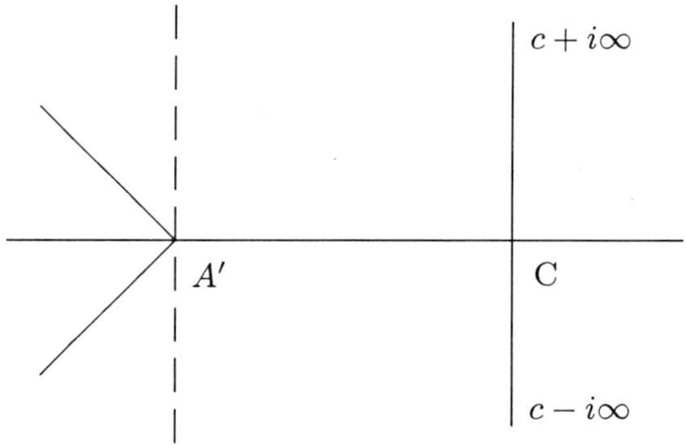

Fig 1.4.1

From complex Laplace transform, we get the Borel-Laplace formalism. We use the conformal transformation $z = \frac{1}{x}$ (automorphism of $\mathbb{C}^* = \mathbb{C} - \{0\}$). We set $g(z) = f(x)$. We have (cf. Fig. 1.4.3):

$$\varphi(t) = \frac{1}{2\pi i} \int_{\gamma'} g(z) e^{tz} dz,$$

$$\varphi(t) = \frac{1}{2\pi i} \int_{\gamma} f(x) e^{t/x} \frac{dx}{x^2} = \mathcal{B}f(t).$$

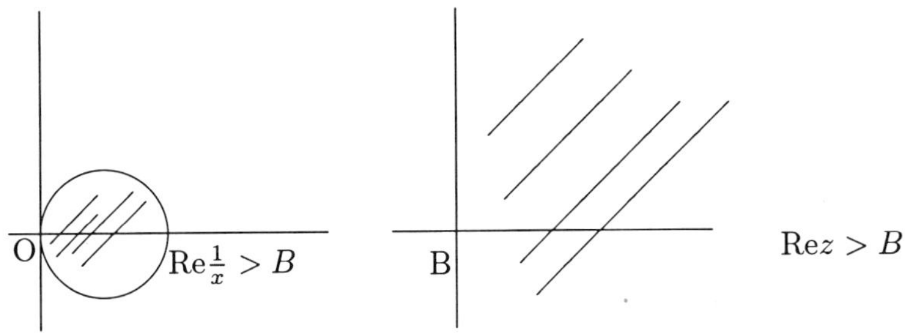

Fig 1.4.2

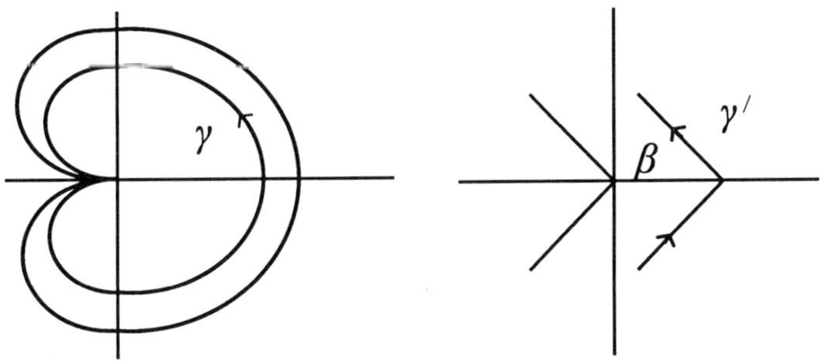

Fig 1.4.3

By definition, $\mathcal{B}$ is the (actual) Borel-transform. We will also call Laplace transform, the transform

$$f(x) = \mathcal{L}\varphi(x) = \int_{\mathbb{R}+} \varphi(t)e^{-t/x}dt.$$

The formal counterparts are:

$$\hat{\mathcal{B}}(x^n) = \frac{t^{n-1}}{(n-1)!}, \qquad \hat{\mathcal{L}}(t^n) = n!x^{n+1}$$

and

$$\hat{\mathcal{B}}(1) = \delta \text{ (Dirac at } t = 0), \qquad \hat{\mathcal{L}}(\delta) = 1.$$

1.5. *Ramis-Sibuya theorem and k-summability*

Theorem 1.4. *We assume that* $\{V_1, ..., V_m\}$ *is an open covering of the punctured disc* $D^* = \{x \in \mathbb{C}, 0 < |x| < r\}$ *by open sectors, such that*

the three by three intersections are void. Let $f_1, ..., f_m$ be a collection of holomorphic functions satisfying the following conditions:
(i) each $f_j \in O(V_j)$ is bounded on V_j ($j = 1, 2, ..., m$);
(ii) $f_{j,j+1} = f_{j+1} - f_j \in \mathcal{A}^{<0}(V_{j,j+1})$ ($j = 1, 2, ..., m$; "$m+1 = 1$"; $V_{j,j+1} = V_j \cap V_{j+1}$).

Then there exists a unique formal power series $\hat{f} \in \mathbb{C}[[x]]$ such that the f_j's admit $\hat{f}$ as asymptotic expansion (in the classical sense) on the V_j's.

Moreover, let $k > 0$. If (ii) is replaced by
(ii)' $f_{j,j+1} \in \mathcal{A}^{\leq -k}(V_{j,j+1})$,
then $\hat{f} \in \mathbb{C}[[x]]_{\frac{1}{k}}$ and $f_j \sim \hat{f}$ in Gevrey $\frac{1}{k}$ sense on V_j ($j = 1, 2, ..., m$).

Corollary 1.1. *If $f_j \in \mathcal{A}^{<0}(V_j)$ and $f_{ij} \in \mathcal{A}^{\leq -k}(V_{i,j})$, then $f_j \in \mathcal{A}^{\leq -k}(V_j)$, that is the f_j's admit an exponential decay of order k on the V_j's.*

This result is named "Ramis - Sibuya Theorem". The statement with hypothesis (ii) is due to Malgrange. The corollary is due to Sibuya. The statement with hypothesis (ii)' is due to Ramis. The proof that we will sketch below is due independently to Ramis and Sibuya. (Malgrange's argument was quite different.)

Lemma 1.3. *Let $\{V_1, ..., V_m\}$ be a covering of the punctured disc D_R^* ($R > 0$) by open sectors. (We suppose that 3 by 3 intersections are void.) Let $\varphi_{j,j+1} \in \mathcal{A}^{\leq -k}(V_{j,j+1})$ ("$m+1 = 1$") be a given family of functions ($j = 1, 2, ..., m$). Then (reducing $R > 0$ if necessary) there exists a family of functions $\{\varphi_j\}_{j=1,2,...,m}$, such that:*
(i) $\varphi_j \in \mathcal{A}(V_j)$ ($j = 1, 2, ..., m$).
(ii) $\varphi_{j+1} - \varphi_j = \varphi_{j,j+1}$ ($j = 1, 2, ..., m$; "$m+1 = 1$").

Moreover, if $\varphi_{j,j+1} \in \mathcal{A}^{\leq -k}(V_{j,j+1})$ ($j = 1, 2, ..., m$), then
(i)' $\varphi_j \in \mathcal{A}_{\frac{1}{k}}(V_j)$ ($j = 1, 2, ..., m$).

Proof. By linearity, we can suppose that $\varphi_{j,j+1} \equiv 0$ except for $j = 1$, $j + 1 = 2$. Then we set $\varphi_{1,2} = g$, and $V_{1,2} = V_1 \cap V_2$; $V = \{x \in \mathbb{C}, \alpha < \arg x < \beta, |x| < R\}$. We have $\beta < \alpha + 2\pi$. We will consider "ramified sectors" (using the Riemann surface of the Logarithm). In particular we introduce the ramified sector $\tilde{V} = \{x \in \tilde{D}^*, \arg x \in [\alpha, \beta + 2\pi]\}$. (cf. Fig 1.5.1)

We choose $0 < \tau < R$ and denote $\gamma = [0, \tau e^{i\delta}]$ the corresponding segment, for a fixed direction of argument $\delta \in [\alpha, \beta]$. For $\alpha < \delta < \delta' < \beta$,

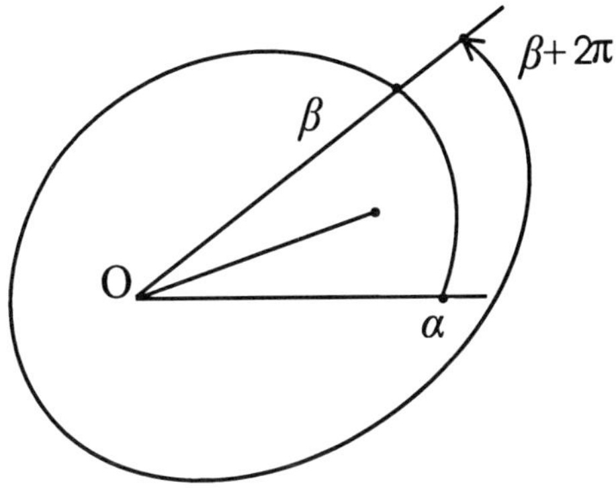

Fig 1.5.1

we set $\gamma' = [0, \tau e^{i\delta'}]$. Let ρ' be the circle arc joining $\tau e^{i\delta}$ to $\tau e^{i\delta'}$. We introduce

$$h_\gamma(x) = \frac{1}{2i\pi} \int_\gamma \frac{g(\xi)}{x - \xi} d\xi,$$

and $h_{\gamma'}$, $h_{\rho'}$ by analog formulae.

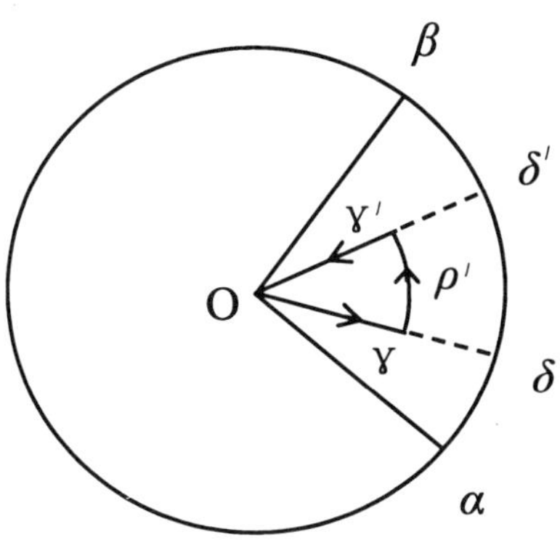

Fig 1.5.2

By Cauchy integral formula, we get

$$h_{\gamma'} - h_{\rho'} - h_\gamma \equiv 0$$

for x not belonging to the "triangular" domain delimited by γ, ρ', γ'. The function $h_\gamma(x)$ is defined on $\mathbb{C} - \gamma$, in particular for $\arg x \in (\gamma, \gamma + 2\pi)$. For x such that $\arg x \in (\gamma, \gamma + 2\pi)$, we can replace the contour of integration γ

by $\rho' + \gamma'$, without changing $h_\gamma(x)$:

$$h_\gamma(x) = h_{\rho'}(x) + h_{\gamma'}(x).$$

But $h_{\rho'}(x) + h_{\gamma'}(x)$ is defined and holomorphic for x such that $|x| < r$ and $\arg x \in (\gamma', \gamma' + 2\pi)$.

Therefore we get an analytic extension of $h_\gamma(x)$ to the ramified sector $\{x \in \tilde{D}^* : |x| < r, \arg x \in (\delta, \delta' + 2\pi)\}$. Moving δ' and therefore γ' towards $\arg x = \beta$, we get an analytic extension to the ramified sector $\{|x| < r, \arg x \in (\delta, \beta + 2\pi)\}$. Then we can suppose $\alpha < \delta'' < \delta < \beta$ and move δ'' towards α.

By a similar argument, we get an analytic extension of h_γ to the ramified sector: $\{|x| < r, \arg x \in (\alpha, \beta + 2\pi)\}$. We denote $\tilde{h}$ this extension. Let x be such that $\alpha < Algx < \beta$. We choose δ', δ'' such that $\alpha < \delta'' < \arg x < \delta' < \beta$.

We consider the closed simple contour $\gamma'' + \rho'' - \rho - \gamma''$ (oriented positively). By the Cauchy integral formula, we get

$$g(x) = \frac{1}{2i\pi} \int_{\gamma'' + \rho'' - \gamma' - \rho'} \frac{g(\xi)}{\xi - x} d\xi = h_{\gamma'' + \rho''}(x) - h_{\gamma' + \rho'}(x)$$

and $g(x) = \tilde{h}(e^{2i\pi} x) - \tilde{h}(x)$.

The function g is the variation of $\tilde{h}$. In order to end the proof of Lemma 1.3, it remains to prove asymptotic estimates for $\tilde{h}$. We can choose freely γ with $\arg x \in (\alpha, \beta)$ and prove such estimates for h_γ and a sector W corresponding to $\arg x \in [\delta + \varepsilon, \delta + 2\pi - \varepsilon]$ ($\varepsilon > 0$).

We write

$$\frac{1}{x - \xi} = -\sum_{p=0}^{n-1} \frac{x^p}{\xi^{p+1}} - \frac{x^n}{\xi^{n+1}} \frac{1}{x - x/\xi},$$

therefore

$$h_\gamma(x) = h(x) = \frac{1}{2i\pi} \int_\gamma (\sum_{p=0}^{n-1} \frac{x^p}{\xi^{p+1}}) g(\xi) d\xi + \frac{1}{2i\pi} \int_\gamma \frac{x^n g(\xi)}{\xi^{n+1} - x\xi^n} d\xi.$$

We set

$$a_p = \frac{1}{2i\pi} \int_\gamma \frac{g(\xi)}{\xi^{p+1}} d\xi$$

and we have

$$|h(x) - \sum_{p=0}^{n-1} a_p x^p| = |x|^n |\frac{1}{2\pi}\frac{1}{2i\pi}\int_\gamma \frac{g(\xi)}{\xi^{n+1} - x\xi^n} d\xi$$
$$\leq |x|^n \int_0^r \frac{|g(te^{i\delta})|}{t^n(te^{i\delta} - x)} dt.$$

Because $|\arg x - \delta| \geq \varepsilon$, we have $|te^{i\delta} - x| \geq |x| \sin\varepsilon$, therefore

$$|h(x) - \sum_{p=0}^{n-1} a_p x^p| \leq \frac{|x|^{n-1}}{\sin\varepsilon}\int_0^r \frac{|g(te^{i\delta})|}{t^n} dt.$$

The function g being infinitely flat at 0, we have

$$0 \leq \sup_{t\in(0,1]} |\frac{|g(te^{i\delta})|}{t^n}| = \ell_n < +\infty;$$

therefore

$$|h(x) - \sum_{p=0}^{n-1} a_p x^p| \leq \frac{|x|^{n-1}}{\sin\varepsilon} r\ell_n.$$

It follows that

$$|h(x) - \sum_{p=0}^{n-1} a_p x^p| \leq (\frac{r\ell_{n+1}}{\sin\varepsilon} + |a_n|)|x|^n = M_n|x|^n.$$

If, moreover, $g \in \mathcal{A}^{\leq -1}(V)$, then $|g(te^{i\delta})| \leq ke^{-\frac{B}{t}}$ for some K, $B > 0$, and it follows:

$$\int_0^r \frac{|g(te^{i\delta})|}{t^n} dt \leq K\int_0^r t^{-n}e^{-\frac{B}{t}} dt.$$

We have

$$\int_0^r t^{-n}e^{-\frac{B}{t}} dt = \int_{\frac{1}{r}}^{+\infty} B^{-n+1}\mu^{n-2}e^{-\mu} d\mu,$$

$$\int_0^r t^{-n}e^{-\frac{B}{t}} dt \leq \frac{1}{B^{n-1}}\Gamma(n-2) = \frac{1}{B^{n-1}}(n-3)!.$$

Our estimates follow easily; we can do similar computations for any $k > 0$.

This ends the proof of Lemma 1.3. This lemma implies easily Theorem 1.4.

We apply Lemma 1.3 with $\varphi_{i,i+1} = f_{i,i+1} = f_{i+1} - f_i$. We get a family of functions $\{\varphi_i\}_{i=1,\ldots m}$, $\varphi_i \in \mathcal{A}(V_i)$ (or $\mathcal{A}_{\frac{1}{k}}(V_i)$) such that $\varphi_{i,i+1} = \varphi_{i+1} -$

φ_i. Then we get $\varphi_{i+1} - \varphi_i = f_{i+1} - f_i$, and $\varphi_{i+1} - f_{i+1} = \varphi_i - f_i$ on $V_{i,i+1} = V_i \cap V_{i+1}$. We set $\psi_i = \varphi_i - f_i$ on V_i. We have $\psi_i = \psi_{i+1}$ on $V_i \cap V_{i+1}$, therefore the ψ_i's $(i = 1, ...m)$ define a holomorphic function ψ_i on $D^* = \bigcup_{i=1,...m} V_i$.

By hypothesis the f_i's are bounded, therefore ψ is bounded on D^* and (Riemann's extension theorem) can be uniquely extended to a function ψ holomorphic on $D : \psi \in \mathbb{C}\{x\}$. Then $f_i = \varphi_i + \psi$ and $f_i \in \mathcal{A}(V_i)$ $(\mathcal{A}_{\frac{1}{k}}(V_i))$, which ends the proof of Theorem 1.4.

We can now go back to the study of k-summability. We recall that $\hat{f} \in \mathbb{C}[[x]]$ is k-summable $(k > 0)$, if it is k-summable in every direction d but perhaps a finite number (the so-called singular directions of $\hat{f}$: $d_1, ...d_\ell$). The finite set $\Sigma(\hat{f}) = \{d_1, ..., d_\ell\}$ is the singular support of $\hat{f}$; $\Sigma(\hat{f}) = \emptyset \Leftrightarrow \hat{f} \in \mathbb{C}\{x\}$, *i.e.* convergent.

It's easy to see that, when d moves between two consecutive singular directions d_i and d_{i+1}, then the sums f_d of $\hat{f}$ in the direction d give together analytically and define a holomorphic function f_i on a "sector" V_i[10] (cf. Fig 1.5.3):

$$\arg x \in (\arg d_i - \frac{\pi}{2k}, \arg d_{i+1} + \frac{\pi}{2k}).$$

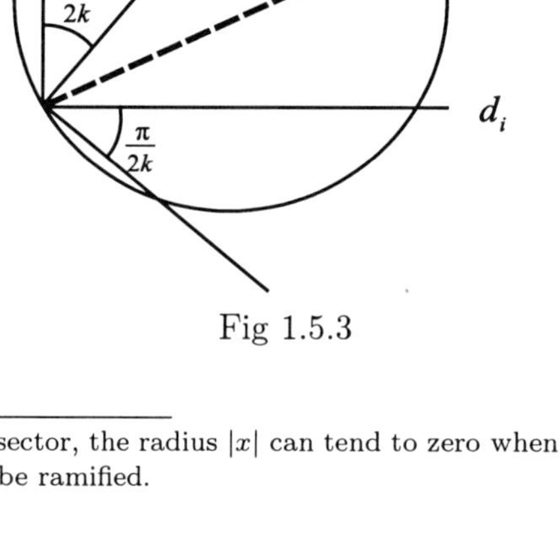

Fig 1.5.3

[10] it is an eye-shaped sector, the radius $|x|$ can tend to zero when $\arg x \to \arg d_i - \frac{\pi}{2k}$ or $\arg d_{i+1} + \frac{\pi}{2k}$; it can be ramified.

The functions f_i and f_{i+1}, $(i = 1, ..., \ell - 1)$ have the same asymptotic expansion at the origin: f_i, $f_{i+1} \sim \hat{f}$, in Gevrey$\frac{1}{k}$ sense. Therefore $f_{i+1} - f_i = f_{i,i+1}$ is asymptotic to zero on $V_{i,i+1} = V_i \cap V_{i+1}$ in Gevrey $\frac{1}{k}$ sense (*i.e.* is exponentially decreasing with an order k). The eye-shaped "sector" $V_{i,i+1}$ admits an opening exactly equal to $\frac{\pi}{k}$.

There is a converse to this property which (even if it seems surprising) will give a very efficient way to prove that a power series derived from a problem of dynamical system is k-summable.

Theorem 1.5. *We assume that $\{V_1, ..., V_m\}$ is an open covering of the punctured disc D^* by open sector, such that the 3 by 3 intersections are void. Let $\{f_1, ..., f_m\}$ be a collection of holomorphic functions satisfying the following conditions:*
(i) each $f_j \in \mathcal{O}(V_j)$ is bounded on V_j $(j = 1, ..., m)$;
(ii) $f_{j,j+1}$ defined in $V_{j,j+1} = V_j \cap V_{j+1}$ can be extended in an holomorphic function defined on an eyeshaped "sector" $V'_{j,j+1}$ with opening $\frac{\pi}{k}$ such that this extension (denoted also $f_{j,j+1}$) satisfies $f_{j,j+1} \in \mathcal{A}^{\leq -k}(\mathbb{V}'_{j,j+1})$.

Then there exists a unique formal power series $\hat{f}$ such that $f_j \sim \hat{f}$ $(i = 1, ..., m)$. Moreover $\hat{f}$ is k-summable and the f_i's correspond to sums of $\hat{f}$ in different directions. If we denote by Σ the finite set of the bisecting lines of the sectors $V_{i,i+1}(i = 1, ..., m)$, then $\Sigma(\hat{f}) \subset \Sigma$.

Proof. We leave it to the reader (it is easy, using Theorem 1.4) and our definition of k-summability.

In the case of dynamical systems, it is in many cases possible to find directly the f_i's using dynamical methods. Then it remains to estimate the differences $f_{i+1} - f_i = f_{i,i+1}$ in order to prove an exponential decay (of some order). It is important to notice that it is easy to linearize this last problem, therefore the method is very powerful for non-linear problems.

Examples. We will describe in some details two k-summable series. We will begin with our favorite example: the Euler series. The next example will be a power series coming from the asymptotic expansion of the Airy function at infinity[11].

Euler series. We start from $\hat{f}(x) = \sum\limits_{n=0}^{+\infty} (-1)^n n! x^{n+1}$. Here $k = 1$. Its formal Borel transform is $\hat{B}\hat{f}(t) = \frac{1}{1+t} = 1 - t + t^2 - \cdots$. By 1-summation

[11]For more details and other examples from special functions theory, cf. [Mar Ram].

in the direction $\mathbb{R}^+$, we get

$$f_{\mathbb{R}^+}(x) = \int_0^{+\infty} \frac{e^{-t/x}}{1+t} dt.$$

More generally, the integral

$$\int_d \frac{e^{-t/x}}{1+t} dt$$

exists if $d \neq \mathbb{R}^-$ and defines the sum f_d in the direction d.

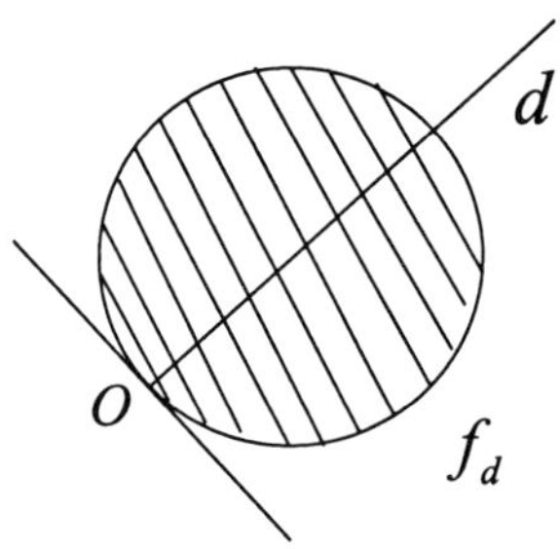

Fig 1.5.4

The sum f_d is defined in a disc (cf. Fig 1.5.4) and Gevrey 1 asymptotic to $\hat{f}$ in every subsector of this disc. When we move d, the holomorphic functions f_d glue together analytically and define an holomorphic function $\hat{f}$ on an eye-shaped "sector" $\widetilde{V}$ whose opening is 3π ($-\pi < \arg x < \pi$); $\widetilde{f}$ is asymptotic to $\hat{f}$ on this "sector" in Gevrey 1 sense.

We denote $d = e^{i\theta}\mathbb{R}^+$. The allowed summation directions correspond to $\theta \in]-\pi, \pi[$. The function $\tilde{f}$ is multivalued: it has only one value f on $\operatorname{Re} x > 0\}$ and two values f^+, f^- on $\{\operatorname{Re} x < 0\}$. The "branches" f^+ and f^- correspond respectively to f_{d^+} and f_{d^-} where $d^+ = e^{i\theta^+}\mathbb{R}^+$, $d^- = e^{i\theta^-}\mathbb{R}^+$, $\theta^+ = \pi + \alpha$, $\theta^- = \pi - \alpha$, where $\alpha > 0$ is "small".

We can easily compare f^+ and f^-:

$$f^-(x) - f^+(x) = \int_\gamma \frac{e^{-k/x}}{1+k} dt, \qquad \text{where} \quad \gamma = d^- - d^+.$$

We can "deform" γ in a bounded contour γ' without changing the integral (Cauchy theorem and elementary estimates). Then we get

$$\int_{\gamma'} \frac{e^{-t/x}}{1+t} dt = 2i\pi \operatorname{Res}(\frac{e^{-t/x}}{1+t}; t = -1) = 2i\pi e^{1/x}$$

So, $f^-(x)-f^+(x)=2i\pi e^{1/x}$ which is, as we waited, exponentially small on the half-plane $\{\mathrm{Re}\, x<0\}$: $f^- - f^+ \in \mathcal{A}^{\leq -1}(\{\mathrm{Re}\, x<0\})$. The functions $\tilde{f}$, f, f^+, f^- are *actual* solutions of the linear *non homogeneous* differential equation (Euler equation):

$$x^2y' + y = x.$$

The function $f^-(x) - f^+(x) = 2i\pi e^{1/x}$ is an actual solution of the corresponding *homogeneous* differential equation:

$$x^2y' + y = 0.$$

We can now describe the *monodromy* of the problem, that is what happens to the solutions when we turn around the origin by *analytic continuation.*

We start from f on $\{\mathrm{Re}\, x>0\}$ and we extend it by analytic continuation along a loop turning around 0 in the positive sense (we replace $\mathbb{R}^+ = e^{i0}\mathbb{R}^+$ by $d = e^{i\theta}\mathbb{R}^+$ with $\theta \in]0,\pi[$); firstly we get f^-. When d crosses $\mathbb{R}^-$ (θ crosses π), f^- *jumps* to $f^+ = f^- - 2i\pi e^{1/x}$. Therefore in this region $f^+ + 2i\pi e^{1/x}$ is the analytic continuation of f. It works for $\theta^+ \in]-\pi, 0[$. Then we get an *extremely important* phenomena (an example of the Stokes phenomena): near the direction $\mathbb{R}^-$ (the singular direction of $\hat{f}$: $\Sigma(\hat{f}) = \{\mathbb{R}^-\}$), the two functions f^+ and f^- admit the *same asymptotics* (*i.e.* $\hat{f}$); they differ by $2i\pi e^{1/x}$ which is *infinitely flat.* The function $f^+ + 2i\pi e^{1/x}$ admits the asymptotics $\hat{f}$ on $\{\mathrm{Re}\, x<0\}$, *but* when we cross the line $-i\mathbb{R}^+ = e^{-i\pi/2}\mathbb{R}^+$, f^+ keeps the same Asymptotics $\hat{f}$ and $2i\pi e^{1/x}$ jumps, turning its asymptotics (*i.e.* 0), and becomes infinitely big (exponentially big) on $\{\mathrm{Re}\, x>0\}$. On the line $-i\mathbb{R}^+$ it is bounded and *oscillates.* The line $-i\mathbb{R}^+$ is (I think *improperly*[12]) named *Stokes line* by many authors. Turning backwards we get a similar phenomena on the other Stokes line: $i\mathbb{R}^+ = e^{i\pi/2}\mathbb{R}^+$.

After one turn around the origin (in the positive sense) by *analytic continuation*, the function f is *replaced* by the function $f + 2i\pi e^{1/x}$ (it is an *affine* transformation in the affine space of solutions of Euler ordinary differential equation). We observe that $f + 2i\pi e^{1/x}$ is *not asymptotic* to $\hat{f}$. Its *dominant* part is $2i\pi e^{1/x}$ which exposes when $x \to 0$ in the half plane $\{\mathrm{Re}\, x>0\}$.

---

[12]The true Stokes phenomena happens when we cross the *singular line* $\mathbb{R}^-$. At the beginning we cannot notice it by asymptotics and we observe it only when we cross $-i\mathbb{R}^+$.

Airy equation (cf. [17]) This second example is the description of the solutions of Airy equation

$$y'' - zy = 0 \qquad (1)$$

near $z = \infty$, using asymptotics. Stokes discovered Stokes phenomena studying these solutions.

We consider Airy ordinary differential equation on the Riemann sphere $\mathbb{P}^1(\mathbb{C}) = \mathbb{C} \cup \{\infty\}$. Its *only singular* point on this Riemann sphere is clearly $z = \infty$. A first consequence is that any solution on a small disc centered at $z_0 \in \mathbb{C}$ extends analytically to all $\mathbb{C}$: it is *uniform* (single valued) and if we turn around ∞ along a loop by analytic continuation, we will go back to the *same solution*: the (actual) *monodromy* transformation is the *identity.*

Airy ordinary differential equation (1.5.1) admits a basis of *convergent* power series solutions at $z = 0$ (cf. 1.2). The sums of these series are *entire* functions(they are holomorphic on $\mathbb{C}^*$).

We recall

$$Ai(z) = \frac{1}{2\pi i}\int_{-i\infty}^{+i\infty} e^{\zeta z - \frac{\zeta^3}{3}}\, d\zeta = \frac{1}{\pi}\int_0^{+\infty} \cos(\frac{t^3}{3} + zt)dt.$$

We introduce the power series expansion

$$\hat{A}(z) = \sum_{n \geq 0} c_n (-z)^{-\frac{3n}{2}}$$

where

$$c_n = \frac{\Gamma(n+\frac{1}{6})\Gamma(n+\frac{5}{6})}{n!}(\frac{3}{4})^n \quad (n \in \mathbb{N}^*); \qquad c_0 = 1.$$

Let $t = x^{\frac{1}{\nu}}$ with $\nu \in \mathbb{N}^*$. We set, by definition, $\mathbb{C}[[t]]_s = \mathbb{C}[[x^{\frac{1}{\nu}}]]_s$. If $\hat{f} \in \mathbb{C}[[x^{\frac{1}{\nu}}]]_s$, we will say that it is of order νs in x in a generalized sense.

Theorem 1.6. *(i) The power series $\hat{A}(z) \in \mathbb{C}[[z^{-\frac{1}{2}}]]$ is Gevrey of order $\frac{2}{3}$ in $\frac{1}{z}$ and $\frac{3}{2}$-summable (in $\frac{1}{z}$) or 3-summable in t $(= z^{-\frac{1}{2}})$ in every direction except* $\arg t = -\pi$ *or* $\pm\frac{\pi}{3}$ *(mod. 2π).*

(ii) The sum of $\frac{1}{4\pi^{\frac{3}{2}}} z^{-1/4} e^{-2/3z^{3/2}} \hat{A}(z)$ in the direction $\mathbb{R}^+$ is equal to the classical Airy function $Ai(z)$.

We set $\hat{B}(z) = \sum_{n \geq 0} c_n z^{-\frac{3n}{2}} = \hat{A}(-z)$. Then

$$z^{-1/4} e^{-2/3z^{3/2}} \hat{A}(z), \qquad z^{-1/4} e^{2/3z^{3/2}} \hat{B}(z)$$

form a formal system of fundamental solutions of Airy ordinary differential equation at $z = \infty$. By $\frac{3}{2}$-summation in a general (*i.e.* non singular) direction we get a system of actual fundamental solutions. The jumps when we move the direction correspond to Stokes phenomena (cf. 2.1 below)

2. Applications to ordinary differential equations

2.1. *Linear ordinary differential equations, index theorem and Newton polygons*

We will study in this part spaces of solutions of linear analytic differential equations. We will give only the main results (for detailed proofs, cf. [25], [26]).

The situation is local and we are interested in solutions of an operator $D = a_m(x)(\frac{d}{dx})^m + \cdots + a_0(x)$ (*where* $a_0, \cdots, a_m \in \mathbb{C}\{x\}$) in some (topological) vector spaces. An important result is the finiteness of the dimension of a solution space. Traditionally we can try to compute this dimension using a fixed point method. This is elementary but it can be technically difficult. Here we will use a different approach based on finite index operators (Fredholm Operators). It is less precise: it gives a lower estimate of the dimension (and not in general the dimension itself), but it is very easy to apply (due in particular to stability by compact perturbations).

Definition 2.1. Let E, F be two complex vector spaces. A linear map $u : E \to F$ has a finite index $\chi(u)$ if $\dim_{\mathbb{C}} \ker u$ and $\dim_{\mathbb{C}} \operatorname{coker} u = \dim_{\mathbb{C}} F/\operatorname{im} u$ are finite. Then $\chi(u) = \dim \ker u - \dim \operatorname{coker} u$.

Example 2.1.

$$D = x : \mathbb{C}[[x]] \to \mathbb{C}[[x]], \qquad \chi(x) = -1;$$
$$D = \frac{d}{dx} : \mathbb{C}[[x]] \to \mathbb{C}[[x]], \qquad \chi(\frac{d}{dx}) = 1.$$

Proposition 2.1. *Let E, F, G three complex vector spaces and $u : E \to F$, $v : F \to G$ two linear operators. We suppose that u and v have a finite index. Then the operator $v \circ u : E \to F$ has also a finite index and $\chi(v \circ u) = \chi(v) + \chi(u)$.*

Example 2.2.

$$\chi(x^j(\frac{d}{dx})^i) = i - j \qquad (i, j \in N)$$
$$x^j(\frac{d}{dx})^i : \mathbb{C}[[x]] \to \mathbb{C}[[x]].$$

Lemma 2.1. *(F. Riesz) Let E be a complex Banach space. Let $K \in L(E, E)$ be a compact linear endomorphism of E. Then the operator $id_E + K$ is a finite index endomorphism of E and $\chi(id_E + K) = 0$.*

Theorem 2.1. *Let E, F be two complex Banach spaces. Let $K \in L(E, F)$ be a compact operator. Let $u : E \to F$ be a continuous operator admitting a finite index. Then $u + K \in L(E, F)$ has a finite index and $\chi(u+K) = \chi(u)$.*

Be careful: In general $dim \ker(u + K) \neq dim \ker u$.

Lemma 2.2. *Let E_1, F_1 and E_2, F_2 be Banach spaces or Fréchet spaces, or DFS spaces$^{(1)}$ (same type for each pair, but the two pairs are allowed to admit different types)*[13].

If we have a commutative diagram

$$\begin{array}{ccc} E_1 & \xrightarrow{\mu_1} & F_1 \\ v\downarrow & & \downarrow w \\ E_2 & \xrightarrow{\mu_2} & F_2 \end{array}$$

where u_1, u_2, v, w are linear continuous maps, u_1 are finite index maps, v and w are injective and w is dense, then:

(i) We have an exact sequence

$$0 \to \ker \overline{\mu}_2 \to E_2/E_1 \xrightarrow{\overline{\mu}_2} F_2/F_1$$

and $\ker \bar{u}_2$ has finite dimension with $\dim \ker \bar{u}_2 = \chi(u_2) - \chi(u_1)$. This implies $\chi(u_2) \geq \chi(u_1)$.

(ii) The pair (v, w) is a quasi-isomorphism (i.e. it induces isomorphisms on the kernels and cokernels of u_1 and u_2) if and only if $\chi(u_1) = \chi(u_2)$.

This lemma generalizes some results of Dieudonné and Schwartz and was communicated to the author by B. Malgrange.

We will now explain how to compute indices of a differential operator D using the Newton polygon $N(D)$ of D. The Newton polygon in this form is due to the author [26]. (There are a lot of Newton polygons associated to differential equations are more general functional equations in the literature: Fine, Adams, Komatsu $\cdots$)

Let $D = a_m(\frac{d}{dx})^m + \cdots + a_0$ be a holomorphic (or meromorphic) linear differential operator: $a_0, \cdots a_m \in \mathbb{C}\{x\}$ (*or* $\mathbb{C}\{x\}[x^{-1}]$). We write the

[13] DFS spaces are dual of FS (Fréchet Schwartz) spaces or equivalently DF spaces of Schwartz type.

expansion of $a_i(x)$ at the origin:

$$a_i(x) = \sum_{j \in \mathbb{Z}} a_{ij} x^j;$$

$a_{ij} \in \mathbb{C}$ and $a_{ij} = 0$ if $j \leq n_i \in \mathbb{Z}$.

By definition the valuation $v(a_i)$ of a_i is the smallest integer j such that $a_{ij} \neq 0$.

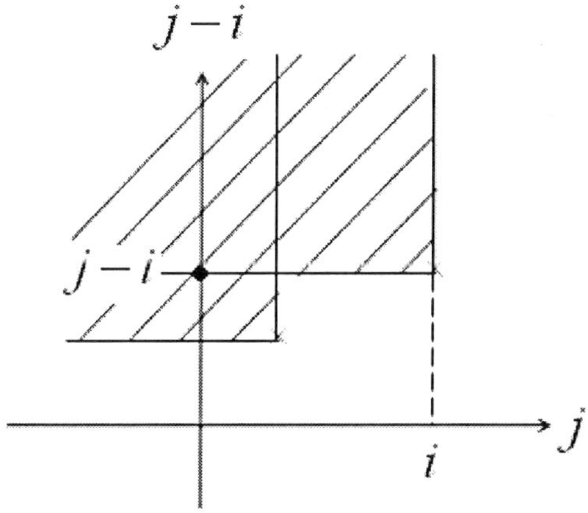

Fig 2.1.1

We can write D as an infinite sum of elementary operators $a_{ij} x^j (\frac{d}{dx})^i$ with $a_{ij} \neq 0$. If $a_{ij} \neq 0$, we put a dot at the point of coordinates $(i, j-i)$ (cf. Fig 2.1.1.). For each dot (a, b), we translate the second quadrant at the point (a, b) and we draw the convex hull of all the quadrants (translated at all the allowed values of $(i, j-i)$): cf. Fig 2.1.2. We get the Newton polygon $N(D)$ of D.

The lowest point of $N(D)$ corresponding to a fixed value of i is $(i, v(a_i) - i)$. The theory of index of O.D.E. begins with a result of Malgrange:

Theorem 2.2. *(Malgrange) Let* $D = a_m(\frac{d}{dx})^m + \cdots + a_0 \in \mathbb{C}\{x\}[\frac{d}{dx}]$. *Then:*
(i) $D : \mathbb{C}[[x]] \to \mathbb{C}[[x]]$ *has a finite index:*

$$\chi_{formal} = \inf_{i=0,\cdots,m} [i - v(a_i)]$$

(ii) $D : \mathbb{C}\{x\} \to \{x\}$ *has a finite index:*

$$\chi_{an} = m - v(a_m)$$

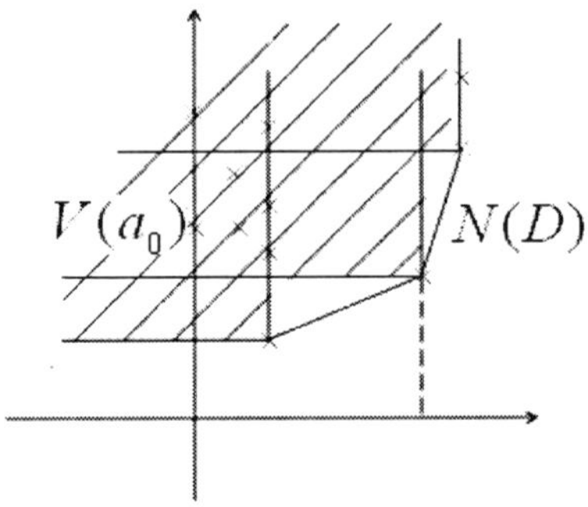

Fig 2.1.2

(iii) We consider $D : \mathbb{C}[[x]]/\mathbb{C}\{x\} \xrightarrow{D} \mathbb{C}[[x]]/\mathbb{C}\{x\}$*. Then* D *is surjective, has a finite index, and*

$$dim \ker D = \chi(D) = \chi_{formal} - \chi_{an}$$

By definition $\chi_{formal} - \chi_{an} \geq 0$ is the *irregularity index* of D.

Corollary 2.1. *The following conditions are equivalent:*
(i) D *is a Fuchsian operator (regular singular);*
(ii) $N(D)$ *is a rectangle (i.e. it has no strictly positive slopes);*
(iii) The irregularity index of D *is zero.*

Remark 2.1. The formal index corresponds to the lowest part (horizontal slope) of $N(D)$ (with a change of sign). The analytic index corresponds to the lowest point of the vertical slope of $N(D)$ (with a change of sign) which is the dot marked for the symbol $a_{m,v(a_m)}x^{v(a_m)}(\frac{d}{dx})^m$ of D.

The problem of irregularity of an operator was studied by many authors: J. Moser, Levelt, Gérard-Levelt, Malgrange, N. Katz $\cdots$

In [26], I introduced an interpolation between Malgrange analytic and formal indexes. The idea is to use the interpolation between $\mathbb{C}\{x\}$ and $\mathbb{C}[[x]]$ by Gevrey spaces $\mathbb{C}[[x]]_s$ of formal power series:

$$\mathbb{C}\{x\} = \mathbb{C}[[x]]_0 \subset \mathbb{C}[[x]]_s \subset \mathbb{C}[[x]] = \mathbb{C}[[x]]_\infty;$$

$\mathbb{C}[[x]]_s$ increases with $s \in (0, +\infty)$.

Theorem 2.3. *(Ramis) Let* $s \in [0, +\infty]$*. We set* $k = \frac{1}{s}$*. Then:*
(i) $D : \mathbb{C}[[x]]_s \to \mathbb{C}[[x]]_s$ *has a finite index:*

$$\chi_s(D) = i(k) - v(a_i(k)) \quad (\textit{cf. Fig } 2.1.3)$$

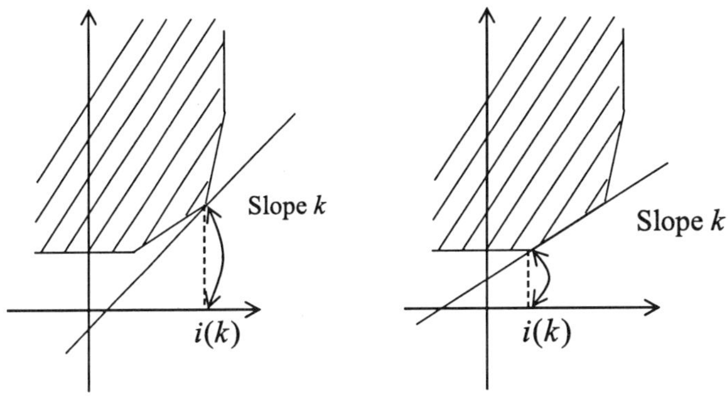

Fig 2.1.3

(ii) $\dim\ker(D : \mathbb{C}[[x]]_s/\mathbb{C}\{x\} \to \mathbb{C}[[x]]_s/\mathbb{C}\{x\})$ *decreases with* s *and is locally constant except for some "jump points" corresponding to the finite set of slopes of the Newton polygon* $N(D)$.

If we fix k, then we can consider all the lines of slope k cutting the closed set $N(D)$ (*i.e.* such that the intersection is not void). The smallest intersection is in general reduced to a point $(i(k), v(a_{i(k)}) - i(k))$. We have an exceptional situation when k is one of the slopes of the Newton polygon $N(D)$; in such a case the smallest intersection is a segment (the corresponding slope of $N(D)$) and we choose for $i(k)$ the smallest possible value.

Proof of Theorem 2.3. We will give only the main ideas. When $D \in \mathbb{C}[x][\frac{d}{dx}]$ this result is due to O. Perron who proved it by a delicate computation using a fixed point method. Our proof is the following: We write

$$D = a_{i(k),v(a_{i(k)})}x^{v(a_{i(k)})}(\frac{d}{dx})^{i(k)} + D_1 = D_0 + D_1$$

in the "general case" and $D = D_0 + D_1$, where D_0 is the sum of the $a_{i,j}x^j(\frac{d}{dx})^i$ where (i,j) belongs to the segment of slope k of $N(D)$ in the exceptional case. Then we write $\mathbb{C}[[x]]_{\frac{1}{k}}$ as an union (inductive limit) of Banach spaces of type $l^1(\mathbb{C}[[x]]_{\frac{1}{k}}$ is a DFS space). The next step is to interpret D as a compact perturbation of D_0 (*i.e.* D_1 as a compact operator) between some Banach spaces (much precisely we do that for a family of pairs of Banach spaces). It remains to compute $\chi(D_0)$ "by hand" (using a variant of Example 2.1.2 and if necessary Proposition 2.1.1 and a simple computation with Euler operator), to get $\chi(D) = \chi(D_0 + D_1) = \chi(D_0)$

using theorem 2.1.1, and to conclude by an inductive limit argument based on Lemma 2.1.2. This proves (i).

Using Lemma 2.1.2 and (i), we get a small exact sequence

$$0 \to H_s \to \mathbb{C}[[x]]_s/\mathbb{C}\{x\} \xrightarrow{D} \mathbb{C}[[x]]_s/\mathbb{C}\{x\} \to 0$$

where

$$H_s = \ker(D : \mathbb{C}[[x]]_s/\mathbb{C}\{x\} \to \mathbb{C}[[x]]_s/\mathbb{C}\{x\})$$

The index $\chi_s(D) = \dim_{\mathbb{C}} \ker H_s$ clearly decreases with s, remains locally constant except perhaps when $s = \frac{1}{k}$ corresponds to a slope k of $N(D)$. It is semicontinuous on the left. We denote by $k_1 > k_2 > \cdots k_\ell$ the strictly positive slopes (finite) of $N(D)$. We set $s_r = \frac{1}{k_r}$ $(r = 1, \cdots, \ell)$.

Then $H_s = H_{s_\ell}$ if $s \geq s_\ell$; if $s \in [s_r, s_{r+1})$ then $H_s = H_{s_r}$; if $s \in [0, s_1)$, then $H_s = (0)$. If $\hat{f} \in \mathbb{C}[[x]]$ is a solution of $Dy = g \in \mathbb{C}\{x\}$, then $\hat{f} \in \mathbb{C}[[x]]_s$ with $s \leq s_l$; moreover, if $s \in [s_r, s_{r+1})$ then we have exactly $\hat{f} \in \mathbb{C}[[x]]_{s_r}$. Hence, we get the following result:

Theorem 2.4. *(Ramis) If D is a linear differential operator analytic in a neighborhood of the origin and if $\hat{f} \in \mathbb{C}[[x]]$ is a formal power series solution of the inhomogeneous O.D.E. $Dy = g$, where $g \in \mathbb{C}\{x\}$ is convergent, then $\hat{f}$ is Gevrey of optimal Gevrey order $s = 1 + \frac{1}{k}$, where k is one of the strictly positive (finite) slopes of the Newton polygon $N(D)$ of D.*

A formal power series $\hat{f}$ is said *of optimal Gevrey order s* if $\hat{f} \in \mathbb{C}[[x]]_s$ and $\hat{f} \notin \mathbb{C}[[x]]_{s'}$ for any $s' < s$. Maillet proved (1904) that a formal power series solution $y = \hat{f} \in \mathbb{C}[[x]]$ of a (not necessarily linear) analytic O.D.E.

$$G(x, y, y', \cdots, y^{(m)})$$

is Gevrey of some order s, but his estimate for s is quite bad. Maillet theorem was rediscovered in the 50s by K. Mahler. Recently Malgrange extended theorem 2.4 to the nonlinear case getting a new version of Maillet theorem with optimal Gevrey estimates; he uses a Newton polygon along the choosen formal solution.

2.2. *Fundamental existence theorems*

Our aim is to "represent" a formal power series $\hat{f} \in \mathbb{C}[[x]]$ solution of an analytic O.D.E.:

$$G(x, y, y', \cdots, y^{(m)}) = 0$$

by an actual solution $y = f(x)$ of the equation, holomorphic on a sector and asymptotic to $\hat{f}$. As we explained before, the story begins with H. Poincaré. Many authors worked on this problem (Malmquist, Birkhoff, Hukuhara, $\cdots$), but the complete solution is quite recent [30]. Here we will mainly study the linear case in relation with Gevrey estimates, k-summability and Stokes phenomena.

It is more easy to work with differential systems

$$(\Delta) \qquad Y' = A(x)Y$$

where Y is an unknown function of the complex variable x, taking its value in $\mathbb{C}^n$, and A a given (n, n) meromorphic matrix in a (small) neighborhood of the origin.

It is easy to derive a system from an O.D.E of order n in using the classical trick:

$$Y = (y, y', \cdots, y^{(m-1)}).$$

Conversely we can derive (non uniquely) an O.D.E of order n: $D_y = 0$, from a system (Δ) using a "cyclic vector method" [26]. The Newton polygon $N(D)$ is independent of the choices and we can set $N(\Delta) = N(D)$.

At the end of the century, Fabry got a fundamental system of formal solutions for an analytic linear O.D.E at a singular point. For systems the result is due (independently) to Hukuhara and Turrittin.

Theorem 2.5. *(Hukuhara - Turrittin) Let*

$$(\Delta) \qquad Y' = A(x)Y$$

be a (germ of) meromorphic differential system at the origin. Then it admits a formal fundamental matrix solution

$$\hat{F} = \hat{H}(t)x^L e^{Q(\frac{1}{t})}$$

where: $t^\nu = x$ $(\nu \in N^*)$, $L \in \mathrm{End}\,(m; C)$ *is a constant matrix,* $x^L = e^{(\mathrm{Log}\, x)L}$, $\hat{H} \in GL(m, C[[t]][t^{-1}])$ *is a formal invertible matrix and* $Q = (q_1, \cdots, q_m)$ *is a diagonal matrix where* $q_i \in \frac{1}{t}\mathbb{C}[\frac{1}{t}]$ $(i = 1, \cdots, m)$ *(q_i can be zero).*

Example 2.3. For the system of rank $m = 2$ associated to Airy equation $(Y = (y, y'); y'' - zy = 0)$, we have

$$\nu = 2, \qquad t^2 = x,$$

$$q_1(t) = -\frac{2}{3}t^3 = -\frac{2}{3}x^{\frac{3}{2}}, \qquad q_2(t) = \frac{2}{3}t^3 = \frac{2}{3}x^{\frac{3}{2}}.$$

In Theorem 2.5, the integer ν is the LCM of the slopes of the Newton polygon $N(\Delta)$ (which belongs to $\mathbb{Q}^+$). The ramification $x = t^\nu$ to (Δ) is a system (Δ_1) in t; the slopes of (Δ_1) are the slopes of (Δ) multiplied by ν ; they are integers (the degree in t of the polynomials q_i).

It is very important to notice that in general the entries of the matrix $\hat{H}(t)$ are divergent series (like in the Airy example $\cdots$)

In a sector V, we can represent the "symbol" x^L by an actual holomorphic function: we write $x^L = e^{L(\operatorname{Log} x)}$ and we choose a determination of the logarithm over V.

In the linear case we have the following fundamental existence theorem (Birkhoff fundamental existence theorem) in Poincaré's style.

Theorem 2.6. *(Birkhoff) Let*

$$(\Delta) \qquad Y' = A(x)Y$$

be a (germ of) meromorphic system. For each direction d, there exists a sector V bisected by d and a holomorphic matrix H in V, asymptotic in V to $\hat{H}$ (in Poincare's sense) such that $F = Hx^Le^Q$ is a fundamental actual solution of (Δ).

We suppose that when we write the expression of F, we have chosen a branch of $\operatorname{Log} x$ on V and therefore a determination of x^L and a determination of $t = x^{\frac{1}{\nu}} = e^{(\frac{1}{\nu})\operatorname{Log} x}$.

The reader can find a proof of Theorem 2.6 in W. Wasow's book [33]. The proof is quite delicate. It is possible to improve this result: if k_ℓ is the smallest strictly positive slope of $N(\Delta)$, then we can choose for V a sector of opening $\frac{\pi}{k_\ell}$. Moreover, for some choices of d it can be possible to choose a bigger sector [29]. We will give some indications below.

Let $q \in \frac{1}{t}\mathbb{C}[\frac{1}{t}] : q(t) = \rho t^{-k} + \cdots (\rho \in \mathbb{C},\ k \in N^*)$. The "dominant part" of $e^{q(t)}$ is $exp(\rho t^{-k})$. The directions (in the t-plane) such that $\rho t^{-k} \in iR$ are oscillation lines for $exp(\rho t^{-k})$. They are (in the classical terminology) Stokes lines. Sometimes they are called anti-Stokes lines (which is more in the spirit of Stokes' work$\cdots$).

We suppose that the Newton polygon $N(\Delta)$ of our system admits only one strictly positive slope $k > 0$. Then $q_i = \rho_i t^{-k}$ $(i = 1, \cdots, m)$. The ρ_i's are solutions of an algebraic equation (the indicial equation associated to the slope k). The simplest case is when the m values of ρ_i are all distinct. In this case, we can consider the system

$$(\operatorname{End}\Delta) \qquad X' = AX - XA = [A, X]$$

where X is an unknown (m, m) square matrix. The exponentials appearing in its formal fundamental solution are the

$$q_{ij} = q_i - q_j = (\rho_1 - \rho_j)t^{-k} \qquad (i, j = 1, \cdots, m),$$

therefore k is the only strictly positive slope of $N(\mathrm{End}\,\Delta)$.

For each $\rho_i \in \{\rho_1, \cdots, \rho_m\}$ different from 0, we get $2k$ Stokes lines. (For Airy case, we get $k = 3$, $m = 2$, therefore 12 Stokes lines in the $t = x^{\frac{1}{2}}$ -plane. For Euler case, we get $k = 1$, therefore 2 Stokes lines.) The set of non zero values of ρ_i has $m' < m$ elements. We have m' families of Stokes lines ($2m'k$ Stokes lines).

Theorem 2.7. *(Birkhoff, Horn) Let*

$$(\Delta) \qquad Y' = A(x)Y$$

be a (germ of) meromorphic system. We suppose that $N(\Delta)$ admits only one strictly positive slope $k > 0$ and that the m values ρ_i are all distinct. To each non zero value of $\rho_{ij} = \rho_i - \rho_j$ corresponds a family of $2k$ Stokes lines of $N(\mathrm{End}\,\Delta)$.

Then, if a sector V contains at most one Stokes lines for each family ($\rho_{ij} \neq 0$; $i, j = 1, \cdots, n$), there exists a (m, m) square holomorphic matrix H on V, H asymptotic to $\hat{H}$ (in Poincaré's sense) such that $F = Hx^Le^Q$ is a fundamental solution of Δ in V.

Idea of Proof. We start from a smaller subsector $W \subset V$ such that we can apply Birkhoff fundamental existence theorem (Theorem 2.6). We get a fundamental solution $F_1 = H_1x^Le^Q$, $H_1 \sim \hat{H}$ on W. Then we can extend analytically H_1 and therefore F_1 on a bigger sector (We are dealing with linear equations). When we cross the first Stokes line which does not belongs to W, it can happen that the existence of H_1 is no longer asymptotic to $\hat{H}$. If this is the case we modify F_1, multiplying it by a well chosen constant matrix $C \in \mathrm{End}\,(m; \mathbb{C})$ on the right.

We end the proof iterating this process. When we have crossed a Stokes line of each family we have loosed all the freeness for a convenient choice of C and it is impossible to cross a new line in the extension process.

This corresponds to a result of uniqueness.

Theorem 2.8. *In the situation of Theorem 2.7, if the sector V contains exactly one Stokes line for each family, then the fundamental solution F is the unique solution asymptotic to $\hat{f}$ (i.e, $H \sim \hat{H}$) on the sector V.*

In the setting of O.D.E., Birkhoff fundamental existence theorem follows from Borel - Ritt theorem and the following result (cf. [33]).

Theorem 2.9. *Let $D \in \mathbb{C}\{x\}[\frac{d}{dx}]$ be a linear differential operator. Then we have $\mathcal{A}^{<0}(V) \xrightarrow{D} \mathcal{A}^{<0}(V) \to 0$. When $g \in \mathcal{A}^{<0}(V)$ ($g \sim 0$), we can solve $Df = g$ with $f \in \mathcal{A}^{<0}(V)$ ($f \sim 0$).*

There is a Gevrey analog of this result [29].

Theorem 2.10. *(Ramis - Sibuya) Let $k > 0$, let $D \in \mathbb{C}\{x\}[\frac{d}{dx}]$. We denote by $k_1 > 0$ the biggest slope of $N(D)$ and we set $k' = \sup(k, k_1)$. Then we have $\mathcal{A}^{<-k}(V) \xrightarrow{D} \mathcal{A}^{<-k}(V) \to 0$ for every sector V with opening $< \frac{\pi}{k'}$.*

From this result we can derive a Gevrey analog of the fundamental existence theorem.

Theorem 2.11. *We consider a non homogeneous equation $Dy = g$, where $g \in \mathbb{C}\{x\}$ is convergent. We suppose that there exists $\hat{f} \in \mathbb{C}[[x]]_s$ such that $D\hat{f} = g$. Let V a sector such that its opening is $< \frac{\pi}{k}$ and $< \frac{\pi}{k_1}$ (where $k_1 > 0$ is the biggest slope of $N(D)$). Then there exists $f \in \mathcal{A}_s(V)$ such that $Df = g$ and $f \sim \hat{f}$ in Gevrey's sense on V.*

Proof. There exists $h \in \mathcal{A}_s(V)$, $h \sim \hat{f}$ in Gevrey's sense (Borel-Ritt-Gevrey Theorem). Then $Dh - g \sim 0$ in Gevrey's sense.

It is important to recall the method of the proof. The first step is to get a quasi-solution of our O.D.E., that is a solution up to some exponentially small quantity (of order k). The second step is to modify our quasi-solution in order to get a true solution by the choice of some exponentially small correction.

Theorem 2.12. *(Ramis) Let $D \in \mathbb{C}\{x\}[\frac{d}{dx}]$. We suppose that $N(D)$ admits only one strictly positive slope $k > 0$. Let $g \in \mathbb{C}\{x\}$. If $\hat{f} \in \mathbb{C}[[x]]$ is a formal power series solution of $Dy = g$, then $\hat{f}$ is k-summable: $\hat{f} \in \mathbb{C}\{x\}_{\frac{1}{k}}$.*

Proof. We use a "geometric" argument.

We consider the exponentials appearing in the formal fundamental solutions of $Dy = 0$. The corresponding lines of maximal decay of these exponentials are named singular directions of D (they are the bisecting lines of consecutive Stokes lines of a given family). We denote $\Sigma(D)$ the singular set of D (the set of singular directions).

Let $d \notin \Sigma(D)$ be a direction. We can choose $\varepsilon > 0$ such that if V is bisected by d and the opening of V is $> \frac{\pi}{k} - \varepsilon$, we can move V in a sector

V' such that the opening of $V \cup V' > \frac{\pi}{k}$ and such that $V \cup V'$ contains exactly a Stokes line of each family.

Then these exists solution of $Dy = g, f \in \mathcal{A}_s(v)$, $f_1 \in \mathcal{A}_s(v')$, $f \sim \hat{f}$ on V, $f_1 \sim_s \hat{f}$ on V'. On $V \cup V'$, $f - f_1 \sim 0$ and $D(f - f_1) = Df - Df_1 = g - g = 0$. Using Theorem 2.8 (unicity theorem) and $D(0) = 0$, we get $f - f_1 = 0$ on $V \cup V'$. Gluing together f and f_1, we get a solution h of $Dy = g$ on $V \cup V'$. We have clearly $h \sim \hat{f}$ on $V \cup V'$ and h is k-summable in the direction d_1 bisecting $V \cup V'$. The result (*i.e.* $\hat{f} \in \mathbb{C}\{x\}_{\frac{1}{k},d}$) follows easily with convenient choices of V' (moving on the other side): $\hat{f}$ is k-summable in every non singular direction (*i.e.* $\in \Sigma(D)$).

Another proof. We use the fundamental existence and uniqueness theorems in Poincaré's sense (precise form). We get a finite set of sectors V_i of existence and unicity limited by two Stokes lines and such that opening of $V_i > \frac{\pi}{k}$. For two consecutive such sectors V_i, V_{i+1}, we have opening of $V_i \cap V_{i+1} = \frac{\pi}{k}$ (we use the Riemann Surface of $\operatorname{Log} x$ if necessary). If we denote f_i the unique solution of $Dy = g$ such that $f_i \in A(V_i)$ and $f \sim f_i$ (Poincaré's sense), for each strict subsector $W \prec V_i$ with opening $W = \frac{\pi}{k}$, we can apply the unicity theorem to W (it contains exactly one Stokes line for each family). By Theorem 2.11, there exists $f \in \mathcal{A}_s(V)$ such that $Df = g$ and $f \sim \hat{f}$ on W. By unicity $f = f_i$ on W and f_i is k-summable in the direction d bisecting W(moving a little bit W): $f_i \in \mathcal{A}_s(V_i)$.

There is a similar version for systems.

Theorem 2.13. *(Ramis) Let*

$$(\Delta) \qquad Y' = A(x)Y$$

be a (germ of) meromorphic differential system at the origin. We suppose that $N(\Delta)$ has only one strictly positive slope $k > 0$. Then in the formal fundamental solution $\hat{F} = \hat{H} x^L e^Q$ of Hukuhara-Turrittin, the entries of $\hat{H}$ are k-summable[14].

We will end with some indications on the nonlinear case. The results are similar but the proofs more difficult [29].

Theorem 2.14. *(Ramis - Sibuya) We consider an analytic O.D.E.*

$$G(x, y, y', \cdots, y^{(m)}) = 0.$$

[14] In a generalized sense: $k\nu$-summable in $t = x^{\frac{1}{\nu}}$.

We suppose that $y = \hat{f}(x) \in \mathbb{C}[[x]]$ *is a formal solution of this equation. Then:*

(i) For each direction d*, there exists a sufficiently small sector* V *bisected by* d *and an* $f \in A(V)$ *such that* f *is a solution of the O.D.E. and* $f \sim \hat{f}$ *in* V *(Poincaré's sense).*

(ii) If $\hat{f} \in \mathbb{C}[[x]]_s$*, we can choose in (i) the function* f *such that* $f \in \mathcal{A}_s(V)$ *and* $f \underset{s}{\sim} \hat{f}$ *in* V *(choosing* V *smaller if necessary). There are also some* k*-summability results in some cases.*

3. Multisummability

The k-summability of formal power series solutions of analytic O.D.E. was conjectured by E. Borel (in Borel - Laplace style version) at the beginning of the XX-th century. He and his followers (Maillet, Leroy, $\cdots$, Turrittin) tried to prove this conjecture. The results of the preceding chapter comfort this conjecture. Unfortunately it is *false*, and in some sense *evidently false*.

Starting again from our favorite example, Euler power series:

$$\hat{f}_1(x) = \sum_{n=0}^{+\infty} (-1)^n n! x^{n+1},$$

we replace x by x^2 and get

$$\hat{f}_2(x) = f_1(x^2) = \sum_{n=0}^{+\infty} (-1)^n n! x^{2n+2}.$$

Now we consider $\hat{f} = \hat{f}_1 + \hat{f}_2$.

Following E. Picard, there is an analogy between algebraic numbers and solutions of linear differential equations. The sum of two algebraic numbers is an algebraic number. Here $\hat{f}_1$and $\hat{f}_2$ are solutions of two (different) linear differential equations, therefore $\hat{f}$ is also a solution of a linear (algebraic) differential equation.

Theorem 3.1. *(Ramis; cf. [29]) The power series expansion* $\hat{f}(x) = \sum_{n=0}^{+\infty} (-1)^n n!(x^n + x^{2n+2})$ *is Gevrey 1 and solution of a linear algebraic differential equation, but there exist no value of* $k \in \mathbb{R}^+$ *such that* $\hat{f}$ *is* k*-summable.*

We have $\hat{f}_2 \in \mathbb{C}[[x]]_{\frac{1}{2}}$ $(k = 2)$ (and $\frac{1}{2}$ is the optimal value). We have $\hat{f}_1 \in \mathbb{C}[[x]]_1$ and 1 is the optimal Gevrey estimate, therefore $\hat{f} \in \mathbb{C}[[x]]_1$

and 1 is also the optimal Gevrey estimate. The power series $\hat{f}_2$ is *not* 1-summable: $\hat{B}\hat{f}_2$ is an entire function but its growth at infinity is too strong ($\hat{f}$ is Gevrey 1 but "too convergent"). Finally $\hat{f}$ is not 1-summable. If $k > 1$, $\hat{f} \notin \mathbb{C}[[x]]_{\frac{1}{k}}$. Therefore $\hat{f}$ is not k-summable.

Theorem 3.2. *(Ramis; tauberian theorem) Let $k' > k > 0$. Then* $\mathbb{C}[[x]]_{\frac{1}{k'}} \cap \mathbb{C}\{x\}_{\frac{1}{k}} = \mathbb{C}\{x\}$.

Roughly speaking: k-summable + less divergent than supposed implies *convergent.*

Corollary 3.1. *Let k_1, $k_2 > 0$ with $k_1 \neq k_2$, Then*

$$\mathbb{C}\{x\}_{k_1} \cap \mathbb{C}\{x\}_{k_2} = \mathbb{C}\{x\}.$$

There is some analogy with prime numbers (Bezout). It remains to prove that $\hat{f}$ is solution of a linear O.D.E.. Let

$D = (\frac{d}{dx})^5[x^5(2-x)(\frac{d}{dx})^2 - x^2(-2x^3+5x^2+4)\frac{d}{dx} - 2(x^2-x+2)]$. The reader can check that $D\hat{f} \equiv 0$. I got D from D_1 and D_2 ($D_1\hat{f}_1 = 0$, $D_2\hat{f}_2 = 0$) using LCM algorithm in *non commutative* Euclidean rings (due O. Ore).

Finally our tauberian theorem appears as an essential obstruction to k-summability of solutions in the general case. Fortunately it is in some sense the *only obstruction.*

Theorem 3.3. *(Ramis) Let*

$$(\Delta) \qquad Y' = AY$$

be a (germ at the origin of) meromorphic differential system. Let

$$(\mathrm{End}\,\Delta) \qquad X' = [A, X]$$

be the associated system with unknown matrix function X. We denote by $k_1 > \cdots > k_\ell > 0$ the strictly positive slopes of the Newton polygon $N(\mathrm{End}\,\Delta)$. Let

$$\hat{F} = \hat{H}(t)x^L e^Q \quad (t^\nu = x; \nu \in \mathbb{N}^*)$$

be a formal fundamental solution of (Δ).

Then there exists a unique (up evident multiplications by holomorphic invertible matrices) decomposition:

$$\hat{H} = \hat{H}_1 \cdots \hat{H}_\ell,$$

where $\hat{H}_i$ is k_i-summable ($i = 1, \cdots, \ell$) (in x, i.e. $k_i\nu$-summable in t). Moreover, if we denote by $S_d\hat{H}_i$ the k_i-sum of $\hat{H}_i$ in a[15] *direction d and if we choose a branch of* Log x *in a sector bisected by d, we can define $S_d\hat{F} = S_d\hat{H}_1 \cdots S_d\hat{H}_\ell x^L e^Q$ in a (sufficiently small) sector bisected by d. Then $F = S_d\hat{F}$ is an actual solution of the system* (Δ) *for every non singular direction $d \notin \Sigma(\mathrm{End}\,\Delta)$.*

Finally the situation is not too bad. In order to check some good theory of summation, or "exact asymptotics" for formal solutions of linear O.D.E., it remains to check some good abstract properties of the summations S_d in Theorem 3.3 and moreover to get *explicit algorithms* of computation for the sums $S_d\hat{f}$. This can be done as we will explain now. Moreover at the end we will get a "miracle": this exact asymptotics built for solutions of *linear* equations will work *also* for arbitrary solutions of arbitrary analytic *non linear* O.D.E.

When d is a singular direction for $\hat{f}$, $S_d\hat{f}$ will not exist but $S_{d_1}\hat{f}$ will exist for d_1 near of d. We will denote $S_{d_1} = S_d^+$ for such a d_1 "after" d and $S_{d_1} = S_d^-$ for such a d_1 "before" d (turning counterclockwise).

Question 3.1. Find a sub-differential algebra M of[16] $\mathbb{C}[[x]]$ ($\mathbb{C}\{x\} \subset M$) and "summation-maps" $S_d^\pm : M \to$ Holomorphic functions on germs of sectors bisected by d with some "good properties":

(i) the maps $S_d^\pm$ are injective homomorphisms of differential algebras;

(ii) $S_d^\pm$ restricted to $M \cap \mathbb{C}\{x\}_{1/k}$ is the k-summation operator ($k \in \mathbb{R}^+$); $S_d^\pm$ restricted to $\mathbb{C}\{x\}$ is the classical sum for convergent series;

(iii) for a given $\hat{f} \in M$, $S_d^+\hat{f} = S_d^-\hat{f}$ except perhaps for a finite set $\Sigma(\hat{f})$ of directions.

(iv) if $d \notin \Sigma(\hat{f})$, $\hat{f} \in M$, then $S_d\hat{f} = S_d^\pm(\hat{f})$ is asymptotic to $\hat{f}$ on a (sufficiently small) sector bisected by d.

Problem 3.1 admits a (unique[17]) positive answer asking that the differential algebra M *contains all the formal* power series solutions of *all* the *analytic linear* O.D.E.. Looking at Theorem 3.3, we can *easily* guess the solution, but it is *very difficult* to prove that this solution (the evident "blend" of k-summable series for all the $k \in \mathbb{R}^+$) works.

---

[15] Non singular for $\hat{H}_i$, *i.e.* for all its entries.

[16] I write $\mathbb{C}[[x]]$ for sake of simplicity, in fact we need all the entire ramifications $\bigcup_{m>0, m\in\mathbb{N}} \mathbb{C}[[x^{1/m}]]$.

[17] If we ask that it extends the k-sums already obtained.

As we saw above, when we apply Hukuhara-Turrittin theorem about existence of formal fundamental solutions, we need in general a *ramified* variable t ($t^\nu = x$; $\nu \in \mathbb{N}^*$). Moreover it will appear in multisummability theorem a new difficulty related to the fact that, for $k < \frac{1}{2}$ (*i.e.* $\pi/k > 2\pi$), a sector of opening π/k *overlap* itself. Therefore it is natural to work with

$$\bigcup_{\nu \in \mathbb{N}^*} \mathbb{C}[[x^{1/\nu}]].$$

We denote $\mathbb{C}[[x^{1/\nu}]]_{1/k}$ the $\mathbb{C}$-differential algebra of Gevrey series of Gevrey order $= \frac{1}{k\nu}$ in the variable $t = x^{1/\nu}$, we denote $\bigoplus_{k>0} \mathbb{C}\{x^{1/\nu}\}_{1/k}$ the $\mathbb{C}$-*vector space* generated by $\bigcup_{k>0} \mathbb{C}\{x^{1/\nu}\}_{1/k}$ (for a fixed $\nu \in \mathbb{N}^*$). Then we set:

$$M' = \bigcup_{\nu \in \mathbb{N}^*} \bigoplus_{k>0} \mathbb{C}\{x^{1/\nu}\}_{1/k}.$$

It is a $\mathbb{C}$-vector space. We have clearly

$$\bigoplus_{k>0} \mathbb{C}\{x^{1/\nu}\}_{1/k} \subset M' \cap \mathbb{C}[[x]].$$

But be careful the *inclusion is strict* [1], [2].

Theorem 3.4. *(Balser) The $\mathbb{C}$-vector space M' is a* differential algebra.

Then $M = M' \cap \mathbb{C}[[x]]$ is a $\mathbb{C}$-differential algebra. It gives a positive answer to Problem 3.1. It is not evident: we have to prove that the maps $S_d^{\pm}$ are well defined and satisfy some properties.

The key point is the following: let $\hat{f} \in M' \cap \mathbb{C}[[x]]$. We can write it (Theorem 3.4) as a sum of (perhaps *ramified*) $k_1, \cdots, k_q$-summable series,

$$\hat{f} = \hat{f}_1 + \cdots + \hat{f}_q;\ \hat{f}_i \in \mathbb{C}\{x^{1/\nu}\}_{1/k_i} \quad (i = 1, \cdots, q).$$

Then we can set: $S_d^{\pm}\hat{f} = S_d^{\pm}\hat{f}_1 + \cdots + S_d^{\pm}\hat{f}_q$ (the sum $S_d^{\pm}\hat{f}_i$ is the sum in the sense of k_i-summability).

Theorem 3.5. *(Balser) The sum $S_d^{\pm}\hat{f}$ is independent of the decomposition $\hat{f} = \hat{f}_1 + \cdots + \hat{f}_q$ in M'.*

Therefore the operators $S_d^{\pm}$ are well defined on M. It is easy to check properties (i), (ii), (iii), (iv). By definition M is the differential algebra of *multisummable* series. For $\hat{f} \in M$, if $\hat{f}$ admits a decomposition. Using $k_1, \cdots, k_q$-summable series, we will say that it is $(k_1, \cdots, k_q)$-summable.

History. After Ramis example $\hat{f} = \hat{f}_1 + \hat{f}_2$ (superposition of two copies of Euler series) which we cannot sum by k-summability it was clear that it was necessary to introduce a *more general* summability method. It seemed reasonable to ask

$$S\hat{f} = S\hat{f}_1 + S\hat{f}_2$$

(the right member being defined by k-summability, $k = 1,\ 1/2$). However, at the beginning, it was (*very* $\cdots$) *far* to be clear that such a S was well defined and even in that case that it was an injective homomorphism of $\mathbb{C}$-differential algebra ! Therefore we adopted an *indirect way*.

• The first formulation is due to Martinet - Ramis [16]. It is based on Ecalle acceleration theory.[18] Roughly speaking, we used finitely many ($\hat{f}$ being *fixed*) operators $\mathcal{B}_k$, $\mathcal{L}_k$ and, because we cannot apply $\mathcal{L}_k$ to functions with a priori a too big growth at infinity, we needed acceleration kernel operators.

From a theoretical viewpoint this formulation is satisfying. It gives *explicit* methods of summation. However these methods use *acceleration kernels* which are complicated "special functions" (related to generalized hypergeometric G-functions of Mejer). Starting from Martinet-Ramis formulation, Balser found an *equivalent formulation* using only the operators $\hat{\mathcal{B}}_k$, $\mathcal{L}_k$ (for arbitrary k).

• Second equivalent formulation of multisummability (Balser). We start from $\hat{f} \in M \subset \mathbb{C}[[x]]$ and its "levels" $k_1, \cdots, k_q$; $k_1 > \cdots .k_q > 0$ (the "slopes" appearing in the decomposition in k-summable series). We set

$$\frac{1}{\chi_i} = \frac{1}{k_i} - \frac{1}{k_{i-1}}; \quad \frac{1}{k_1} = \frac{1}{\chi_1}, \quad (i = 2, \cdots, q)$$

which defines $\chi_1, \cdots, \chi_q > 0$. Then we set

$$S_d\hat{f} = \mathcal{L}_{\chi_1} * \mathcal{L}_{\chi_2} * \cdots * \mathcal{L}_{\chi_q} * S\hat{\mathcal{B}}_{\chi_1} \cdots \hat{\mathcal{B}}_{\chi_q},$$

where d is a non singular direction (for $\hat{f}$), S is the summation operator for *convergent* power series and $*$ is the analytic continuation along the direction d. This new formulation is simpler and more easy to use. (For solutions of linear O.D.E. there exists a computer algebra implementation

[18] The idea of acceleration theory in already in G. H. Hardy work (and in the thesis of his student I. J. Good).

in the ISOLDE package). There is a precise version of the decomposition of $\hat{f} \in M$.

Theorem 3.6. *(Balser) Let $\hat{f} \in \mathbb{C}[[x]]$ be a multisummable power series expansion (i.e. $\hat{f} \in M$). We suppose that $\hat{f}$ is $(k_1, \cdots, k_q)$-summable and we define $\chi_1, \cdots, \chi_q$ as above. Then, if $\chi_1, \cdots, \chi_q > \frac{1}{2}$, we have*

$$\hat{f} = \hat{f}_1 + \cdots + \hat{f}_q$$

with $\hat{f}_i \in \mathbb{C}\{x\}_{1/k_i}$ (unramified).

If the condition $\chi_1, \cdots, \chi_q > \frac{1}{2}$ is *not* verified, we use a ramification $x = t^m$ such that it is satisfied for $m\chi_1, \cdots, m\chi_q$ and we replace x by t in Theorem 3.6.

The multisummability theory allows us to sum in a natural ("branched") way *any solution* of *any* linear analytic O.D.E..

Theorem 3.7. *(Ramis, Balser-Braaksma-Ramis-Sibuya, Malgrange - Ramis) Any formal power series solution of any linear analytic O.D.E. is multisummable, and more precisely $(k_1, \cdots, k_q)$-summable, where $(k_1, \cdots, k_q)$ are the positive slopes of the Newton polygon of the equation.*

The first proof is due to Ramis in relation with Theorem 3.3 (Theorem 3.3 implies multisummability of the entries of $\hat{H}$). Other independent proofs appeared later [3], [15]. Finally multisummability theory, mainly devised to solve *linear* equations, works also for *non-linear* equations. The following theorem was *conjectured* independently by Ecalle and Ramis. The first proof is due to Braaksma, later appeared other independent (quite different) proofs: Ramis-Sibuya [30], Balser [1], [2].

Theorem 3.8. *(Braaksma, Ramis-Sibuya, Balser) Any formal power series solution of any linear or non-linear analytic O.D.E. is multisummable.*

Later Malgrange gave a precise version: the solution $\hat{f}$ is $(k_1, \cdots, k_q)$-summable, $k_1, \cdots, k_q$ being the positive slopes of the Newton-polygon of the equation *along* the solution $\hat{f}$ (in the non-linear case this NP depends on the solution), cf. also [31].

4. Gevrey asymptotic and singular perturbations

Gevrey asymptotic expansions and multisummability theory are tools perfectly adapted to the study of *singular analytic dynamical systems* (more precisely to their *local* study). These tools work for solutions but also for other classical and important problems as *normal forms.* These are important results in this direction for the case of (germs of) analytic differmorphisms tangent to identity: $\varphi(z) = z + \cdots$ (G.D. Birkhoff, Malgrange, Ecalle, Voronin, more recently Mardesic and Rousseau ...) and for the case of *resonant* (germs of) analytic O.D.E in the plane (Martinet, Ramis, and more recently L. Teyssier, Voronin, ...) or in bigger dimensions (L. Stolovitch, B. Braaksma ...). In such situations we have *convergent* normal forms and Gevrey, or more precisely k-summable coordinate transformations reducing a systems to "its" normal form. Comparison between the natural sums gives rise to *non-linear Stokes phenomena* [12].

Resonant differential equations in the plane are local perturbation of $x^{k+1}dy + \lambda y dx = 0$ (saddle-nodes) or $pxdy + qydx = 0$ ($p,\ q \in \mathbb{N}^+$). When we perturb a nilpotent equation as $d(y^2 - x^3) = 2ydy - 2x^2dx = 0$, the situation is more dedicate. We have "prenormal forms" in Takens style. Some are convergent (a recent result of Zoladek). But the "true" normal forms (due to F. Loray) are *divergent.* M. Canalis-Durand and R. Schäfke recently proved that they are in fact Gevrey and even summable [7]. The same authors and E. Paul are working on generalizations.

Another potential field of applications is the field of *singular perturbations* of analytic dynamical systems. Gevrey asymptotics theory is the asymptotic theory *exactly adapted* to this field which is extremely important for the applications.

This appears clearly in the recent *complete solution* by A. Fruchard and R. Schäfke of the famous problem of Ackerberg-O'Malley resonance [9], [11] or in the recent solution of the important Wasow's conjecture for *turning points* by Catherine Stenger [32]. In such problems it is clear that the *good theory* is *Gevrey asymptotics* (not Poincaré's Asymptotics which is *too unprecise*, or multisummability which is *"too precise"* $\cdots$, except for some particular situations).

4.1. *Delay in bifurcations*

We begin with discrete dynamical systems. We study the iteration of a map $f : \mathbb{R} \to \mathbb{R}$ or $\mathbb{C} \to \mathbb{C}$, i.e the family of maps

$$F^{(n)} = F \circ \cdots \circ F$$

(composition n times; $n \in \mathbb{N}$). For sake of simplicity we will suppose that F is polynomial; $F \in \mathbb{R}[x]$ or $\mathbb{C}[x]$. More generally we can consider a *family* of maps $\{F_\lambda\}_{\lambda\in\mathbb{R}}$. We can "slightly" modify the problem, looking at a "small perturbation" of this situation. Let $\varepsilon > 0$ a "small" real number[19]. We replace the family of iterations: $(x, \lambda) \mapsto (F_\lambda(x) = f(x, \lambda), \lambda)$ by

$$(x, \lambda) \mapsto F_\epsilon(x, \lambda) = (f(x, \lambda), \lambda + \varepsilon)$$

(we change a little bit the real (or complex) parameter λ at each iteration map). We will investigate what happens when we perturb a *bifurcation* of the family by a *slow drift*.

A typical example is the Feigenbaum family:

$$f(x, \lambda) = \lambda x(1 - x).$$

We need a hypothesis: we will suppose that there exists an analytic (in λ) curve of fixed points $\lambda \mapsto C_0(\lambda)$ for the unperturbed family.

This problem has been studied by C. Baesens (by a Gevrey approach) and, independently, A. Fruchard (using Neishstadt ideas in the spirit of the summation at the smallest term).

An invariant curve for the perturbed family is $\lambda \mapsto U(\lambda, \varepsilon)$ with

$$F_\varepsilon(U(\lambda, \varepsilon), \lambda) = U(\lambda + \varepsilon, \varepsilon).$$

The two authors prove the existence of such a curve if $\partial_x F_\varepsilon(\lambda, c_0(\lambda)) \neq 1$ *formally*: these exists a *unique*

$$\hat{U}(\lambda, \varepsilon) = c_0(\lambda) + \sum_{n\geq 1} c_n(\lambda)\varepsilon^n$$

satisfying our relation. In general $\hat{U}$ is divergent but[20] Gevrey 1 in ε (the c_n's being holomorphic and uniformly bounded in n on a same disc in λ).

By an incomplete Laplace transform we can get a "quasi-invariant quasi curve" (an exponentially narrow "invariant domain" at a distance of order ε of c_0).

In Fig 4.1.1, the dotted lines correspond to the bifurcation of equilibriums for the unperturbed family. The black line corresponds to the same bifurcation for the drift case. We observe a *delay in the bifurcation*, Feigenbaum drifted case is very interesting. The reader can easily draw the corresponding drifted "cascade" of bifurcations. This drifted case appears in some laser problems in physics. The explanation of the delay phenomenon

[19] In some works on this subject, ε is "infinitely small" (Non Standard Analysis).
[20] C. Baesens.

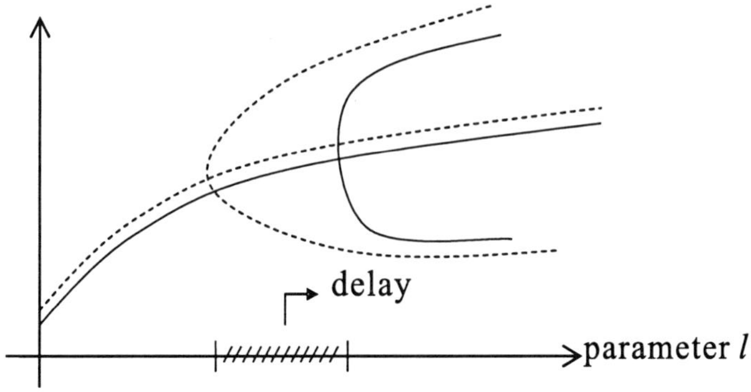

Fig 4.1.1

is essentially the same in the discrete and continuous case. We will limit ourselves to the continuous case.

We consider a family of vector fields on a domain of $\mathbb{C}^p$ (or $\mathbb{R}^p$). The parameter λ is complex (or real):

$$\dot{x} = \frac{dx}{dt} = f_\lambda(x) \qquad (x \in \mathbb{C}^p)$$

is the corresponding family of autonomous O.D.E (t real or complex is the "time"). We will always suppose that f_λ is analytic in x and λ.

We now consider a small perturbation (a *non autonomous* one), $\varepsilon > 0$ being a *fixed* small real;

$$\begin{cases} \dot{x} = \dfrac{dx}{dt} = f_\lambda(x) = f(\lambda, x) \\ \dot{\lambda} = \dfrac{d\lambda}{dt} = \varepsilon \end{cases}$$

We can suppose that $f(0,0) = 0$. The simplest case is when $\dfrac{\partial f}{\partial x}(0,0) \in GL(n;\mathbb{C})$ (*i.e.* when the linear part is *invertible* near the origin). Eliminating the time t, we get

$$\varepsilon \frac{dx}{d\lambda} = f(\lambda, x).$$

The operator $\varepsilon \dfrac{d}{d\lambda}$ is a *singular perturbation*; if $\varepsilon = 0$, we get 0 which is not differential. The discrete case $\lambda \mapsto \lambda + \varepsilon$ corresponds to $\exp(\varepsilon \dfrac{d}{d\lambda})$ which can be considered as a singularly perturbed operator, differential of *infinite order*.

For $\varepsilon = 0$, $\varepsilon \frac{dx}{d\lambda} = f(\lambda, x)$ become $0 = f(\lambda, x)$. This is an analytic *implicit* equation. With our invertibility hypothesis, it admits (in a neighborhood of $\lambda = 0$) a unique solution $c_0(\lambda)$ which is analytic:

$$c_0(\lambda) = \sum_{n>0} a_n \lambda^n.$$

We will suppose that the so-called *slow curve* $\lambda \mapsto c_0(\lambda)$ is tangent to the field (*i.e.* $c_0 \neq 0$)

$$f(x, \lambda)\frac{\partial}{\partial x} + \varepsilon \frac{\partial}{\partial \lambda}.$$

During the spring 1989, in relation with a phenomena of delay of bifurcation observed experimentally for such families, the French mathematician J. Martinet conjectured the following.

Conjecture 4.1. *(Martinet) With the above notation and hypothesis, there exists a unique formal curve $\hat{c}(\varepsilon, \lambda) = \sum_{n \geq 0} c_n(\lambda)\varepsilon^n$ invariant by the field, where the c_n's are holomorphic and uniformly bounded in a same disc. Moreover, $\hat{c}$ is Gevrey 1 in ε (uniformly in a sufficiently small ε).*

We suppose that when λ varies there is a stable equilibrium for $\lambda > 0$ ($\lambda \in \mathbb{R}$) and that we loss the stability for $\lambda = 0$: we have an unstable equilibrium for $\lambda > 0$ (typically a Hopf bifurcation at $\lambda = 0$ with the birth of a limit cycle growing when λ growth, $\lambda > 0$). In such a case we can observe a delay in the bifurcation for the perturbed case: the bifurcation is observed for a *strictly positive* value of the parameter λ.

J. Martinet gave an explanation of this phenomena based upon his conjecture. *Gevrey estimates* are *essential* in this approach.

From Conjecture 4.1 we get a *"quasi-invariant quasi-curve"*: we use an incomplete Laplace transform of $\hat{c}$ (in the variable ε). We get a *"quasi-invariant curve"* (quasi-curve because of the cutting choice ...). It is important to notice that in general $\hat{c}$ is divergent in ε. Roughly speaking we have an invariant region which is exponentially narrow (in ε). The distance between c_0 and this quasi-curve is of the order of ε ($\sim c_1\varepsilon$), the concentration is of the order of some $e^{-h/\varepsilon}$ (typically $\varepsilon = \frac{1}{10}$, $c_1\varepsilon \sim \frac{1}{5}$, $e^{-h/\varepsilon} = e^{-\frac{1}{20}}$).

We can apply a linearization of our O.D.E "along the quasi-invariant curve" (Poincaré variational equation). Up to *exponentially small errors* we are reduced[21] to the case of a curve which is analytic and invariant by the field. This case is similar to the following case: while a slow curve is invariant by the field c_0, it is easily reduced to $c_0(\lambda) \equiv 0$.

Then the explanation of the delay of bifurcation is evident. We will see that on a simple example. We consider the simplest example of our situation:

$$\begin{cases} \dot{x} = \mu x \\ \dot{\mu} = \varepsilon \end{cases} \tag{4.1.1}$$

Here $\lambda = \mu$. For $\mu \neq 0$ *fixed*, we consider the O.D.E:

$$\dot{x} = \mu x \tag{4.1.2}$$

The point $x = 0$ is an equilibrium for (4.1.2). It is a stable equilibrium for $\mu < 0$ and an unstable one for $\mu > 0$. There is a *bifurcation* for $\mu = 0$. We consider the unique solution of (4.1.1) corresponding to the initial condition $x = x_0$, $\mu = \mu_0 < 0$. We see immediately that it can be computed "in closed form" (we have $\dfrac{dx}{d\mu} = \dfrac{\mu}{\varepsilon}x$) :

$$x(\mu) = x_0 e^{\frac{\mu^2 - \mu_0^2}{2\varepsilon}} \tag{4.1.3}$$

(From 4.1.3 we get $x(\varepsilon, t) = x_0 e^{\frac{(\mu_0 + \varepsilon t)^2 - \mu_0^2}{2\varepsilon}}$.)

We see that for $\mu \in]-\mu_0, \mu_0[$, $x(\varepsilon)$ is infinitely small in ε. Then our solution goes down nearly vertically from the initial condition (x_0, μ_0) to a point exponentially near $(0, \mu_0)$, follows the "horizontal" real axis until μ arrives exponentially near of $-\mu_0 > 0$, then jumps almost vertically to the "final condition" $(x_0, -\mu_0)$. When the solution follows the horizontal axis it remains exponentially near of this axis (cf. Fig 4.1.2).

It remains to prove Martinet conjecture (conjecture 4.1). It was proved by Y. Sibuya almost at the same time it was stated by J. Martinet (and completely independently from any consideration of delay in bifurcation!)

Theorem 4.1. *(Sibuya) Let $n \in \mathbb{N}^*$. We consider a differential equation*

$$\varepsilon^\sigma \frac{dy}{dx} = F(x, \varepsilon, y) = f(x, \varepsilon) + A(x, \varepsilon)y + \sum_{|p| \geq 2} f_p(x, \varepsilon) y^p$$

where $\sigma \in \mathbb{N}^*$, ε, $x \in \mathbb{C}$, $y \in \mathbb{C}^n$. We suppose that F is analytic in (x, ε, y) in a neighborhood of $(0, 0, 0)$ and that $f(0, 0) = 0$. We suppose also that the matrix $A(0, 0)$ is invertible.

[21] More precisely the two cases are similar.

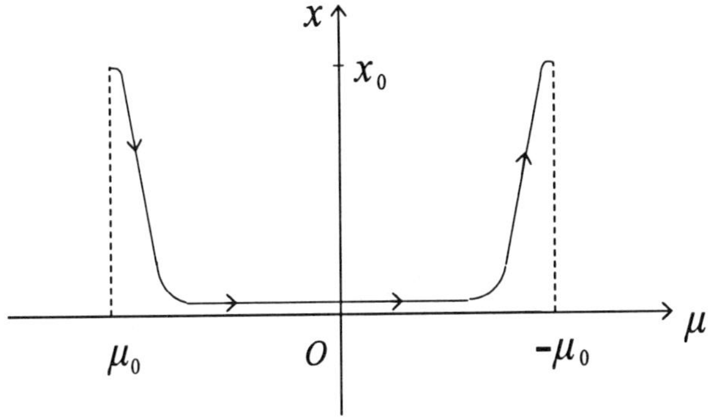

Fig 4.1.2

Then the equation admits a unique formal (in ε) solution $y = \hat{f}(x, \varepsilon) = \sum_{n \geq 0} a_n(x)\varepsilon^n$, where the a_n's are holomorphic and bounded in a same disc of the x-plane. Moreover (reducing this disc if necessary), $\hat{f}$ is Gevrey $\dfrac{1}{\sigma}$ in ε:

$$||a_n|| < c(n!)^{1/\sigma} B^n \qquad (C,\ A > 0;\ ||a_n|| = \sup |a_n(x)|)$$

Sibuya's proof of his theorem was delicate. It used cohomological estimates and Ramis-Sibuya theorem. Later M. Canalis-Durand gave a direct proof using a Gevrey method for P.D.E ("Méthode des contours successions"). Later R. Schäfke gave a very simple proof using the same trick (Nagumo norms) and a majorant equation (which is a singular O.D.E). (The idea is to "replace" $\varepsilon \dfrac{d}{dx}$ by $z^2 \dfrac{d}{dz}$.)

Now there is a natural question: what happens if in Sibuya theorem, or Matinet conjecture .... we suppress the invertibility condition? An example of such a situation was studied *before* Martinet work, it is the so-called *duck phenomena* ("le phénomène canard").

We will describe this phenomena, more precisely how it was discovered by J.L. Callot, F. and M. Diener and E. Benoit following an idea of the French (Alsacian ...) mathematician G. Reeb at the beginning of the eighties [4]. Later we will see that it is related to a more sophisticated version of Sibuya theorem [6]. But that was discovered a long time after!

Finally we have an unified theory involving a lot of Gevrey asymptotics[22].

References

1. W. Balser, *From Divergent Power Series to Analytic Functions. Theory ans application of multisummable power series*, Springer Lecture Notes in Maths 1582 (1994).
2. W. Balser, *Formal Power Series and Linear Systems of Meromorphic Ordinary Differential Equations*, Springer Universitext (2000).
3. W. Balser, B.L.J. Braaksma, J.P. Ramis, Y. Sibuya, Multisummability of formal power series solutions of linear ordinary differential equation, *Asymptotic Analysis* **5** (1991), p. 27-45.
4. E. Bénoît, J.L. Callot, F. Diener, M. Diener, Chasse au canard, *Collectanea mathematica* **32** (1981), p. 37-119.
5. B. Braaksma, Multisummability and Stokes Multipliers of linear meromorphic differential equations, *J. Differential Equations* **92** (1991), p. 45-75.
6. M. Canalis-Durand, J. P. Ramis, R. Schäfke, Y, Sibuya, Gevrey solution of singularly perturbed differential equations. *J. Reine Angew. Math.* **518** (2000), p. 95-129.
7. M. Canalis-Durand, R. Schäfke, On the normal form of a system of differential equations with nilpotent linear past, *C.R. Acad. Sci. Paris* **336** (2003), p. 129-134.
8. R. Feymann, *QED Quantum Electrodynamics, The strange Theory of Light and Matte*, Princeton University Press (1985).
9. A. Fruchard, R. Schäfke, Overstability and Resonance, *Ann. Inst. Fourier* **53** (2003), p. 227-264.
10. A. Fruchard, R. Schäfke, Bifurcation delay and difference equations, *Nonlinearity* **16** (2003), p. 2199-2220.
11. A. Fruchard, R. Schäfke, Classification of resonant equations, preprint IRMA (010) 2003, Strasbourg.
12. Yu.S. Il'Yaskenkv (Editor), Nonlinear stokes phenomena, AMS, advances in Soviet Mathematics, 14, 1991.
13. W. Magnus, F. Oberhettinger, R. P. Soni, *Formulas and theorems for the special functions of Mathematical Physics*, Springer-Verlg Berlin (1966).
14. B. Malgrange, Sommation des séries divergentes, *Expositiones Mathematicae*,**13** (1995), p. 147-176.
15. B. Malgrange, J. P. Ramis, Functions Multisummables, *Ann. Inst. Fourier, Grenoble* **41** (1991), p. 353-368.
16. J. Martinet, J. P. Ramis, Elementary Acceleration and Multisummability, *Ann. Inst. Henri Poincare, Physique Théorique* **54** (1991), p. 331-401.

[22]Unfortunately, by lack of time the author could not finish writing this part of his courses; for more details, see [24] and his recent works (common with Juan Moralès-Ruiz): *Galoisian obstructions to integrability of Hamiltonian systems.* I (Mathods Appl. Anal. 8, No.1, 33-95 (2001)); II (Mathods Appl. Anal. 8, No.1, 97-111 (2001)); ...

17. J. Martinet, J. P. Ramis, Theorie de Galois defferentielle et Resommation, *Computer Algebra and differential equations*, E. Tournier Ed., Academis Press (1989), p. 117-214.
18. J. Martinet, J. P. Ramis, Problems de modules pour des equations des equations differentielles non lineaires du premier order, *Publications Math. de 1'I. H. E. S.* **55** (1982), p. 64-164.
19. J. Martinet, J. P. Ramis, Classification analytique des equations differentielles non lineaires resonnantes du premier order, *Ann. Sci. Éc. Norm. Supér.* **16** (1983), p. 571-621.
20. E. Matzinger, Etudes des solutions surstables de l'équation de Van der Pol, *Ann. Fac. Sci. Toulouse*, VI Sér., Math. **10** (2001), p. 713-744.
21. F. W. J. Olver, *Asymptotics and Special Functions*, Academis Press (1974).
22. H. Poincaré, Acta Math. 8 (1882).
23. H. Poincaré, *Mécanique Céleste*, Tome II (1892), reprint Dover Publ. New-York (1957).
24. J. P. Ramis, *Séries divergentes et théories asymptotiques*, Panoramas et Synthses 0, Socit Mathmatique de France(1994).
25. J. P. Ramis, Devissage Gevrey, *Astérisque*, **59-60** (1978), p. 173-204.
26. J. P. Ramis, *Théorèmes d'indices Gevrey pour les équations différentielles ordi-naires*, Memoirs of the American Mathematical Society, 296(1984).
27. J. P. Ramis, Leéries k-sommables et leurs applications, *Complex Analysis, Microlocal Calculus and Relativistic Quantum Theory*, Proceedings, Les Houches 1979, Springer Lecture Notes in Physics, 126(1980), p. 178-199.
28. J. P. Ramis, R. Schäfke, Gevrey separation of fast and slow variable, *Nonlinearity* **9** (1996), p. 353-384.
29. J. P. Ramis, Y. Sibuya, Hukuhara's domains and fundamental existence and uniqueness theorems for asymptotic solutions of Gevrey type, *Asymptotic Analysis* **2** (1989), p. 39-94.
30. J. P. Ramis, Y. Sibuya, A new proof of multisummability of formal solutions of nonlinear meromorphic differential equations, *Ann. Inst. Fourier* **44** (1994), p. 811-848.
31. Y. Sibuya, *Linear differential equations in the complex domain: problems of analytic continuation*, Translation of Mathematical Monographs 82, AMS (1990).
32. C. Stenger, Points tournants de systems d'équations différentielles ordinaires singulièrement perturbées, preprint IRMA (019) 1999.
33. W. Wasow, *Asymptotic Expansions for Ordinary Differential Equations*, Interscience Publishers New-York (1965),reprint Dover Publ. New-York (1987).
34. E.T. Whittaker, G. Watson, *A Course of Mordern Analysis*, Cambridge University Press (1902).
35. C. Zhang, Développement Asymptotiques q-Gevrey et séries Gq-sommables, *Ann. Ins. Fourier* **49** (1999), p. 227-261.

SPIKES FOR SINGULARLY PERTURBED REACTION-DIFFUSION SYSTEMS AND CARRIER'S PROBLEM

MICHAEL J. WARD
Dept. of Mathematics,
University of British Columbia,
Vancouver, B.C., Canada V6T 1Z2
E-mail: ward@math.ubc.ca

1. Introduction

For a two-component reaction-diffusion system, Turing [97] in 1952 used a linear stability analysis to show that a spatially homogeneous steady-state solution that is stable in the absence of diffusion can be destablilized when the ratio of the diffusion coefficients is sufficiently large. This diffusion induced instability of the homogeneous state leads to various types of stable spatial patterns. Spike patterns are those where one or both of the chemical species concentrate, or localize, at certain points in the domain. There has been considerable progress over the past thirty years in developing weakly nonlinear theories to analyze small amplitude patterns that emerge from an initially spatially homogeneous solution. However, there are many open problems that relate to the stability and dynamics of fully nonlinear, but localized structures, in reaction-diffusion systems. In the survey chapter on pattern formation by Knobloch (see [61]), he remarks that "The question of stability of finite amplitude structures, be they periodic or localized, and their bifurcation is a major topic that requires new insights". In this survey we make an effort in this direction by giving some results for the existence, stability, and dynamics, of spike patterns to a few reaction-diffusion systems, including the Gierer-Meinhardt (GM) model [36], the Gray-Scott (GS) model [37], and the Schnakenburg (SC) model [90]. Specific open problems are also mentioned.

The GM model has been used to model a range of localization processes in nature, including morphogenesis and the formation of sea-shell patterns (cf. [70], [71]). In dimensionless form, the GM model is given by

$$a_t = \varepsilon^2 \Delta a - a + \frac{a^p}{h^q}\,, \quad x \in \Omega\,; \quad \partial_n a = 0\,, \quad x \in \partial\Omega\,, \tag{1.1a}$$

$$\tau h_t = D\Delta h - h + \frac{a^m}{\varepsilon^N h^s}\,, \quad x \in \Omega\,; \quad \partial_n h = 0\,, \quad x \in \partial\Omega\,. \tag{1.1b}$$

Here Ω is a bounded simply-connected domain in $\mathbb{R}^N$, a and h are the activator and inhibitor concentrations, $\varepsilon^2 \ll 1$ and D are the two diffusivities, τ is a reaction-time constant, and ∂_n denotes the outward normal derivative. The exponents (p, q, m, s) in the GM model are assumed to satisfy

$$p > 1, \qquad q > 0, \qquad m > 0, \qquad s \geq 0, \qquad \zeta \equiv \frac{mq}{p-1} - (s+1) > 0\,. \tag{1.2}$$

For $N > 3$, we assume that $p < p_c = (N+2)/(N-2)$, where p_c is the critical Sobolev exponent.

The Gray-Scott system, introduced in [37] and popularized in the numerical study of [86], models an irreversible reaction involving two reactants in a gel reactor, where the reactor is maintained in contact with a reservoir of one of the two chemicals in the reaction. In the nondimensional variables of [74], this system can be written in one spatial dimension as

$$v_t = \varepsilon^2 v_{xx} - v + Auv^2\,, \quad |x| < 1\,; \quad v_x(\pm 1, t) = 0\,, \tag{1.3a}$$

$$\tau u_t = Du_{xx} + (1-u) - uv^2\,, \quad |x| < 1\,; \quad u_x(\pm 1, t) = 0\,. \tag{1.3b}$$

Here $D > 0$, $\tau > 1$, $\varepsilon \ll 1$, and $A > 0$. Two ranges of the feed-rate parameter A are of interest: the high feed-rate regime where $A = O(1)$ and the low feed-rate regime where $A = O(\varepsilon^{1/2})$. In this latter regime, we introduce the new variables

$$A = \varepsilon^{1/2} \tilde{A}\,, \qquad v = \varepsilon^{-1/2} \nu\,. \tag{1.4}$$

In terms of (1.4), (1.5) is transformed to

$$\nu_t = \varepsilon^2 \nu_{xx} - \nu + \tilde{A} u \nu^2\,, \quad |x| < 1\,; \quad \nu_x(\pm 1, t) = 0\,, \tag{1.5a}$$

$$\tau u_t = Du_{xx} + (1-u) - \varepsilon^{-1} u \nu^2\,, \quad |x| < 1\,; \quad u_x(\pm 1, t) = 0\,. \tag{1.5b}$$

The Schnakenberg model [90] is another well-known two-component reaction-diffusion system. In non-dimensional form, and in one spatial dimension, this system is given by (cf. [107]),

$$u_t = \varepsilon^2 u_{xx} - u + vu^2 \,, \quad |x| < 1 \,; \quad u_x(\pm 1, t) = 0 \,, \tag{1.6a}$$

$$\tau v_t = D v_{xx} + \frac{1}{2} - \frac{b}{\varepsilon} v u^2 \,, \quad |x| < 1 \,; \quad v_x(\pm 1, t) = 0 \,. \tag{1.6b}$$

In this survey we will only consider the semi-strong interaction limit where $\varepsilon \ll 1$ and $D = O(1)$. In this limit, only one of the two chemical species in (1.1), (1.5), and (1.6), becomes localized. For each of these problems, we will find that the localizing chemical can be well-approximated by dilations of the radially symmetric ground-state solution $w(\rho)$, with $\rho = |y|$, where $w(\rho)$ is the unique solution to

$$w^{''} + \frac{(N-1)}{\rho} w^{'} - w + w^p = 0 \,, \quad \rho \geq 0 \,, \tag{1.7a}$$

$$w(0) > 0 \,, \quad w^{'}(0) = 0 \,, \quad w \sim c\rho^{-(N-1)/2} e^{-\rho} \,, \quad \text{as} \quad \rho \to \infty \,. \tag{1.7b}$$

where $p > 1$ and c is a positive constant. There exists a unique solution to (1.7) when $N = 1$ and $N = 2$, while for $N \geq 3$ we require that $p < p_c$ (cf. [65]. The one-dimensional problem with $N = 1$, plays an important role in our survey. In this case, with $\rho = y$, we readily calculate that

$$w(y) = \left(\frac{p+1}{2}\right)^{1/(p-1)} \left(\cosh\left[\frac{(p-1)y}{2}\right]\right)^{-2/(p-1)} \,, \quad c = (2p+2)^{1/(p-1)} \,. \tag{1.8}$$

For the special case $p = 2$ and $N = 1$, we get $w(y) = \frac{3}{2}\text{sech}^2(y/2)$.

The topics in this survey are organized into sections of increasing problem complexity. In Sec. 2 we consider scalar nonlinear problems. We first focus on constructing k-spike equilibria to nonlinear boundary value problems of the type originally considered in [13]. Then, we describe some results for spikes in scalar quasilinear elliptic problems. In Sec. 3 we study the stability and dynamics of spikes for certain scalar, but nonlocal, problems. One such problem arises from the shadow limit associated with (1.1). In Sec. 4 we survey some bifurcation and dynamical phenomena that occur for one-spike solutions to the reaction-diffusion system (1.1). Finally, in Sec. 5 we give some equilibrium and stability results for multi-spike patterns of the reaction-diffusion systems (1.1), (1.5), (1.3), and (1.6).

2. Spike Equilibria for Scalar Problems

In this section we begin by constructing asymptotic solutions for

$$\varepsilon^2 u^{''} + Q(u) = 0 \,, \qquad -1 < x < 1 \,, \tag{2.1}$$

with various boundary conditions at $x = \pm 1$. The assumptions on $Q(u)$ are that $Q(u)$ is smooth and that it has only two zeroes, s and s_b, with $s < s_b$, where $Q^{'}(s) < 0$, and $Q^{'}(s_b) > 0$. In addition we assume that there exists a $u_m > s_b$ such that $\int_s^{u_m} Q(\eta)\, d\eta = 0$. In Fig. 1 we plot such a $Q(u)$, and in Fig. 2 we plot the corresponding phase-plane u_z versus u, where $z = x/\varepsilon$, showing a saddle point at $u = s$ and a separatrix structure.

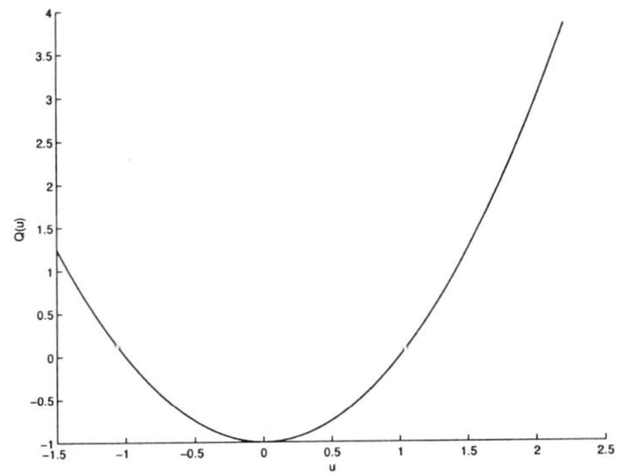

Figure 1. Plot of $Q(u)$ versus u with $Q(s) = 0$, $Q^{'}(s) < 0$, $Q(s_b) = 0$, and $Q^{'}(s_b) > 0$. There, exists a $u_m > s_b$ such that $\int_s^{u_m} Q(\eta)\, d\eta = 0$. Here $s = -1$ and $s_b = 1$.

The separatrix corresponds to a homoclinic solution $w(y)$ to (2.1) on $-\infty < y < \infty$, with $w(\pm\infty) = s$. Up to an arbitrary phase-shift, $w(y)$ satisfies

$$w^{''} + Q(w) = 0\,, \quad -\infty < y < \infty\,, \tag{2.2a}$$

$$w^{'}(0) = 0\,, \quad w(0) > 0\,; \quad w(y) \sim s + ce^{-\sigma y}\,, \quad |y| \to \infty\,. \tag{2.2b}$$

Here, $\sigma > 0$ and $c > 0$ are defined by

$$\sigma = \left[-Q^{'}(s)\right]^{1/2}, \quad \log c = \log(u_m - s) + \int_{u_m}^{s} \left[\frac{-\sigma}{\left[-2\,V(\eta)\right]^{1/2}} + \frac{1}{\eta - s} \right] d\eta\,, \tag{2.2c}$$

where $V(\eta) \equiv \int_{u_m}^{\eta} Q(\theta)\, d\theta$. This homoclinic solution leads to the existence of spike layer solutions for (2.1), where the spikes are localized near certain points in the domain. Depending on the boundary conditions for (2.1), there may be boundary layers near one or both endpoints that correspond to pieces of the homoclinic orbit.

A phase-plane analysis of (2.1) for various boundary conditions was made in [84] and [88]. The main observation is that by using the stretching

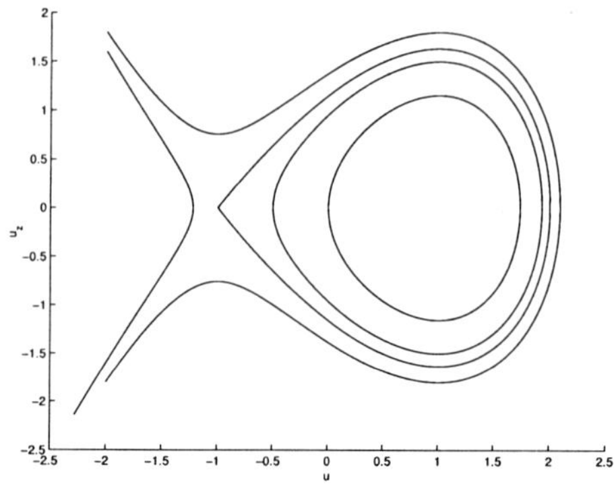

Figure 2. Plot of the phase-plane u_z versus u, with $z = x/\varepsilon$, for a $Q(u)$ that admits spike-type solutions.

$y = x/\varepsilon$ in (2.1), it is clear that boundary and internal layer solutions for (2.1) correspond to trajectories in the phase-plane that are close to the separatrix in Fig. 2 and are away from the saddle point region near $u = s$. Since the solution spends a long time near the saddle point and a comparably shorter time to make a complete or partial circuit around the separatrix, it is clear that $u \sim s$ on $-1 \leq x \leq 1$ except for $O(\varepsilon)$ transitions, or spike layers, located near the endpoints or at some interior points in the domain. This geometrical picture is very useful for obtaining qualitative information regarding the admissible spike configurations. In particular, it shows that for the boundary data $u = s$ at $x = \pm 1$, solutions with two or more interior spikes are impossible for (2.1) since any trajectory must necessarily lie outside the closed region bounded by the separatrix. Despite the simplicity of the geometrical picture afforded by the phase-plane, it does not readily yield detailed quantitative information regarding questions such as determining the locations of the spikes for arbitrary boundary data, the possibility of bifurcation behavior, and determining the number of solutions that exist for a fixed but small ε. Furthermore, the phase-plane does not generalize to the singularly perturbed quasilinear elliptic problems in two or more spatial dimensions. Therefore, it is desirable to develop an asymptotic method to construct spike layer solutions to (2.1) that can be extended to the multi-dimensional case.

The first example illustrating the difficulty in applying the method of matched asymptotic expansions to treat (2.1) was given by Carrier and Pearson (see pages 202-205 of [13]) for the special nonlinearity $Q(u) =$

$u^2 - 1$, and with $u(\pm 1) = 0$. For this reason, we refer to (2.1) as Carrier's problem. Near each endpoint there are two possible boundary layer solutions. Superimposed on these boundary layer solutions, they tried to construct a solution with one interior spike. However, as shown in [13], a routine application of the method of matched asymptotic expansions fails to determine the interior spike location x_0. In the vicinity of $x = x_0$, it is easy to see that u has the form

$$u(x) \sim 2w\left[\sqrt{2}\varepsilon^{-1}(x - x_0)\right] - 1\,, \tag{2.3}$$

where $w(y)$ is given in (1.8) with $p = 2$. This failure in determining x_0 is not restricted to the choice $Q(u) = u^2 - 1$ but is typical for the class of problems (2.1). An extension of the method of matched asymptotic expansions was used in [66] to determine the spike locations for $Q(u) = u^2 - 1$. There it was shown that the failure of a routine application of the method of matched asymptotic expansions in determining the spike locations was a result of ignoring exponentially small terms in the expansion of the solution. By extending this method to retain and match the dominant exponentially small terms, it was shown how to find the correct spike layer locations for $Q(u) = u^2 - 1$. Another analytical approach to determine the spike locations for $Q(u) = u^2 - 1$ was given in [51]. They employed a variational principle, with trial functions from the matched asymptotic expansion solution, and determined the spike locations by making the variation stationary with respect to the spike layer locations. More recently, for the nonlinearity $Q(u) = u^2 - 1$ a rigorous shooting method has been developed in [85] for constructing spike layer solutions, and for determining the number of such solutions for a fixed ε with $\varepsilon \ll 1$. Another rigorous approach based on Green's functions is given in [52]. One limitation of these approaches are that they rely heavily on analytical formulae that are available only for $Q(u) = u^2 - 1$, and they are inherently methods that will work only in one space dimension.

In [100], an analytical method called the *projection method*, was used to determine the spike layer locations for (2.1) for various boundary conditions. As shown in [100], the indeterminacy in constructing interior spike solutions for (2.1) is a result of the occurrence of exponentially small eigenvalues in the spectrum of the linearized operator. In this sense, the linearized problem is exponentially ill-conditioned. The projection method, which combines traditional matched asymptotic analysis with spectral theory, exploits this exponential ill-conditioning by imposing limiting solvability conditions on the solution to the linearized equation. In this sense the projection method has much in common with a Lyapunov-Schmidt approach to study the linearized equation. Although this method is formal, it does not require any detailed knowledge of the explicit form of the homoclinic orbit and so can treat an arbitrary $Q(u)$ of the form shown in Fig. 1. In addition, it can be extended to treat related problems in several space dimensions.

In addition to Carrier's problem, there are other classes of singularly perturbed boundary-value problems where a straightforward application of the method of matched asymptotic expansions fails to determine the solution uniquely as a result of neglecting important exponentially small terms. Such problems include boundary layer resonance phenomena for turning point problems [67], heteroclinic interfaces in reaction-diffusion systems, certain viscous shock waves for convection-diffusion problems, the Cahn-Hillard equation etc. These problem are all exponentially ill-conditioned and the projection method can be used to eliminate the indeterminacy. A survey of these other problems is given in [102].

2.1. *Interior Spike Solutions: No Boundary Layers*

We now outline the projection method for constructing a solution to

$$\varepsilon^2 u'' + Q(u) = 0\,, \qquad -1 < x < 1\,, \tag{2.4a}$$

$$\varepsilon u'(1) + \kappa_r \left[u(1) - s\right] = 0\,; \qquad \varepsilon u'(-1) - \kappa_l \left[u(-1) - s\right] = 0\,, \tag{2.4b}$$

with one interior spike and with no boundary layers near $x = \pm 1$. Here $\kappa_r \geq 0$ and $\kappa_l \geq 0$ and s is the saddle-point location for $Q(u)$. In the phase-plane u_z versus u, where $z = x/\varepsilon$, these special boundary conditions geometrically represent straight line segments originating from the saddle-point $u = s$. Each of these lines is either locally inside or outside the loop bounded by the separatrix. This geometrical fact will have an implication regarding the location of a spike for (2.4).

The first step in the projection method is to look for an approximate solution to (2.4) in the form

$$u(x) = w\left[\varepsilon^{-1}(x - x_0)\right] + R(x)\,, \tag{2.5}$$

where x_0, with $|x_0| < 1$, is to be found, and the error $R(x)$ is assumed to be small. We substitute (2.5) into (2.4) and linearize. Assuming that x_0 is not $O(\varepsilon)$ close to ± 1, we can use the far-field behavior of $w(y)$ in (2.2b) to estimate the residuals in satisfying the boundary conditions in (2.4b). In this way, we obtain that R satisfies

$$LR \equiv \varepsilon^2 R'' + Q'(w)R = 0\,, \qquad -1 < x < 1\,, \tag{2.6a}$$

$$\varepsilon R'(1) + \kappa_r R(1) \sim c e^{-\sigma y_r}(\sigma - \kappa_r)\,, \tag{2.6b}$$

$$\varepsilon R'(-1) - \kappa_l R(-1) \sim c e^{\sigma y_l}(\kappa_l - \sigma)\,. \tag{2.6c}$$

Here we have defined $y_r \equiv \varepsilon^{-1}(1 - x_0)$ and $y_l = -\varepsilon^{-1}(1 + x_0)$.

Next, we consider the associated eigenvalue problem

$$L\phi \equiv \varepsilon^2 \phi'' + Q'(w)\,\phi = \lambda\phi\,, \quad -1 < x < 1\,, \tag{2.7a}$$

$$\varepsilon\phi'(1) + \kappa_r\,\phi(1) = 0\,; \quad \varepsilon\phi'(-1) - \kappa_l\phi(-1) = 0\,. \tag{2.7b}$$

For $\varepsilon > 0$ fixed, there exists a countably infinite number of eigenvalues λ_j of (2.7) for $j = -1, 0, 1, \ldots$. The corresponding eigenfunctions form a complete set with respect to the space of square integrable functions. However, since $\varepsilon \ll 1$, there is an eigenvalue of (2.7), labeled by λ_0, which tends to zero as $\varepsilon \to 0$, and the corresponding (unnormalized) eigenfunction ϕ_0 is nearly proportional to $w'(y)$, where $y = \varepsilon^{-1}(x - x_0)$. Since w is a spike, the derivative w' has exactly one zero-crossing. Hence, by standard Sturm-Liouville oscillation theory, we expect that λ_0 is the second largest eigenvalue in the spectrum. The principle, or equivalently, largest eigenvalue, labeled by λ_{-1}, tends to a positive value as $\varepsilon \to 0$. This is discussed more fully in Sec. 3 when we analyze the stability of spike layer solutions.

Roughly speaking, the exponential ill-conditioning results from a slight break of a translation invariance that occurs for Carrier's problem posed in the infinite line. On the infinite line, where there is translation invariance, w' is an exact eigenfunction corresponding to a zero eigenvalue. For our finite-domain problem where the spike is localized near $x = x_0$ the translation invariance is broken only through the interaction of the far-field behavior of $w(y)$ with the boundaries at $x = \pm 1$. Near the boundaries, $w - s$ is exponentially small, and thus this interaction is exponentially weak. As we now show, this weak interaction perturbs the zero eigenvalue of translation by exponentially small terms.

We now calculate λ_0. Applying Green's identity to ϕ_0 and w', we get

$$\frac{\lambda_0}{\varepsilon}\left(\phi_0, w'\right) = -\phi_0(1)\left[\varepsilon w''(y_r) - \kappa_r w'(y_r)\right] + \phi_0(-1)\left[\varepsilon w''(y_l) - \kappa_l w'(y_l)\right]\,, \tag{2.8}$$

where the inner product (f,g) is defined by $(f,g) \equiv \int_{-1}^{1} fg\,dx$. Using the far-field behavior for $w(y)$ as $|y| \to \infty$ from (2.2b), (2.8) reduces to

$$\lambda_0\left(\phi_0, w'\right) \sim -\varepsilon\phi_0(1)c\sigma\left(\sigma - \kappa_r\right)e^{-\sigma\kappa_r} + \varepsilon\phi_0(-1)c\sigma\left[\sigma - \kappa_l\right]e^{\sigma y_l}\,. \tag{2.9}$$

To determine $\phi_0(\pm 1)$, we use boundary layer theory to calculate ϕ_0 as

$$\phi_0(x) = w^{'}\left[\varepsilon^{-1}(x - x_0)\right] + \phi_r[\varepsilon^{-1}(1 - x)] + \phi_l[\varepsilon^{-1}(x + 1)], \qquad (2.10)$$

where $\phi_l(\eta) \to 0$ and $\phi_r(\eta) \to 0$ as $\eta \to +\infty$. Since $Q^{'}(u) \sim Q^{'}(s) = -\sigma^2$ near $x = \pm 1$, we calculate from (2.1) that $\phi_l(\eta) = b_l e^{-\sigma\eta}$ and $\phi_r(\eta) = b_r e^{-\sigma\eta}$ for some coefficients b_l and b_r to be determined. We then substitute (2.10) into the boundary conditions (2.7b), and we use the far-field behavior (2.2b) to calculate $w^{'}$ at $x = \pm 1$. In this way we determine b_l and b_r, and from this we calculate

$$\phi_0(1) \sim -\frac{2c\sigma^2}{\kappa_r + \sigma} e^{-\sigma\varepsilon^{-1}(1-x_0)}, \quad \phi_0(-1) \sim \frac{2c\sigma^2}{\kappa_l + \sigma} e^{-\sigma\varepsilon^{-1}(1+x_0)}. \qquad (2.11)$$

Substituting (2.11) into (2.9), and using (2.10) to asymptotically evaluate $\left(\phi_0, w^{'}\right) \sim \varepsilon \int_{-\infty}^{\infty} w^{'2}(y)\, dy$, we obtain the following result:

Proposition 2.1. *For $\varepsilon \ll 1$, the spectrum of the linearized problem (2.6) around a one-spike solution has an exponentially small eigenvalue λ_0 with*

$$\lambda_0 \sim -\frac{2c^2\sigma^3}{\int_{-\infty}^{\infty} w^{'2}\, dy}\left[\left(\frac{\kappa_r - \sigma}{\kappa_r + \sigma}\right) e^{-2\sigma y_r} + \left(\frac{\kappa_l - \sigma}{\kappa_l + \sigma}\right) e^{2\sigma y_l}\right], \qquad (2.12)$$

where $y_r \equiv \varepsilon^{-1}(1 - x_0)$ and $y_l = -\varepsilon^{-1}(1 + x_0)$. The corresponding eigenfunction ϕ_0 is given by (2.10).

Next, we select the correct value of x_0 by imposing a limiting solvability condition on the solution to the linearized problem. We expand R in (2.6) in terms of the normalized eigenfunctions ϕ_j of (2.7). Using Green's second identity, we readily derive that

$$R = \sum_{j=-1}^{\infty} \frac{C_j}{\lambda_j}\phi_j, \quad C_j = -c\varepsilon\left[e^{-\sigma y_r}(\sigma - \kappa_r)\phi_j(1) - e^{\sigma y_l}(\kappa_l - \sigma)\phi_j(-1)\right]. \qquad (2.13)$$

Since $\lambda_0 \to 0$ as $\varepsilon \to 0$, a necessary condition for (2.6) to have a solution in this limit is that $C_0 \to 0$ as $\varepsilon \to 0$. This limiting solvability condition, which ensures that the projection of the error R onto the eigenspace associated with the exponentially small eigenvalue is zero, yields an equation for x_0. Setting $C_0 = 0$, and using (2.11) for $\phi_0(\pm 1)$, we obtain that

$$-e^{-2\sigma y_r}\left(\frac{\sigma - \kappa_r}{\sigma + \kappa_r}\right) \sim e^{2\sigma y_l}\left(\frac{\kappa_l - \sigma}{\sigma + \kappa_l}\right). \qquad (2.14)$$

Solving (2.14) for x_0, we obtain the following result:

Proposition 2.2. *Let* $\varepsilon \ll 1$ *and suppose that* $(\sigma - \kappa_r)(\sigma - \kappa_l) > 0$. *Then, a solution to (2.4) with one interior spike satisfies* $u \sim w\left[\varepsilon^{-1}(x - x_0)\right]$, *where* $w(y)$ *is the homoclinic solution satisfying (2.2), and* x_0 *satisfies*

$$x_0 \sim -\frac{\varepsilon}{4\sigma} \log \left[\left(\frac{\kappa_r - \sigma}{\kappa_l - \sigma} \right) \left(\frac{\kappa_l + \sigma}{\kappa_r + \sigma} \right) \right] . \tag{2.15}$$

We now make several remarks. From (2.15), we see that an interior one-spike solution does not exist when $(\sigma - \kappa_r)(\sigma - \kappa_l) < 0$. This condition is readily seen from the phase-plane shown in Fig. 2. When $\kappa_r = \kappa_l$, we get $x_0 = 0$ as expected. Secondly, Eq. (2.15) also suggests a sensitivity of the solution to small changes in the data. In particular, if $\kappa_l = 0$ and $\kappa_r = \sigma - e^{-\nu/\varepsilon}$, for some $\nu > 0$, the spike layer location x_0 is perturbed by $O(1)$ as $\varepsilon \to 0$. In fact, as shown in Sec. 2.3 below, other types of extreme sensitivity to exponentially small perturbations in Carrier's problem are possible as a result of the exponential ill-conditioning. Finally, the exponential ill-conditioning leads to significant numerical difficulties when trying to compute spike layer solutions to (2.4) using finite difference schemes. As a result of the large condition number of the linearized problem, small residuals in Newton's method do not generally imply that the iterations have converged to a close approximate solution. In particular, suppose that $Q(u) = -u + u^2$, $\kappa_r = \kappa_l = 0$, and $\varepsilon = 0.05$. Then, we calculate that $\sigma = 1$, $c = 6$, and $\int_{-\infty}^{\infty} w'^2 \, dy = 6/5$. Therefore, from (2.12), we obtain

$$\lambda_0 \sim 120 e^{-2/\varepsilon} = 5.1 \times 10^{-16} . \tag{2.16}$$

This shows that the condition number of the system is on the order $O(10^{16})$, and hence even with 16 digit floating point computations, the solution will be sensitive to roundoff error. With extended precision computations, which retain 32 digits in the computation, the negative effects of this ill-conditioning can be alleviated somewhat, but only for slightly smaller values of ε. Exponential ill-conditioning, and the numerical consequences, is discussed further in [67], [94], [93], and [102]. Numerical difficulties for the related problem of Euler buckling for large applied loads, modeled by $u'' + \lambda \sin u = 0$ for $\lambda \gg 1$, has been studied recently in [31].

Next, we follow Sec. 3 of [100] and show how to construct multi-spike solutions to (2.4) having n interior spikes at some unknown x_j for $j = 0, \ldots, n-1$, where $-1 < x_0 < x_1 < \ldots < x_{n-1} < 1$. In the remainder of this section we label the inter-spike separations d_j by $d_j = x_j - x_{j-1}$ for $j = 0, \ldots, n$, where we have labeled $x_{-1} = -1$, and $x_n = 1$. We assume that the spikes are well-separated in the sense that $d_j \gg O(\varepsilon)$ for $j = 0, \ldots, n$. We look for an n-spike solution to (2.4) in the form

$$u \sim u_c(x) \equiv \sum_{j=0}^{n-1} w\left[\varepsilon^{-1}(x - x_j)\right] - (n-1)\, s\,. \tag{2.17}$$

We substitute $u = u_c + R$ into (2.4), where $R \ll 1$. In analogy with (2.6), we obtain that R satisfies

$$\varepsilon^2 R^{''} + Q^{'}(u_c)\, R = E \equiv \sum_{i=0}^{n-1} Q\left[w(y_i)\right] - Q(u_c)\,, \tag{2.18a}$$

$$\varepsilon\, R^{'}(1) + \kappa_r\, R(1) \sim -c\,(\kappa_r - \sigma)e^{-\sigma\varepsilon^{-1}d_n}\,, \tag{2.18b}$$

$$\varepsilon\, R^{'}(-1) - \kappa_l\, R(-1) \sim -c\,(\sigma - \kappa_l)e^{-\sigma\varepsilon^{-1}d_0}\,. \tag{2.18c}$$

Here we have defined y_i by $y_i \equiv \varepsilon^{-1}(x - x_i)$, for $i = 0, \ldots, n-1$.

The eigenvalue problem associated with (2.18) is (2.7), where $Q^{'}(w)$ in (2.7a) is replaced by $Q^{'}(u_c)$. When the spikes are well-separated, there are n eigenvalues, labeled by λ_j for $j = -n, -n+1, \ldots, -1$ that tend to a common positive value as $\varepsilon \to 0$. These are the largest eigenvalues in the spectrum of the linearization. The next largest group of eigenvalues are the exponentially small eigenvalues λ_j for $j = 0, \ldots, n-1$ that result from the broken translation invariance. The corresponding eigenfunction near the core of the jth spike is $\phi_j \sim w^{'}(y_j)$ for $j = 0, \ldots, n-1$. The solution to (2.18) is expanded in terms of all of the normalized eigenfunctions ϕ_j as $R = \sum_{j=-n}^{\infty} C_j \lambda_j^{-1} \phi_j$. Using Green's identity we derive

$$C_j = (E, \phi_j) - \varepsilon\left[(\varepsilon\, R^{'}(1) + \kappa_r\, R(1))\phi_j(1) - (\varepsilon\, R^{'}(-1) - \kappa_l\, R(-1))\phi_j(-1)\right]\,. \tag{2.19}$$

Here $(f, g) \equiv \int_{-1}^{1} fg\, dx$. A necessary condition for there to be a solution to (2.18) in the limit $\varepsilon \to 0$, is that the projection of R onto the subspace spanned by ϕ_j, for $j = 0, \ldots, n-1$, vanishes. Setting $C_j = 0$ for $j = 0, \ldots, n-1$, we obtain n coupled algebraic equations for x_j given by

$$(E, \phi_j) \sim B_j \equiv \varepsilon\, c\left[(\sigma - \kappa_r)e^{-\sigma\varepsilon^{-1}d_n}\phi_j(1) + (\sigma - \kappa_l)e^{-\sigma\varepsilon^{-1}d_0}\phi_j(-1)\right]\,, \tag{2.20}$$

for $j = 0, \ldots, n-1$.

To determine explicit equations for x_j, we must evaluate the inner product term and the boundary term B_j in (2.20). For well-separated spikes, the residual E is exponentially small on $-1 < x < 1$. Since $\phi_j \sim w^{'}(y_j)$

near $x = x_j$, it follows that the dominant contribution to the inner product term (E, ϕ_j) arises from the region near $x = x_j$. To determine this contribution we must retain the dominant exponentially small terms in E near $x = x_j$. These terms arise from the nearest neighbor spikes located at $x = x_{j+1}$ and $x = x_{j-1}$ when $j = 1, \ldots, n-2$. The spikes at x_0 and x_{n-1}, which are closest to the boundary, have only one nearest neighbor. A detailed calculation as given in §3 of [100] shows that

$$(E, \phi_j) \sim \begin{cases} 2\varepsilon\, c^2\sigma^2\, e^{-\sigma\varepsilon^{-1}d_1}\,, & j = 0\,, \\ 2\varepsilon\, c^2\sigma^2 \left(e^{-\sigma\varepsilon^{-1}d_{j+1}} - e^{-\sigma\varepsilon^{-1}d_j}\right), & j = 1, \ldots, n-2\,, \\ -2\varepsilon\, c^2\sigma^2 e^{-\sigma\varepsilon^{-1}d_{n-1}} & j = n-1\,. \end{cases} \tag{2.21}$$

Next, we asymptotically calculate the boundary term B_j in (2.20). The dominant terms in B_j for $j = 0, \ldots, n-1$ arise when $j = 0$ or $j = n-1$. To compute B_0 and B_{n-1} we calculate $\phi_0(-1)$ and $\phi_{n-1}(1)$ in a similar way as was done above for a one-spike solution. We readily obtain that

$$\phi_0(-1) \sim \frac{2\sigma^2 c}{\kappa_l + \sigma} e^{-\sigma\varepsilon^{-1}d_0}\,, \qquad \phi_{n-1}(1) \sim -\frac{2\sigma^2 c}{\kappa_r + \sigma} e^{-\sigma\varepsilon^{-1}d_n}\,. \tag{2.22}$$

From (2.20), we get that

$$B_0 \sim 2\sigma^2 c^2\, \varepsilon \left(\frac{\sigma - \kappa_l}{\sigma + \kappa_l}\right) e^{-2\sigma\varepsilon^{-1}d_0}\,, \quad B_{n-1} \sim -2\sigma^2 c^2\, \varepsilon \left(\frac{\sigma - \kappa_r}{\sigma + \kappa_r}\right) e^{-2\sigma\varepsilon^{-1}d_n}\,. \tag{2.23}$$

For $j = 1, \ldots, n-2$, B_j is of an exponential order smaller than (E, ϕ_j). Therefore, in (2.20) we can set $B_j = 0$ for $j = 1, \ldots, n-2$.

Substituting (2.23) and (2.21) into (2.20), we get that (2.20) reduces to

$$e^{-\sigma\varepsilon^{-1}(d_1 - 2d_0)} \sim \left(\frac{\sigma - \kappa_l}{\sigma + \kappa_l}\right), \quad e^{-\sigma\varepsilon^{-1}(d_{n-1} - 2d_n)} \sim \left(\frac{\sigma - \kappa_r}{\sigma + \kappa_r}\right). \tag{2.24a}$$

$$e^{-\sigma\varepsilon^{-1}d_{j+1}} \sim e^{-\sigma\varepsilon^{-1}d_j}\,, \qquad j = 1, \ldots, n-2\,. \tag{2.24b}$$

It is clear from (2.24a) that there are no n-spike solutions to (2.4) when either $(\sigma - \kappa_l) < 0$ or $(\sigma - \kappa_r) < 0$. This is also readily seen from the phase-plane shown in Fig. 2. Only when $(\sigma - \kappa_l) > 0$ and $(\sigma - \kappa_r) > 0$ are both satisfied is the trajectory inside the region bounded by the separatrix, which then allows for n-spike solutions with $n \geq 2$. By solving (2.24) explicitly, we obtain the following result for the spike locations:

Proposition 2.3. *Let $\varepsilon \ll 1$ and suppose that $(\sigma - \kappa_r) > 0$ and $(\sigma - \kappa_l) > 0$.*

Then, for $n \geq 2$, there is an n-spike solution to (2.4) of the form

$$u \sim u_c(x) \equiv \sum_{j=0}^{n-1} w\left[\varepsilon^{-1}(x - x_j)\right] - (n-1)\, s\,, \tag{2.25}$$

where x_j is given by

$$x_j \sim -1 + \frac{(2j+1)}{n} + \frac{\varepsilon}{2}\left[\eta_l - \frac{1}{2n}(\eta_r + \eta_l)(1+2j)\right], \quad j = 0, \ldots, n-1\,. \tag{2.26}$$

Here η_l and η_r are defined by

$$\eta_l \equiv \sigma^{-1} \log\left(\frac{\sigma - \kappa_l}{\sigma + \kappa_l}\right), \quad \eta_r \equiv \sigma^{-1} \log\left(\frac{\sigma - \kappa_r}{\sigma + \kappa_r}\right). \tag{2.27}$$

For the Neumann problem where $\kappa_l = \kappa_r = 0$, (2.26) shows that $x_j = -1 + (2j-1)/n$, for $j = 0, \ldots, n-1$, so that $d_j = 2/n$ for $j = 1, \ldots, n-1$ and $d_0 = d_n = 1/n$. This symmetric spike spacing for the Neumann problem also follows readily from the "time-map" associated with the phase-plane.

2.2. *Interior Spike Solutions: Boundary Layers*

We now construct n-spike solutions to (2.1) for the boundary data

$$u(-1) = u_L\,, \quad u(1) = u_R\,, \tag{2.28}$$

where $s < u_L, u_R < u_m$ (see Fig. 1). A composite expansion for an n-spike solution has the form

$$u \sim u_c(x) \equiv \sum_{j=0}^{n-1} w\left[\varepsilon^{-1}(x - x_j)\right] - (n-1)\, s + \left[w_L\left(\varepsilon^{-1}(x+1)\right) - s\right] + \left[w_R\left(\varepsilon^{-1}(1-x)\right) - s\right]. \tag{2.29}$$

The boundary layer terms $w_L(\xi)$ and $w_R(\xi)$ for $\xi \geq 0$ at the left and right endpoints, respectively, are portions of the homoclinic orbit, which satisfy

$$w_L'' + Q(w_L) = 0\,, \quad w_L(0) = u_L\,, \quad w_L(\xi) \sim s + c_L e^{-\sigma\xi} \quad \text{as } \xi \to \infty\,, \tag{2.30a}$$

$$w_R'' + Q(w_R) = 0\,, \quad w_R(0) = u_R\,, \quad w_R(\xi) \sim s + c_R e^{-\sigma\xi} \quad \text{as } \xi \to \infty\,, \tag{2.30b}$$

for some constants $c_L > 0$ and $c_R > 0$. From the phase-plane, it is clear that there are two solutions to each of (2.30a) and (2.30b). One solution decreases monotonically to s, while the other first increases to u_m and then decreases to s. Therefore, there are two values for c_L and for c_R.

We substitute $u = u_c + R$ into (2.1), where $R \ll 1$. In analogy with (2.18), we obtain that R satisfies

$$\varepsilon^2 R'' + Q'(u_c) R = E \equiv \sum_{i=0}^{n-1} Q\left[w(y_i)\right] - Q(u_c) + Q[w_R(-y_n)] + Q\left[w_L(y_{-1})\right] , \tag{2.31a}$$

$$R(-1) \sim -ce^{-\sigma\varepsilon^{-1}d_0} ; \quad R(1) \sim -ce^{-\sigma\varepsilon^{-1}d_n} . \tag{2.31b}$$

Here $y_i \equiv \varepsilon^{-1}(x - x_i)$, for $i = -1, \ldots, n$, where $x_{-1} \equiv -1$ and $x_n \equiv 1$. By eliminating the projection of R onto the eigenspace associated with the exponentially small eigenvalues, the spike layer locations satisfy

$$(E, \phi_j) \sim B_j \equiv \varepsilon^2 c \left(e^{-\sigma\varepsilon^{-1}d_n} \phi_j'(1) - e^{-\sigma\varepsilon^{-1}d_0} \phi_j'(-1) \right) , \quad j = 0, \ldots, n-1 , \tag{2.32}$$

where $\phi_j \sim w'(y_j)$ near the jth spike.

The analysis of (2.32) differs in two main respects from the analysis in Sec. 2.1. Firstly, to evaluate (E, ϕ_j), we note that the interior spikes closest to the endpoints, corresponding to $j = 0$ and $j = n - 1$, now have two nearest neighbors instead of only one. Specifically, for the spike at $x = x_0$, the nearest neighbors are the spike at $x = x_1$ and the boundary layer solution w_L. A similar situation occurs for the spike nearest the right endpoint. Secondly, it can be shown that for each $j = 0, \ldots, n - 1$ the boundary term B_j in (2.32) is asymptotically smaller than (E, ϕ_j). In contrast, for the problem in Sec. 2.1, the contribution from B_0 and B_{n-1} had to be retained. Therefore, we can set $B_j = 0$ for $j = 0, \ldots, n - 1$ in (2.32), to obtain that the layer locations asymptotically satisfy $(E, \phi_j) \sim 0$. By calculating (E, ϕ_j) asymptotically as in Sec. 7.1 of [100], we obtain, in analogy with (2.24), that the inter-spike separations d_j satisfy

$$ce^{-\sigma\varepsilon^{-1}d_1} \sim c_L e^{-\sigma\varepsilon^{-1}d_0} ; \quad c_R e^{-\sigma\varepsilon^{-1}d_n} \sim ce^{-\sigma\varepsilon^{-1}d_{n-1}} . \tag{2.33a}$$

$$e^{-\sigma\varepsilon^{-1}d_{j+1}} \sim e^{-\sigma\varepsilon^{-1}d_j} , \quad j = 1, \ldots, n - 2 . \tag{2.33b}$$

By solving (2.33) explicitly, we obtain the following result:

Proposition 2.4. *Let $\varepsilon \ll 1$ and consider (2.1) with the Dirichlet conditions $u(-1) = u_L$ and $u(1) = u_R$, where $s < u_L, u_R < u_m$. Then, there is an n-spike solution of the form*

$$u \sim u_c(x) \equiv \sum_{j=0}^{n-1} w\left[\varepsilon^{-1}(x - x_j)\right] - (n - 1)\, s + \left[w_L\left(\varepsilon^{-1}(x + 1)\right) - s\right] + \left[w_R\left(\varepsilon^{-1}(1 - x)\right) - s\right] , \tag{2.34}$$

where w_L and w_R are boundary layer functions satisfying (2.30). For this problem, there are are four possible sets $(x_0, \ldots, x_n)$ of spike-layer locations, which for $j = 0, \ldots, n-1$ satisfy

$$x_j \sim -1 + \frac{2}{n+1}(1+j) + \frac{\varepsilon}{\sigma(n+1)}\left[\log\left(\frac{c^{1-n}c_L^n}{c_R}\right) + j\log\left(\frac{c^2}{c_R c_L}\right)\right]. \tag{2.35}$$

There are two values for c_L and for c_R defined in (2.30).

Therefore, for Dirichlet conditions we get $d_j = 2/(n+1) + O(\varepsilon)$ for $j = 0, \ldots, n$, whereas for the problem considered in Sec. 2.1, we found that $d_j = 2/n + O(\varepsilon)$ for $j = 1, \ldots, n-1$, with $d_0 = 1/n + O(\varepsilon)$, $d_n = 1/n + O(\varepsilon)$. Although (2.35) was was derived only for the case $n > 1$, the result (2.35) turns out to be correct even for one spike where $n = 1$.

We now give an explicit example to illustrate this result. We consider Carrier's original problem with $Q(u) = u^2 - 1$, $u_L = 0$, and $u_R = 0$. For this example, we can calculate w, w_L and w_R explicitly, to obtain that

$$\sigma = \sqrt{2}, \quad c = 12, \quad c_L = 12e^{-\kappa_1\sqrt{2}}, \quad c_R = 12e^{-\kappa_2\sqrt{2}}, \tag{2.36}$$

where κ_1 and κ_2 can be either $\pm\sqrt{2}\cosh^{-1}(\sqrt{3}) = \pm\sqrt{2}\log(\sqrt{3}+\sqrt{2})$. Then, from (2.35), we obtain four different sets of spike layer locations for $j = 0, \ldots, n-1$ given by

$$x_j \sim -1 + \frac{2(j+1)}{n+1} \pm \sqrt{2}\varepsilon\log(\sqrt{3}+\sqrt{2}), \tag{2.37a}$$

$$x_j \sim -1 + \frac{2(j+1)}{n+1} \pm \sqrt{2}\varepsilon\log(\sqrt{3}+\sqrt{2}) \mp \frac{2\varepsilon(j+1)}{n+1}\sqrt{2}\log(\sqrt{3}+\sqrt{2}). \tag{2.37b}$$

This example was given at the end of Sec. 7.1 of [100]. A rigorous proof of this result has only been obtained recently in Theorem 3 of [85]. For $n = 2$, the two solutions corresponding to (2.37a) are shown in Fig. 3.

Finally, we consider a situation where there is a boundary layer at only one of the endpoints. We consider (2.1) with $u(-1) = u_L$ and $u'(1) = 0$, and we look for an n-spike solution where there is no boundary layer at $x = 1$. Using the projection method, we readily obtain that the inter-spike separations d_j satisfy $ce^{-\sigma\varepsilon^{-1}d_1} \sim c_L e^{-\sigma\varepsilon^{-1}d_0}$, and $e^{-\sigma\varepsilon^{-1}d_{n-1}} \sim e^{-2\sigma\varepsilon^{-1}d_n}$, together with (2.33b). By solving this system explicitly, we obtain the following result:

Proposition 2.5. *Let $\varepsilon \ll 1$ and consider (2.1) with the boundary conditions $u(-1) = u_L$ and $u'(1) = 0$. Then, there is an n-spike solution to this*

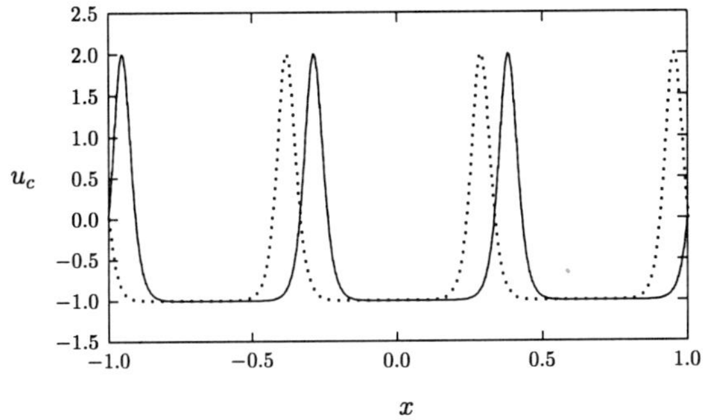

Figure 3. Plot of u_c, given in (2.34), versus x for a two-spike solution for $Q(u) = u^2 - 1$ with $\varepsilon = 0.03$. The two curves correspond to the two choices in (2.37a).

problem with no boundary layer near $x = 1$ *of the form*

$$u \sim u_c(x) \equiv \sum_{j=0}^{n-1} w\left[\varepsilon^{-1}(x - x_j)\right] - (n-1)\,s + \left[w_L\left(\varepsilon^{-1}(x+1)\right) - s\right]. \tag{2.38}$$

Here w_L *is the boundary layer function satisfying (2.30). For this problem, there are are two possible sets* $(x_0, \ldots, x_n)$ *of spike-layer locations, which for* $j = 0, \ldots, n-1$ *satisfy*

$$x_j \sim -1 + \frac{4(j+1)}{2n+1} + \frac{2\varepsilon}{\sigma(2n+1)}\left[\log\left(\left(\frac{c}{c_L}\right)^{\frac{1}{2}-n}\right) + j\,\log\left(\frac{c}{c_L}\right)\right]. \tag{2.39}$$

There are two values of c_L *defined in (2.30).*

Therefore, for this Neumann/Dirichlet problem, the inter-layer separations satisfy $d_j = 4/(2n+1) + O(\varepsilon)$ for $j = 0, \ldots, n-1$, and $d_n = 2/(2n+1) + O(\varepsilon)$. For the special case where $n = 1$, we get $x_0 = 1/3$.

2.3. *Exponential Sensitivity to the Data*

We now consider two types of perturbations of Carrier's problem, where a seemingly minor modification of (2.1) leads to a dramatic effect on the solution behavior. This extreme sensitivity is a result of the exponential ill-conditioning of (2.1) for interior spike solutions.

We first construct spike solutions to (2.1) with $n \geq 2$ interior spikes, when exponentially small terms are added to certain Dirichlet boundary conditions. Specifically, we consider (2.1) with

$$u(-1) = s + A\,e^{-\sigma\varepsilon^{-1}/n}\,, \quad u(1) = s + A\,e^{-\sigma\varepsilon^{-1}/n}\,. \tag{2.40}$$

Here $A > 0$ and is $O(1)$ as $\varepsilon \to 0$. This problem models the initial formation of boundary layers near the endpoints. For the special case $Q(u) = u^2 - 1$, and for two interior spikes, such a problem was analyzed using variational methods in [51], where it was shown that a bifurcation can occur if A exceeds a threshold. The general case for arbitrary $n \geq 2$, and for arbitrary $Q(u)$ of the form shown in Fig. 1, was treated using the projection method in §6.1 of [100], where the following result was obtained:

Proposition 2.6. *Let $\varepsilon \ll 1$ and consider (2.1) with the boundary conditions (2.40). Consider an n-spike solution of the form (2.25) with $n \geq 2$. These equilibria exhibit a saddle-node bifurcation structure as A is varied. In particular, there are four possible sets $(x_0, \ldots, x_n)$ of spike-layer locations when $A > 2c$, two sets when $A_s(n) < A < 2c$, and no such sets when $A < A_s(n)$. Here $A_s(n)$ is defined by*

$$A_s(n) = 2\,c\left(\frac{n}{n+1}\right)\left(\frac{n-1}{n+1}\right)^{(1-n)/2n}, \quad n = 2, 3, \ldots\,. \tag{2.41}$$

Therefore, by perturbing the boundary conditions by exponentially small terms, new solutions can be created by saddle-node bifurcations. To illustrate this result, let $Q(u) = u^2 - 1$. Then, $c = 12$, and $A_s(2) = 21.06$ from (2.41). Therefore, for $\varepsilon \ll 1$, we predict that there are no two-spike solutions to (2.1) with boundary conditions (2.40) when $0 < A < 21.06$.

The next problem considers the effect of modifying a coefficient in the differential operator in (2.1) by an exponentially small but spatially varying term. In [46] a solution with one interior spike was constructed for

$$\frac{\varepsilon^2}{\kappa}\left(\kappa u'\right)' + Q(u) = 0\,, \quad -1 < x < 1\,; \quad u'(\pm 1) = 0\,, \tag{2.42a}$$

where, for some constants ν and $q > 0$, $\kappa = \kappa(x;\varepsilon)$ has the form

$$\kappa(x;\varepsilon) = 1 + \varepsilon^{\nu} g(x) e^{-\varepsilon^{-1}q}\,. \tag{2.42b}$$

Here $g(x)$ is a smooth function. In [46] the projection method was used to determine the location of a spike for an interior one-spike solution with no boundary layers. The result is summarized as follows:

Proposition 2.7. *Let $\varepsilon \ll 1$ and consider a solution to (2.42) with one interior spike. The solution is given asymptotically by $u \sim w\left[\varepsilon^{-1}(x - x_0)\right]$,* *where $w(y)$ solves (2.2). The spike location x_0 is a root of $h(x_0) = 0$, where*

$$h(x_0) \equiv \frac{2\varepsilon c^2}{\beta_0}\left(e^{-2\sigma\varepsilon^{-1}(1-x_0)} - e^{-2\sigma\varepsilon^{-1}(1+x_0)}\right) - \varepsilon^{\nu+2}e^{-q/\varepsilon}g'(x_0)\,. \quad (2.43)$$

Here $\beta_0 \equiv \int_{-\infty}^{\infty} w'^2(y)\,dy$.

For $0 < q < q_c$, it is possible that (2.43) has multiple roots, which corresponds to multiple interior one-spike solutions. To illustrate this, let $g(x) = x^2/2$, and $Q(u) = -u + u^2$, for which $\sigma = 1$, $c = 6$, and $\beta_0 = 6/5$. A simple calculation shows that $x_0 = 0$ is always a solution, and that we have a pitchfork bifurcation when $q = q_c$, where

$$q_c = 2 + (\nu + 2)\varepsilon\log\varepsilon - \varepsilon\log(240)\,. \quad (2.44)$$

For $0 < q < q_c$, there are three roots to (2.43) on $-1 < x_0 < 1$, whereas for $q > q_c$ there is only one such root. Therefore, by perturbing Carrier's problem by an exponentially small but spatially varying term new interior one-spike solutions may be generated by a pitchfork bifurcation.

2.4. *Related Problems: One Space Dimension*

There have been several related studies of spike solutions in ODE's, including problems with spatially varying coefficients [1], [69], and problems involving two components [53]. In addition, there are some further questions that remain to be explored.

Question 2.1. Can one give a rigorous analytical proof, along the lines of [85], of the asymptotic results in Propositions 2.1–2.7 without using the details of the phase-plane?

Question 2.2. For a given $\varepsilon \ll 1$ fixed, can one determine a bound on the number $N(\varepsilon)$ of internal spikes for Carrier's problem (2.1) with various boundary conditions?

For the case $Q(u) = u^2 - 1$, it was proved in [85] that $N(\varepsilon) < 0.41\varepsilon^{-1} + 1$ for $\varepsilon \ll 1$. What is the result for other nonlinearities $Q(u)$?

Question 2.3. Can one numerically compute multi-spike solutions to (2.1) for $\varepsilon \ll 1$, and with various boundary conditions, in a reliable manner despite the exponential ill-conditioning?

This is a key open problem. Standard numerical methods for (2.1) are subject to a loss of precision as a result of the ill-conditioning. However, for

Carrier's other problem where $Q(u)$ is modified to allow for heteroclinic, rather than homoclinic, solutions, an exact nonlinear WKB transformation was introduced in [87] and used to numerically compute multi-layer solutions for $\varepsilon \ll 1$. The sensitivity of these solutions to exponentially small perturbations in the boundary data was also studied numerically with this approach. A distinct advantage of the WKB transformed problem is that it is well-conditioned as $\varepsilon \to 0$. Our question is whether it is possible to use a related change of variables to remove the ill-conditioning associated with (2.1). In particular, can one compute solutions by re-casting (2.1) in terms of the distance function? If successful, such a method could be extended to numerically treat the quasilinear multi-dimensional problems of Sec. 2.5.

Question 2.4. What is the global bifurcation diagram for multi-spike solutions as ε is increased?

For $\varepsilon \gg 1$, multi-spike solutions are born from a bifurcation at infinity. For a few choices of $Q(u)$, branches of n-spike solutions were computed numerically in [88] as ε is decreased from a large positive value. It was found that each n-spike branch undergoes a saddle-node bifurcation at some $O(1)$ value of ε. Does this generic feature holds for other $Q(u)$? Can such a path following method in ε allow one to compute solutions for $\varepsilon \ll 1$?

Finally, we mention a modification of Carrier's original problem. This problem concerns constructing multi-bump solutions for

$$\varepsilon^2 u'' + u^2 - 1 + 2b(1 - x^2)u = 0\,, \quad -1 < x < 1\,; \quad u(\pm 1) = 0\,. \qquad (2.45)$$

It was proved in [1] that if $\varepsilon \ll 1$, $b > 0$, $n \geq 2$, and if u has minima at x_k, for $k = 1, \ldots, n$, where $-1 < x_1 < \ldots < x_n < 1$, then $|x_k| < M\varepsilon|\log \varepsilon|$ for some $M > 0$ independent of ε. Therefore, (2.45) admits multi-bump solutions clustered near $x = 0$. A similar, but formal, result was obtained in [69]. This result is natural in that (2.45) can be written as a nonlinear Schrodinger equation $\varepsilon^2 u'' - 1 + u^2 - V(x)u = 0$, where the potential $V(x) = 2b(x^2 - 1)$ has a global minimum at $x = 0$ when $b > 0$. The possibility of multi-bump solutions near non-degenerate minima of $V(x)$ is well-known (cf. [99], [22]). We conjecture that a one-spike solution to Carrier's original problem with $b = 0$ will undergo a pitchfork bifurcation at $x_0 = 0$ when b is raised to an exponentially small value. The bifurcation should be similar to that for problem (2.42). This leads to the next question.

Question 2.5. What are the bifurcation properties of k-spike solutions with $k \geq 1$ to (2.45) when b is exponentially small?

2.5. *Spikes for Quasilinear Elliptic PDE*

In this subsection, we construct spike solutions for

$$\varepsilon^2 \Delta u + Q(u) = 0\,, \quad x \in \Omega\,; \quad \partial_n u = 0 \quad x \in \partial\Omega\,, \tag{2.46}$$

where Ω is a bounded, simply-connected, domain in $\mathbb{R}^N$, with $N > 1$. We assume that $Q(s) = 0$, with $Q^{'}(s) < 0$, and that there exists a unique radially symmetric ground-state solution $w(\rho)$, with $\rho = |y|$, that satisfies

$$w^{''} + \frac{(N-1)}{\rho} w^{'} + Q(w) = 0\,, \quad \rho \geq 0\,; \quad w(0) > 0\,, \tag{2.47a}$$

$$w^{'}(0) = 0\,, \quad w \sim s + c\rho^{-(N-1)/2} e^{-\sigma\rho}\,, \quad \text{as} \quad \rho \to \infty\,, \tag{2.47b}$$

where $c > 0$ and $\sigma = \left[-Q^{'}(s)\right]^{1/2}$. An important example is $Q(u) = -u + u^p$ for $1 < p < p_c$, where $p_c = \frac{N+2}{N-2}$ is the Sobolev exponent for $N \geq 3$, and $p_c = \infty$ if $N = 2$. For this case, $w(\rho)$ satisfies (1.7). Equation (2.46) is the multi-dimensional version of Carrier's problem (2.1), where $w(\rho)$ replaces the homoclinic solution $w(y)$.

The study of spike solutions to (2.46) was largely initiated in the pioneering work of Ni and Takagi (cf. [68], [76], [77]). An earlier survey of results for (2.46), and for some related problems, is given in [78].

We now follow [78] and give a rough summary of the results of [76] and [77], characterizing the "least-energy solution" of (2.46) for $Q(u) = -u + u^p$, and with $u > 0$ in Ω. For this problem, the energy functional for (2.46) is

$$J_\varepsilon(u) = \frac{1}{2} \int_\Omega \left(\varepsilon^2 |\nabla u|^2 + u^2\right) dx - \frac{1}{p+1} \int_\Omega u_+^{p+1}\, dx\,, \tag{2.48}$$

where $u_+ = \max(u, 0)$. As argued in [76] and [77], J_ε has a minimum when restricted to the set of solutions of (2.46) with $u > 0$ in Ω. This minimizing solution is called the "least-energy solution". Since an interior spike solution has, asymptotically, twice the energy of a boundary spike solution, the least-energy solution must be a one-spike solution centered at some point on $\partial\Omega$. To determine the actual point $\xi_\varepsilon \in \partial\Omega$ where the spike concentrates, a two-term expansion for J_ε as $\varepsilon \to 0$ is required. For a spike centered at $\xi_\varepsilon \in \partial\Omega$, it was shown in [77] that

$$J_\varepsilon(u) = \varepsilon^N \left(\frac{1}{2} I(w) - C\varepsilon H\left(\xi_\varepsilon\right) + o(\varepsilon)\right)\,, \tag{2.49}$$

where C is a positive constant independent of ε, and $I(w)$ is defined by

$$I(w) = \int_{\mathbb{R}^N} \left[\frac{1}{2}\left(|\nabla w|^2 + w^2\right) - \frac{1}{p+1} w^p\right] dy\,. \tag{2.50}$$

Here $H(\xi_\varepsilon)$ is the mean curvature of $\partial\Omega$ at ξ_ε. By minimizing (2.50), the following result was proved in [76], [77]:

Proposition 2.8. *Consider (2.46) with $u > 0$ in Ω and $Q(u) = -u + u^p$, where $1 < p < p_c$. Then, for $\varepsilon \ll 1$, this problem has a least-energy solution u_ε with exactly one global maximum point at ξ_ε. Moreover, $\xi_\varepsilon \in \partial\Omega$, and $H(\xi_\varepsilon) \to \max_{\partial\Omega} H$ as $\varepsilon \to 0$, where H is the mean curvature of $\partial\Omega$.*

Recently, there has been much work characterizing other types of boundary spike solutions that have a higher energy. The theory for this class of solutions is now rather complete. In Theorem 1.2 of [109] it was proved that (2.46) admits a boundary spike-layer solution that concentrates at any non-degenerate critical point of the mean curvature $H(P)$ with $P \in \partial\Omega$. In [39] a boundary k-spike solution was constructed where the spikes are separated by $O(1)$ as $\varepsilon \to 0$, and where each spike concentrates at a different local maximum point of the mean curvature $H(P)$. In [42] it was shown that for any integer $k > 0$, there exists boundary k-peak solutions to (2.46) where the peaks all cluster near a local minimum point of $H(P)$. This clustering effect is qualitatively similar to the spike-clustering phenomena for the one-dimensional problem (2.45) described in Sec. 2.4. Boundary spike solutions, together with asymptotic estimates for the eigenvalues of the linearization that tend to zero as $\varepsilon \to 0$, are also given in [111] and [4].

Next, we describe some results for interior k-spike solutions to (2.46). These solutions have the form

$$u \sim \sum_{j=1}^{k} w\left[\varepsilon^{-1}|x - \xi_j|\right] , \tag{2.51}$$

for some $\xi_j \in \Omega$ for $j = 1, \ldots, k$ to be found. As for Carrier's problem in Sec. 2.1–2.3, since $w(|y|)$ decays exponentially as $|y| \to \infty$, the linerarization of (2.46) around an interior spike solution is exponentially ill-conditioned.

For an interior one-spike solution, it was shown formally in [101], and proved rigorously in [110], that the spike concentrates at a local maximum of the distance function. The result and method used in [101] is described more precisely below. Geometrically, this one-spike result for (2.46) is asymptotically equivalent to that given in Proposition 2.2 for the one-dimensional problem in the sense that an interior spike for (2.46) concentrates at a point in Ω that is farthest from the boundary. In [41] a solution to (2.46) that has one boundary spike and one interior spike was constructed. It was found that the location of the interior spike is moved an $O(1)$ distance from the center of the largest inscribed sphere for Ω in the direction away from the boundary spike. Such a mixed boundary/interior spike solution is the multi-dimensional equivalent of the result in Proposition 2.5 for Carrier's problem, where for $\varepsilon^2 u'' + Q(u) = 0$ with $u(-1) = u_L$ and $u'(1) = 0$, an interior spike is located not at the midpoint $x_0 = 0$ but instead at $x_0 = 1/3$.

For an interior k-spike solution of (2.46), with $k \geq 1$, the following result with relatively minor technical differences, was given in [59], [40], and [3]:

Proposition 2.9. *Equation (2.46) admits an interior k-spike solution given asymptotically by (2.51), where the concentration points $\xi_{1\varepsilon}, \ldots, \xi_{k\varepsilon}$ tend to local maximum points of $\psi(\xi_1, \ldots, \xi_k)$ as $\varepsilon \to 0$, where*

$$\psi(\xi_1, \ldots, \xi_k) \equiv \min_{i,j,l=1,\ldots,k | j \neq l} \left(dist(\xi_i, \partial\Omega), \frac{1}{2}|\xi_j - \xi_l| \right) . \tag{2.52}$$

Notice that this result is geometrically very similar to the analogous result in (2.24) for a k-spike solution to Carrier's problem with Neumann boundary conditions. The main difference in the multi-dimensional case is that, depending on the topology of Ω, there can be many different choices for the set of spike locations. From Proposition 2.9, it is clear that the spike locations are asymptotically equivalent to a corresponding geometric ball-packing problem. The next result, given in Corollary 1.8 of [3], makes this equivalence precise.

Proposition 2.10. *Let $S_1, \ldots, S_k \in \Omega$ be nonoverlapping spheres of the same radius d, and assume that $S_1, \ldots, S_k$ are packed in such a way that when considered as rigid bodies in a rigid container Ω, the set $\xi_1, \ldots, \xi_k$ of their centers becomes also a rigid body. Then, for $\varepsilon > 0$ sufficiently small, (2.46) has a solution with k spikes that localize at $\xi_1, \ldots, \xi_k$.*

In Fig. 2–6 of [3], many illustrations of this "rigid-body" geometrical construction are shown. In particular, in Fig. 5 of [3], several possibilities are shown for packing eight small spheres of a common radius inside a spherical domain Ω.

Although the basic theory for spike solutions of (2.46) is rather well established, there are two questions that should be explored.

Question 2.6. Formulate a numerical method to compute interior k-spike solutions to (2.46) for $\varepsilon \ll 1$ that overcomes the exponential ill-conditioning. What are the global bifurcation branches of solutions to (2.46) as a function of the topology of Ω and of ε, for both $\varepsilon \ll 1$ and $\varepsilon \gg 1$?

Question 2.7. Using techniques in computational geometry, investigate how the topology of Ω influences the geometrical ball-packing characterization given in Proposition 2.10 for a k-spike solution of (2.46) with k large. Can one estimate the number of solutions for ε small but fixed?

In [101], a multi-dimensional extension of the projection method, as outlined in Sec. 2, was used to determine an interior spike location. We now sketch this method for a one-spike solution of

$$\varepsilon^2 \Delta u + Q(u) = 0\,, \quad x \in \Omega\,; \quad \varepsilon \partial_n u + b(u - s) = 0\,, \quad x \in \partial\Omega\,. \tag{2.53}$$

Here $\Omega \in \mathbb{R}^2$, and $b = b(\xi) \geq 0$, where ξ is arclength along the smooth boundary $\partial\Omega$. We look for an interior one-spike solution of the form $u \sim w\left[\varepsilon^{-1}|x - x_0|\right] + R$, where $R \ll 1$. Using the far-field behavior of w in (2.47b), we find that R satisfies

$$L_\varepsilon R \equiv \varepsilon^2 \Delta R + Q'(w) R = 0\,, \quad x \in \Omega\,, \tag{2.54a}$$

$$\varepsilon\, \partial_n R + bR \sim -c\varepsilon^{1/2} r^{-1/2} e^{-\sigma\varepsilon^{-1} r} \left[b - \sigma \hat{r}\cdot\hat{n}\right]. \tag{2.54b}$$

Here $c > 0$ and $\sigma > 0$ are defined in (2.47b). Moreover, $r = |x - x_0|$, $\hat{r} = (x - x_0)/r$, and $\hat{n}$ is the unit outward normal to $\partial\Omega$.

Next, we consider the eigenvalue problem

$$L_\varepsilon \phi = \lambda \phi\,, \quad x \in \Omega\,; \quad \varepsilon \partial_n \phi + b\phi = 0\,. \tag{2.55}$$

This problem has two exponentially small eigenvalues, where the corresponding eigenfunctions ϕ_j for $j = 1, 2$, have the form

$$\phi_j \sim \partial_{x_j} w\left[\varepsilon^{-1}|x - x_0|\right] + \phi_{Lj}\,, \quad j = 1, 2\,. \tag{2.56}$$

Here ϕ_{Lj} is a boundary layer function localized near $\partial\Omega$ that allows the boundary condition in (2.55) to be satisfied. To estimate λ_j, for $j = 1, 2$, we then use Green's identity for ϕ_j and $\partial_{x_j} w$, to derive

$$\lambda_j \left(\partial_{x_j} w, \phi_j\right) = -\varepsilon \int_{\partial\Omega} \phi_j \left(\varepsilon\, \partial_n + b\right)\left[\partial_{x_j} w\right] d\xi\,, \quad j = 1, 2\,, \tag{2.57}$$

where $(u, v) \equiv \int_\Omega uv\, dx$. In [101], ϕ_{Lj} was calculated from a boundary layer analysis, which then determines ϕ_j on $\partial\Omega$. In this way, we get

$$\lambda_j \sim -\frac{c^2\sigma^3}{\pi\beta} \int_{\partial\Omega} \left(\frac{x_j - x_{0j}}{r}\right)^2 r^{-1} e^{-2\sigma\varepsilon^{-1} r} \frac{\left[b - \sigma\hat{r}\cdot\hat{n}\right]\left[1 + \hat{r}\cdot\hat{n}\right]}{\sigma + b}\, d\xi\,, \tag{2.58}$$

for $j = 1, 2$. Here $\beta \equiv \int_0^\infty w'^2 \rho\, d\rho$. This surface integral defining λ_j is of Laplace type, and so can be calculated asymptotically in terms of the points on $\partial\Omega$ closest to x_0. Therefore, it is clear that λ_j for $j = 1, 2$ is exponentially small as $\varepsilon \to 0$. Thus, the problem of determining x_0 is exponentially ill-conditioned. Notice the close similarity in the derivation of (2.58) with that given in (2.8)–(2.12) for Carrier's problem in a one dimensional domain.

To determine the spike location we expand R in terms of the normalized eigenfunctions of (2.55) as $R = \sum_{j=0}^{\infty} C_j \lambda_j^{-1} \phi_j$, where

$$C_j = \varepsilon \int_{\partial\Omega} \phi_j \left(\varepsilon\, \partial_n + b\right) w \, d\xi \,. \tag{2.59}$$

The coefficients C_j for $j = 1, 2$ can be calculated from the far-field behavior of w in (2.47b) and from the behavior of ϕ_j on $\partial\Omega$ as obtained from the boundary layer analysis described above. Since $\lambda_j \to 0$ exponentially for $j = 1, 2$, the corresponding limiting solvability condition is that $C_j \to 0$ for $j = 1, 2$. In this way, it was found in [101] that the spike location x_0 is given by a root of the vector-valued function $I(x_0)$, where

$$I(x_0) \equiv \int_{\partial\Omega} r^{-1}\, e^{-2\sigma\varepsilon^{-1} r} \left[1 + \hat{r}\cdot\hat{n}\right] \left(\frac{b - \sigma \hat{r}\cdot\hat{n}}{\sigma + b}\right) \hat{r}\, d\xi \,. \tag{2.60}$$

The following result, obtained by balancing "forces" as described in the rigid body characterization of Proposition 2.10 of [3], was obtained in [101]:

Proposition 2.11. *Assume that there exists a unique largest inscribed circle B for Ω, with center at x_{in} and radius r_{in}, that makes exactly two-point contact with $\partial\Omega$ at $x(\xi_i) \in \partial\Omega$ for $i = 1, 2$. Suppose that $b_i - \sigma$ has the same sign at each contact point and that $\kappa_i r_{in} > -1$ for $i = 1, 2$, where $b_i \equiv b(\xi_i)$ and where κ_i is the curvature of $\partial\Omega$ at ξ_i. Then, x_0 lies on the chord joining $x(\xi_1)$ and $x(\xi_2)$. Moreover, for $\varepsilon \to 0$, x_0 satisfies*

$$x_0(\varepsilon) = x_{in} + \frac{\varepsilon}{8\sigma} x_0^1 \hat{n}_2 + O(\varepsilon^2)\,, \tag{2.61a}$$

where

$$x_0^1 = 2\log\left[\left(\frac{b_1 - \sigma}{b_1 + \sigma}\right)\left(\frac{b_2 + \sigma}{b_2 - \sigma}\right)\right] + \log\left(\frac{1 + \kappa_2 r_{in}}{1 + \kappa_1 r_{in}}\right)\,, \qquad \hat{n}_2 \equiv \hat{n}(\xi_2)\,. \tag{2.61b}$$

Notice that if $(b_1 - \sigma)(b_2 - \sigma) < 0$, then (2.60) has no root near x_{in}. This condition is qualitatively similar to the condition given in Proposition 2.2 for Carrier's problem. A similar result for when the largest inscribed circle makes three-point contact with $\partial\Omega$ was given in [101].

There are several rigorous results for an interior spike solution for (2.53). For the Dirichlet problem with $b = \infty$, it was proved in [79] that there is a least-energy solution where a one-spike solution concentrates at the maximum of the distance function. This result can be obtained by letting $b \to \infty$ in (2.61). In [8] an interior spike for 2.53) was analyzed in a half-space when b is near the critical value $b = \sigma$. As $b \to \sigma^+$, the spike was found to approach the boundary. The sensitivity of the spike location for b near σ is certainly suggested from (2.61b).

For the Neumann problem, the result in Proposition 2.10 shows that there is a plethora of interior spike solutions. However, as for Carrier's problem in one spatial dimension, interior multi-spike solutions for the Dirichlet problem with $u = s$ on $\partial\Omega$ should not exist. This leads to the next question.

Question 2.8. What are the bifurcation properties of interior k-spike solutions for (2.53) with $k \geq 2$ spikes under the Dirichlet boundary conditions $u = s + Ae^{-\alpha\varepsilon^{-1}}$ for some $\alpha > 0$ and $A > 0$. When $A = 0$, there should be no such solutions. Do the solutions have a saddle-node bifurcation behavior similar to that described in Proposition 2.6 for Carrier's problem?

Question 2.9. For general Dirichlet data with $u = u_b(\xi) > s$ on $\partial\Omega$ can one construct a solution to $\varepsilon^2\Delta u + Q(u) = 0$ that concentrates on the entire boundary of $\partial\Omega$ and that has $k \geq 1$ interior spikes? We conjecture that, in analogy with Proposition 2.4, the interior spikes now concentrate at local maximum points of

$$\psi(\xi_1, \ldots, \xi_k) \equiv \min_{i,j,l=1,\ldots,k|j\neq l} \left(\text{dist}(\xi_i, \partial\Omega), |\xi_j - \xi_l|\right) . \tag{2.62}$$

A related problem where localization occurs is for the nonlinear Schrodinger equation (cf. [38], [22] and [99])

$$\varepsilon^2\Delta u - V(x)u + u^p = 0\,, \quad x \in \Omega\,; \quad \partial_n u = 0 \quad x \in \partial\Omega\,. \tag{2.63}$$

Here $V(x)$ is a smooth positive potential with $V(x) > V_0 > 0$ in Ω, and p is subcritical. This is the multi-dimensional counterpart of the modified Carrier problem (2.45). It is well-known that there exists spike solutions of (2.63) that localize near non-degenerate local maxima and minima of $V(x)$. Equation (2.63) also admits spike-clustering phenomena where k spikes all cluster near a minimum of $V(x)$ (cf. [99]). It would be interesting to compare the spike phenomena for (2.63) for an exponentially shallow potential with that for the quasilinear problem (2.46). This leads to the next question.

Question 2.10. What are bifurcation properties of spike solutions for (2.63) for a potential of the form $V(x) = 1 + e^{-\sigma/\varepsilon}\hat{V}(x)$, where $\sigma > 0$?

The spike locations should be determined from a competition between the distance function and the localizing effect of the potential $\hat{V}(x)$.

Another problem where localization occurs is in the construction of hot-spot solutions for Bratu's problem

$$\triangle u + \lambda e^u = 0\,, \quad x \in \Omega\,; \quad u = 0\,, \quad x \in \partial\Omega\,. \tag{2.64}$$

Here Ω is a bounded, simply-connected, domain in $\mathbb{R}^2$. The qualitative feature of hot-spot solutions is that $u \to \infty$ as $\lambda \to 0$ in a localized region near some $x = \xi_j$, for $j = 1, \ldots, k$ while $u = O(1)$ as $\lambda \to 0$ away from these points. Using complex analysis, a system of equations for the hot-spot locations ξ_j, for $j = 1, .., k$, was derived in [75]. An alternative method based on singular perturbation theory was used in [101]. The following result characterizes the hot-spot locations:

Proposition 2.12. *For $\varepsilon \to 0$, the hot-spot locations $\xi_1, \ldots, \xi_k$ for (2.64) satisfy the coupled system*

$$\nabla R_d(x;\xi_j)|_{x=\xi_j} + \sum_{i \neq j}^{k} \nabla G_d(x;\xi_i)|_{x=\xi_i} = 0\,. \quad j = 1, \ldots, k\,. \tag{2.65}$$

Here $G_d(x;\xi)$ is the Dirichlet Green's function, with $R_d(x;\xi)$ as its regular part, so that

$$\triangle G_d = -\delta(x-\xi) \quad x \in \Omega\,; \quad G_d = 0\,, \quad x \in \partial\Omega\,, \tag{2.66}$$

and

$$R_d(x,\xi) = G_d(x,\xi) + \frac{1}{2\pi}\log|x-\xi|\,. \tag{2.67}$$

Therefore, the criteria determining the hot-spot locations for (2.64) is very different from that given in Proposition 2.9 for the spike locations of (2.46). This difference arises from a logarithmic, or Coulomb-type, singularity in the far-field behavior of the local hot-spot profile (cf. [101]). For a one hot-spot solution, the hot-spot location satisfies $\nabla R_d(x;\xi_1) = 0$ at $x = \xi_1$. For a convex domain Ω, R_d is convex (cf. [12]), and so there is only one hot-spot location. This criterion for a hot-spot solution with $k = 1$ is actually very similar to the criterion developed in Sec. 4 that determines the location of a spike for the full GM model (1.1).

3. Spikes for Nonlocal Scalar Problems

In this section we begin by examining the stability of the equilibrium spike solutions constructed in Sec. 2. Consider the time-dependent problem

$$u_t = \varepsilon^2 \triangle u - u + u^p \,, \quad x \in \Omega\,; \quad \partial_n u = 0 \quad x \in \partial\Omega\,. \tag{3.1}$$

Here Ω is a bounded domain in R^N, and p is a subcritical exponent.

Let w_ε be an interior one-spike equilibrium solution to (3.1). The center of the spike x_0 satisfies $\text{dist}(x_0, \partial\Omega) = O(1)$ as $\varepsilon \to 0$. By linearizing (3.1), we find that the stability of this solution is determined by the spectrum of

$$L_\varepsilon \phi_\varepsilon \equiv \varepsilon^2 \triangle \phi_\varepsilon - \phi_\varepsilon + p w_\varepsilon^{p-1} \phi_\varepsilon = \lambda^\varepsilon \phi_\varepsilon \,, \quad \partial_n \phi_\varepsilon = 0\,. \tag{3.2}$$

Letting $\varepsilon \to 0$, and defining $y = \varepsilon^{-1}(x - x_0)$, we have that $w_\varepsilon \to w$, where $w(|y|)$ satisfies (1.7). In this way, we obtain

$$L_0 \phi \equiv \triangle \phi - \phi + p w^{p-1} \phi = \lambda \phi \,, \quad \phi \to 0 \quad \text{for} \quad |y| \to \infty\,. \tag{3.3}$$

We refer to L_0 as the local operator, and (3.3) as the *infinite-line local eigenvalue problem.* The consequence of the exponential decay of $w(|y|)$ as $|y| \to \infty$ is that (3.3) is independent of the shape of Ω, of ε, and of x_0. A key result for (3.3), obtained in [68], is the following:

Proposition 3.1. *Consider (3.3) written as $L_0 \phi_l = \nu \phi_l$ for $\phi_l \in H^1(\mathbb{R}^N)$. This problem admits the eigenvalues $\nu_0 > 0$, $\nu_1 = \cdots = \nu_N = 0$, and $\nu_{N+k} < 0$ for $k \geq 1$. The eigenvalue ν_0 is simple, and the corresponding eigenfunction is radially symmetric with constant sign.*

This result was proved in Theorem 2.12 of [68]. Therefore, there is exactly one unstable eigenvalue $\nu_0 > 0$ for (3.3). The eigenfunctions corresponding to the zero eigenvalues are the translation modes $\phi_j = \partial_{y_j} w(|y|)$ for $j = 1, \ldots, N$. Each of these modes has exactly one nodal line.

In the one-dimensional case, the following more precise result for the spectrum of (3.3) was obtained in [27] using hypergeometric functions:

Proposition 3.2. *Let $J = J(p)$ be a positive integer such that $J < (p+1)/(p-1) \leq J+1$. Then, for $\phi_l \in H^1(R)$, the infinite-line local eigenvalue problem $L_0 \phi_l = \nu \phi_l$ has $J+1$ discrete eigenvalues given by*

$$\nu_j = \frac{1}{4}\left[(p+1) - j(p-1)\right]^2 - 1\,, \qquad j = 0, \ldots, J\,. \tag{3.4}$$

The continuous spectrum of L_0 lies in the range $-\infty < \nu < -1$.

This result is Proposition 5.6 of [27]. Notice that $\nu_0 > 0$, $\nu_1 = 0$, and $\nu_j \in (-1, 0)$ for $2 \leq j \leq J$. However, when $p \geq 3$, then $J = 1$, and there are no discrete eigenvalues in the interval $(-1, 0)$. Alternatively, there are many discrete eigenvalues that appear in $(-1, 0)$ as $p \to 1^+$. When $p = 2$, the only discrete eigenvalues are $\nu_0 = 5/4$, $\nu_1 = 0$, and $\nu_2 = -3/4$.

Although, for $\varepsilon \ll 1$, it is clear that the discrete eigenvalues of the infinite-line local eigenvalue problem should be exponentially close to corresponding eigenvalues of (3.2), a rigorous result to this effect is rather technical. The formal analysis in Sec. 2.1 leading to Proposition 2.1 determines the exponentially small change in the translation mode $\nu_1 = 0$ as a result of the finite domain. The underlying idea is that if the infinite-line local eigenvalue problem has a discrete eigenvalue corresponding to an eigenfunction with an exponential decay as $|y| \to \infty$, then this eigenvalue should be perturbed by only exponentially small terms by the finite domain. An analysis incorporating this idea was made in [17], where it was proved (see Theorem 3.1 of [17]) that the first three eigenvalues of (3.2) are exponentially close to those of (3.3) in the sense that

$$\lambda_0^\varepsilon = \frac{1}{4}(p-1)(p+3) + O\left(e^{-2/\varepsilon}\right), \tag{3.5a}$$

$$\lambda_1^\varepsilon = \frac{4c^2}{\left(\int_{-\infty}^{\infty} w'^2(y)\, dy\right)} e^{-2/\varepsilon}\left(1 + O(e^{-1/\varepsilon})\right), \tag{3.5b}$$

$$\lambda_2 = \frac{1}{4}(p-1)(p-5) + O\left(e^{-(3-p)/\varepsilon}\right), \qquad \text{when } p < 3\,. \tag{3.5c}$$

Here c and $w(y)$ are defined in (1.8). Equation (3.5b) is also readily obtained by setting $\kappa_l = \kappa_r = 0$ in (2.12) of Proposition 2.1 in Sec. 2.

Since (3.3) has a strictly positive eigenvalue, there is an eigenvalue of (3.2) that remains positive for $\varepsilon \ll 1$. Therefore, an interior one-spike equilibrium solution is unstable for (3.1). Similarly, an equilibrium solution of (3.1) with k interior and well-separated spikes will be unstable due to the existence of k positive eigenvalues that tend to ν_0 as $\varepsilon \to 0$. The existence of one unstable eigenvalue should also occur for an interior one-spike solution of subcritical quasilinear problems of the general form

$$u_t = \varepsilon^2 \Delta u + Q(u)\,, \quad x \in \Omega\,; \quad \partial_n u = 0 \quad x \in \partial\Omega\,. \tag{3.6}$$

Here $Q(u)$ has the form shown in Fig. 1. This leads to our first question.

Question 3.1. Can one characterize the discrete spectrum of the linearization of (3.6) around an interior one-spike solution for more general $Q(u)$?

Since (3.1) will not have stable equilibrium spike solutions, it is natural to ask whether stability can occur for systems of reaction-diffusion equations that admit spike solutions. The simplest type of coupling in a two-component reaction-diffusion system is to consider the so-called shadow limit where the diffusion coefficient of one of the species is taken to infinity and the reaction-time constant for the same species is set to zero. Several examples to illustrate this limit are given below. Before discussing these systems in any detail, we first illustrate qualitatively the mechanism through which a spike can be stabilized by the shadow problem.

The shadow limiting process on a reaction-diffusion system typically leads to a nonlocal scalar PDE of the form

$$u_t = \varepsilon^2 \Delta u + Q(u\,;\sigma_\varepsilon)\,, \quad x \in \Omega\,; \quad \partial_n u = 0\,, \quad x \in \partial\Omega\,, \tag{3.7a}$$

$$\sigma_\varepsilon \equiv \int_D g(u\,;\varepsilon)\, dx\,. \tag{3.7b}$$

Suppose that (3.7) has a radially symmetric localized equilibrium solution of the form $u = u_q\left(\varepsilon^{-1}|x - x_0|\right)$ where $u_q(|y|) \to s$ exponentially as $|y| \to \infty$, for some constant s. Here, we assume that $x_0 \in D$ with $\text{dist}(x_0, \partial\Omega) = O(1)$ as $\varepsilon \to 0$. Then, the stability of this solution is determined by the spectrum of the finite-domain nonlocal eigenvalue problem

$$M_\varepsilon \phi_\varepsilon \equiv \varepsilon^2 \Delta \phi_\varepsilon + Q_u \phi_\varepsilon + Q_\sigma \int_\Omega g_u \phi_\varepsilon\, dx = \lambda^\varepsilon \phi_\varepsilon\,, \quad x \in \Omega\,, \tag{3.8a}$$

$$\partial_n \phi_\varepsilon = 0\,, \quad x \in \partial\Omega\,. \tag{3.8b}$$

where the coefficients in the differential operator are evaluated at u_q. The stability of the spike on an $O(1)$ time-scale follows if we can show that there are no $O(1)$ eigenvalues that satisfy $\text{Re}(\lambda) > 0$. Since the coefficients in the differential operator depend only on $y = \varepsilon^{-1}|x - x_0|$, we look for localized eigenfunctions $\Phi(y)$, which decay as $|y| \to \infty$. Therefore, it is natural to try to compare the spectrum of (3.8) with that of

$$M_0 \Phi \equiv \Delta\Phi + Q_u \Phi + Q_\sigma \varepsilon^N \int_{\mathbb{R}^N} g_u \Phi\, dy = \lambda \Phi\,, \quad y \in \mathbb{R}^N\,, \quad \Phi \to 0 \quad y \to \infty\,. \tag{3.9}$$

Here the derivatives are with respect to the y variable. This problem is referred to as the infinite-line nonlocal eigenvalue problem.

We first note that the spectrum of (3.9) has N zero eigenvalues with corresponding eigenfunctions $\Phi_j(y) = \partial_{y_j} u_q(|y|)$, for $j = 1, \ldots, N$. For these functions the nonlocal term in (3.9) vanishes identically since g_u is radially symmetric in $|y|$. As a result of the exponential decay of u_q, the discrete eigenvalues of (3.9) should be exponentially close to corresponding eigenvalues of the finite-domain nonlocal eigenvalue problem (3.8). This suggests that there are N eigenvalues of (3.8) that will be exponentially small, and whose eigenfunctions $\phi_{j\varepsilon}$ can be approximated by $\phi_{j\varepsilon} = \partial_{x_j} u_q + \phi_{bj}$, where ϕ_b is a boundary layer function localized near $\partial\Omega$ that allows the no-flux condition (3.8b) to be satisfied. Notice that the boundary layer calculation is in the same spirit as that done in Sec. 2.1 for Carrier's problem.

Secondly, we note that if we neglect the nonlocal term in (3.9), the resulting local eigenvalue problem will have one eigenvalue that is strictly positive corresponding to an eigenfunction Φ_l that is of one sign. Since in general $\int_\Omega g_u \Phi_l \, dy \neq 0$, the nonlocal term in (3.9) will perturb this eigenpair significantly. The key step in the analysis is reduced to determining whether the nonlocal term in (3.9) is sufficiently strong to push this positive eigenvalue associated with the local problem into the left half-plane $\text{Re}(\lambda) < 0$. Since (3.8) only perturbs this eigenvalue by exponentially small terms, it remains strictly in the left half-plane for the finite-domain nonlocal problem. If this spectral condition holds, it would follow that an interior one-spike equilibrium solution is *metastable* in the sense that the eigenvalues in the spectrum of the finite-domain nonlocal problem (3.8) that have the largest real parts are exponentially small as $\varepsilon \to 0$. The corresponding eigenfunctions are closely approximated by the translation modes $\partial_{y_j} u_q(|y|)$, for $j = 1, \ldots, N$.

This rough sketch outlines the mechanism through which the nonlocal term can eliminate one unstable eigenvalue of the corresponding local eigenvalue problem and ensure stability on an $O(1)$ time-scale. Depending on the sign of the exponentially small eigenvalues, an interior one-spike solution may not stable on an exponentially long time-scale. However, these exponentially small eigenvalues will lead to the existence of a metastable time-dependent behavior for an interior one-spike solution.

As mentioned above, the key step in the analysis is to find conditions for which there are no eigenvalues of (3.9) with $\text{Re}(\lambda) > 0$. In general, eigenvalue problems of the type (3.9) and (3.8) are difficult to analyze since they are in general not self-adjoint, and hence complex eigenvalues are possible. To illustrate this possibility, consider the eigenvalue problem (3.8) in one space dimension when $\Omega = [-1, 1]$ and $x_0 = 0$. The resulting problem has the general form

$$\varepsilon^2 \phi_{xx} + (A(x/\varepsilon) - \lambda)\,\phi = \delta B(x/\varepsilon) \int_{-1}^{1} C(x/\varepsilon)\phi \, dx \,, \quad -1 < x < 1 \,, \tag{3.10}$$

with $\phi_x(\pm 1) = 0$. Here δ is a parameter measuring the strength of the nonlocal term. This eigenvalue problem is not self-adjoint unless $B(x) = kC(x)$ for some constant k. For fixed ε, many properties of self-adjoint eigenvalue problems of the class (3.10) were obtained in [35] and [9].

Consider the example of [50] where $\varepsilon = 1$, $A(x) \equiv 0$, and

$$B(x) \equiv 2.5 + \cos(\pi x) + 2\cos(2\pi x) \qquad C(x) \equiv 1.1 - \frac{1}{2}\cos(\pi x) + \frac{1}{2}\cos(2\pi x) \,. \tag{3.11}$$

Moveable eigenvalues are those eigenvalues of the local problem that are perturbed by the nonlocal term. Fixed eigenvalues refer to those eigenvalues of the local problem that remain independent of δ, since their eigenfunctions are orthogonal to $C(x)$. For this example, the only moveable eigenvalues are those for which the eigenfunctions lie in the subspace spanned by

$$\phi = s_0 + s_1 \cos(\pi x) + s_2 \cos(2\pi x) \,, \tag{3.12}$$

for some s_0, s_1, and s_2. Substituting (3.11) and (3.12) into (3.10) where $\varepsilon = 1$, we get the matrix eigenvalue problem $(\Lambda - \delta D)\, \boldsymbol{s} = \lambda \boldsymbol{s}$, where

$$\Lambda \equiv \begin{pmatrix} 0 & 0 & 0 \\ 0 & -\pi^2 & 0 \\ 0 & 0 & -4\pi^2 \end{pmatrix} , \qquad D \equiv \begin{pmatrix} 5.5 & -1.25 & 1.25 \\ 2.2 & -0.5 & 0.5 \\ 2.2 & -0.5 & 0.5 \end{pmatrix} , \qquad \boldsymbol{s} \equiv \begin{pmatrix} s_0 \\ s_1 \\ s_2 \end{pmatrix} . \tag{3.13}$$

The real parts of the eigenvalues as a function of δ are shown in Fig. 4. In this figure, the dotted lines correspond to the fixed eigenvalues $-k^2\pi^2/4$ for $k = 1$ and $k = 3$, corresponding to the eigenfunctions $\phi = \cos(k\pi(x+1)/2)$ for $k = 1, 3$. This simple example shows that nonlocal non self-adjoint eigenvalue problems of the form (3.10) can have complex eigenvalues through the collision of two moveable eigenvalues.

An important class of nonlocal infinite-line eigenvalue problems that arises in determining the stability of spike solutions in several different systems is the following nonlocal non self-adjoint problem:

$$M_0\Phi \equiv L_0\Phi - \alpha w^p \left(\frac{\int_{\mathbb{R}^N} w^{m-1}\Phi\, dy}{\int_{\mathbb{R}^N} w^m\, dy} \right) = \lambda\Phi \,, \quad y \in \mathbb{R}^N \,; \quad \Phi \to 0 \ \ |y| \to \infty \,. \tag{3.14a}$$

Here $w(|y|)$ satisfies (1.7), and L_0 is the local operator

$$L_0\Phi \equiv \Delta\Phi - \Phi + pw^{p-1}\Phi \,. \tag{3.14b}$$

We assume that $m > 1$ and $1 < p < p_c$, where p_c is the critical Sobolev exponent. Notice that $\partial_{y_j} w(|y|)$ lies in the kernel of M_0 for $j = 1, \ldots, N$,

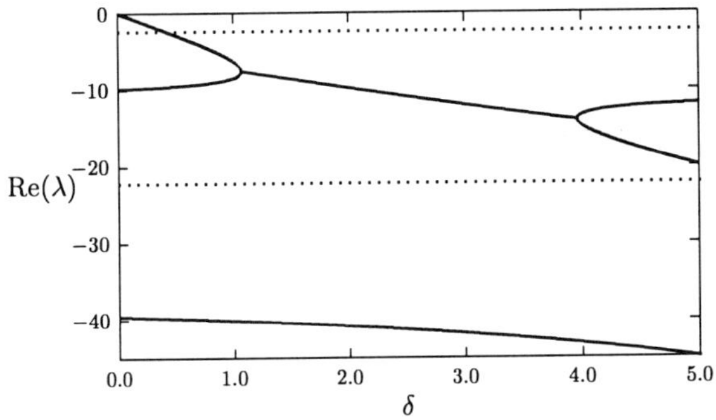

Figure 4. The real parts of the eigenvalues of (3.13) (solid curves) versus δ. Two of them are complex when $1.076 < \delta < 3.970$. The dotted lines are two fixed eigenvalues $\lambda = -\pi^2/4$ and $\lambda = -9\pi^2/4$, not contained in (3.13), which are independent of δ.

and so $\lambda = 0$ is a fixed eigenvalue. There are two key formulae for L_0, obtained by a direct calculation, that are needed below

$$L_0^{-1} w^p = \frac{w}{p-1}\,; \qquad L_0^{-1} w = \frac{w}{p-1} + \frac{1}{2} y \cdot \nabla w\,. \tag{3.15}$$

For the one-dimensional case $N = 1$, where $w(y)$ is given explicitly in (1.8), the following spectral results for (3.14) hold:

Proposition 3.3. *Let $\Phi \in H^1(R)$, and consider any nonzero eigenvalue λ_0 of (3.14). Then, we have the following:*

- *For $0 \le \alpha < p - 1$ we have $Re(\lambda_0) > 0$.*
- *Now suppose $\alpha > p - 1$. Then, if either $m = 2$ and $1 < p \le 5$, or, $m = p + 1$ and $p > 1$, we have $Re(\lambda_0) < 0$.*
- *If $p > 1$ and $m = p$, then we have $Re(\lambda_0) < 0$ when $p - 1 < \alpha \le p$.*

The proof of the first result for $0 \le \alpha < p - 1$ is given in Appendix E of [47]. The proof of the second result for $\alpha > p - 1$ is given in Lemma A and Theorem 1.4 of [108]. The third result is proved in Theorem 1 of [115]. For the multidimensional case where $N > 1$, the following results are known:

Proposition 3.4. *Let $\Phi \in H^1(\mathbb{R}^N)$, and consider any nonzero eigenvalue λ_0 of (3.14). Then, we have the following:*

- *For $0 \le \alpha < p - 1$ we have $Re(\lambda_0) > 0$.*

- *Now suppose $\alpha > p-1$. Then, if either $m = 2$ and $1 < p \leq 1+4/N$, or, $m = p+1$ and $1 < p < p_c$, where p_c is the critical Sobolev exponent, we conclude that $Re(\lambda_0) < 0$.*
- *Let $m = p$. If $2 \leq p \leq 5$ when $N = 2$, or $2 \leq p \leq 3$ when $N = 3$, then we have $Re(\lambda_0) < 0$ when $\alpha = 2p$.*

The proof of the first two results are given in [108], and the proof of the third result for $\alpha = p$ is given in Theorem 5.6 of [91]. Notice that when $m = p+1$, the operator is self-adjoint. Qualitatively, these results show that the nonlocal term may eliminate the unstable eigenvalue of the local operator only when α is large enough.

We now comment on the bounds in these results. The lower bound $\alpha = p - 1$ for stability in the second result of Propositions 3.3 and 3.4 cannot be improved since from (3.15) we readily calculate that $M_0 w = 0$ when $\alpha = p-1$. The upper bound for α in the third result of Proposition 3.3 is not sharp as stated in [115]. The upper bound on p for $m = 2$ in the second result of Proposition 3.3 is indeed sharp as the next result shows.

Proposition 3.5. *Let $\Phi \in H^1(R)$, $m = 2$, and suppose that $p > 5$ in (3.14). Then, there exists an α_m with $\alpha_m > p-1$ such that there are exactly two positive real eigenvalues in the interval $(0, \nu_0)$ for any α with $(p-1) < \alpha < \alpha_m$. In addition, there exists a value α_h, with $\alpha_h > \alpha_m$ such that for $\alpha_m < \alpha < \alpha_h$, there is a pair of complex conjugate eigenvalues in the unstable right half-plane $Re(\lambda) > 0$. When $\alpha = \alpha_h$, there is a pair of complex conjugate eigenvalues on the imaginary axis.*

This result was proved in Proposition 2.7 of [50]. In addition, a detailed numerical study of the spectrum of (3.14) for different values of m and p was given in Sec. 2.2 and Sec. 2.3 of [50].

For any $p \geq 3$ and with $\alpha = 0$, Proposition 3.2 shows that there is only one discrete nonzero eigenvalue of M_0. Proposition 3.5 shows that there are two discrete eigenvalues in the right half-plane for some range of α when $p > 5$. The numerical computations of [50] show that an extra eigenvalue is created out of the edge of the continuous spectrum at a certain value of α. The two discrete moveable eigenvalues then coalesce producing a complex conjugate pair as in the simple example (3.13). This leads to the next question.

Question 3.2. Find other ranges of p, m, and N where any nonzero eigenvalue of (3.14) will have $\mathrm{Re}(\lambda) < 0$? Can one characterize any edge bifurcations for (3.14) from the continuous spectrum? A detailed numerical study for $N > 1$ is an open problem.

The analysis leading to Proposition 3.3–3.5 relies rather heavily on special properties of the nonlinearity $Q(u) = -u+u^p$, most notably the explicit formulae (3.15). This leads to the following question.

Question 3.3. Can one characterize the discrete spectrum of more general problems of the form (3.9) around an interior one-spike solution?

Although the proofs of Propositions 3.3–3.5 are too involved to discuss here in detail, we can still give a qualitative idea on how some of these results are obtained. To do so, we reformulate (3.14) by letting $\psi(y)$ be the solution to

$$L_0\psi \equiv \psi'' - \psi + pw^{p-1}\psi = \lambda\psi + w^p\,; \qquad \psi \to 0 \quad \text{as} \quad |y| \to \infty\,. \tag{3.16}$$

Then, the eigenfunctions of (3.14) can be written as

$$\Phi = \alpha\psi J\,, \qquad J \equiv \frac{\int_{\mathbb{R}^N} w^{m-1}\Phi\, dy}{\int_{\mathbb{R}^N} w^m\, dy}\,. \tag{3.17}$$

We then multiply both sides of (3.17) by w^{m-1} and integrate over $\mathbb{R}^N$. Assuming, that $\int_{\mathbb{R}^N} w^{m-1}\Phi\, dy \neq 0$, we then obtain that the eigenvalues of (3.14) with even eigenfunctions are the roots of $g(\lambda) = 0$, where

$$g(\lambda) \equiv \frac{1}{\alpha} - f(\lambda)\,, \qquad f(\lambda) \equiv \frac{\int_{\mathbb{R}^N} w^{m-1}\left(L_0 - \lambda\right)^{-1} w^p\, dy}{\int_{\mathbb{R}^N} w^m\, dy}\,. \tag{3.18}$$

The function $g(\lambda)$ is analytic in the right half-plane except at the simple pole $\lambda = \nu_0$, where ν_0 is the unique positive eigenvalue of L_0. A simple calculation of the winding number shows that the number M of zeroes of $g(\lambda)$ in $\mathrm{Re}(\lambda) > 0$ is

$$M = 1 + \frac{1}{\pi}\left[\arg g\right]_{\Gamma_I}\,. \tag{3.19}$$

Here $[\arg g]_{\Gamma_I}$ denotes the change in the argument of $g(\lambda)$ along the semi-infinite imaginary axis $\Gamma_I = i\lambda_I$, $0 \leq \lambda_I < \infty$, traversed in the downwards direction. Therefore, to calculate M, we need only determine properties of $g(\lambda)$ on the positive imaginary axis.

We let $\lambda = i\lambda_I$ and we separate real and imaginary parts by writing $g(i\lambda_I) = \tilde{g}_R(\lambda_I) + i\tilde{g}_I(\lambda_I)$. A simple calculation shows that the eigenvalues of (3.14) along the positive imaginary axis are the roots of the coupled system $\tilde{g}_R = \tilde{g}_I = 0$, given by

$$\tilde{g}_R(\lambda_I) \equiv \frac{1}{\alpha} - \tilde{f}_R(\lambda_I) \,, \qquad \tilde{g}_I(\lambda_I) \equiv -\tilde{f}_I(\lambda_I) \,, \tag{3.20a}$$

where $\tilde{f}_R$ and $\tilde{f}_I$ are defined by

$$\tilde{f}_R(\lambda_I) \equiv \frac{\int_{\mathbb{R}^N} w^{m-1} L_0 \left[L_0^2 + \lambda_I^2\right]^{-1} w^p \, dy}{\int_{\mathbb{R}^N} w^m \, dy} \,, \tag{3.20b}$$

$$\tilde{f}_I(\lambda_I) \equiv \frac{\lambda_I \int_{\mathbb{R}^N} w^{m-1} \left[L_0^2 + \lambda_I^2\right]^{-1} w^p \, dy}{\int_{\mathbb{R}^N} w^m \, dy} \,. \tag{3.20c}$$

The following local and global properties of $\tilde{f}_R$ and $\tilde{f}_I$ have been established:

Proposition 3.6. *The function $\tilde{f}_R$ in (3.20b) has the asymptotic behavior*

$$\tilde{f}_R(\lambda_I) \sim \frac{1}{p-1} - \kappa_c \lambda_I^2 + \cdots \,, \quad \text{as } \lambda_I \to 0; \quad \tilde{f}_R(\lambda_I) = 0\left(\lambda_I^{-2}\right) \,, \quad \text{as } \lambda_I \to \infty \,. \tag{3.21}$$

Here $\kappa_c > 0$ if $m = 2$, or, if $m = p+1$ and $1 < p < 1 + 4/N$. When $m = 2$ the function $\tilde{f}_R(\lambda_I)$ is monotone decreasing for $\lambda_I > 0$. The function $\tilde{f}_I$ in (3.20c) has the asymptotic behavior

$$\tilde{f}_I(\lambda_I) \sim \frac{\lambda_I}{p-1} \left[\frac{1}{p-1} - \frac{N}{2m}\right] + O(\lambda_I^3) \,, \quad \text{as} \quad \lambda_I \to 0 \,, \tag{3.22}$$

with $\tilde{f}_I(\lambda_I) = 0\left(\lambda_I^{-1}\right)$ as $\lambda_I \to \infty$. When either, $m = 2$ and $1 < p \leq 1 + 4/N$, or, when $m = p+1$ and $1 < p < p_c$, where p_c is the critical Sobolev exponent, we have the global result that $\tilde{f}_I(\lambda_I) > 0$ for $\lambda_I > 0$.

The local behavior of $\tilde{f}_R$ was derived in Eq. (4.3) of [106]. The condition $\tilde{f}_R' < 0$ for $m = 2$ was derived in the proof of Theorem 2.3 of [106]. The local behavior of $\tilde{f}_I$ was derived in Eq. (4.2) of [106]. The proof that $\tilde{f}_I > 0$ for $m = 2$ and $1 < p \leq 1 + 4/N$ is rather difficult, and was obtained in Theorem 2.3 of [106]. The condition $\tilde{f}_I > 0$ for $m = p+1$ and $1 < p < p_c$ is readily seen by writing $\tilde{f}_I$ as $\tilde{f}_I(\lambda_I) = \lambda_I C(\lambda_I)$, where

$$C(\lambda_I) \equiv \frac{\int_{\mathbb{R}^N} w^p \left[L_0^2 + \lambda_I^2\right]^{-1} w^p \, dy}{\int_{\mathbb{R}^N} w^{p+1} \, dy} \,. \tag{3.23}$$

We readily calculate using (3.15) that

$$C'(\lambda_I) = -2\lambda_I \frac{\int_{\mathbb{R}^N} \left[\left(L_0^2 + \lambda_I^2 \right)^{-1} w^p \right]^2 dy}{\int_{\mathbb{R}^N} w^{p+1}\, dy} < 0\,, \tag{3.24a}$$

$$C(0) = \frac{1}{p-1} \left[\frac{1}{p-1} - \frac{N}{2(p+1)} \right] . \tag{3.24b}$$

Thus, for $1 < p < p_c$, we have that $C(0) > 0$ together with $C'(\lambda_I) < 0$ for $\lambda_I > 0$, and $C(\lambda_I) \to 0^+$ as $\lambda_I \to \infty$. Hence $C(\lambda_I) > 0$ for $\lambda_I > 0$, which establishes that $\tilde{f}_I > 0$ for $\lambda_I > 0$ when $m = p+1$ and $1 < p < p_c$.

Next, we use the properties of g on the imaginary axis to calculate M from (3.19). The following result is readily derived by using (3.20)–(3.22) to calculate $[\arg g]_{\Gamma_I}$:

Proposition 3.7. *Let $\alpha > p-1$. Suppose that at each root of $\tilde{g}_R = 0$, we have that $\tilde{f}_I > 0$. Then, $M = 0$, and there are no eigenvalues of (3.14) in $Re(\lambda) > 0$. Alternatively, suppose that $0 < \alpha < p-1$, and that $\tilde{f}_R$ is monotone decreasing for $\lambda_I > 0$. Then, $M = 1$ and so there is a unique real positive eigenvalue of (3.14).*

Notice that if $\alpha > p-1$, then $\tilde{g}_R(0) < 0$ and $\tilde{g}_I(0) = 0$. As $\lambda_I \to \infty$, we have $\tilde{g}_R \to \alpha^{-1} > 0$ and $\tilde{g}_I \to 0$. Hence, if whenever we have a root of $\tilde{g}_R = 0$ it follows that $\tilde{g}_I < 0$, we conclude that $[\arg g]_{\Gamma_I} = -\pi$, and consequently $M = 0$ from (3.19). Note that $\tilde{g}_I < 0$ is guaranteed whenever $\tilde{f}_I > 0$ for all $\lambda_I > 0$. As seen in Proposition 3.6, this condition is guaranteed for two cases: $m = 2$ and $1 < p \leq 1 + 4/N$, or, $m = p+1$ and $1 < p < p_c$. This criterion then establishes the second statements in Proposition 3.3 and Proposition 3.4. Alternatively, if $0 < \alpha < p-1$ and $\tilde{f}'_R < 0$, then $\tilde{g}_R > 0$ for $\lambda_I > 0$. Consequently, $[\arg g]_{\Gamma_I} = 0$, and hence $M = 1$. This is the first statement in Proposition 3.3 and Proposition 3.4 under a slightly weaker hypothesis. Eliminating the hypothesis that $\tilde{f}_R$ is monotone decreasing, it is readily seen, upon looking for roots of $g(\lambda) = 0$ on the positive real axis, that $M \geq 1$ when $0 < \alpha < p-1$. Finally, we comment on the idea behind Proposition 3.5. For $m = 2$, we have from Proposition 3.6 that $\tilde{f}'_R < 0$ for $\lambda_I > 0$, and hence there exists a unique root to $\tilde{g}_R = 0$ when $\alpha > p-1$. If we can guarantee that $\tilde{g}_I > 0$, or equivalently $\tilde{f}_I < 0$, at this root, then we have $[\arg g]_{\Gamma_I} = +\pi$, and so $M = 2$. For $N = 1$, $m = 2$, and $p > 5$, the local behavior in Proposition 3.6 shows that $\tilde{f}_I < 0$ for $\lambda_I > 0$ sufficiently small. Hence, there is some range of α with $\alpha > p-1$ for which $M = 2$. This is the essence of Proposition 3.5.

The result in Proposition 3.7 gives a simple criterion to determine sufficient conditions for nonzero eigenvalues of (3.14) to satisfy $\text{Re}(\lambda) < 0$. This leads to the following question.

Question 3.4. Can one find other ranges of m, p, and N, to ensure that the positivity condition on $\tilde{f}_I$ given in Proposition 3.7 holds? With this condition any nonzero eigenvalue of (3.14) has $\text{Re}(\lambda) < 0$ when $\alpha > p - 1$.

One of the earliest analyses of metastability for a shadow system was given in [64]. Another activator-inhibitor system that exhibits metastability was given in [5] and [6]. In the next few subsections we give a few explicit examples of the stability and dynamics of spikes for shadow systems.

3.1. *The Shadow Gierer-Meinhardt Model*

Our first example of a shadow system is obtained by letting $D \to \infty$ in the GM model (1.1) to get

$$a_t = \varepsilon^2 \Delta a - a + a^p/h^q \,, \quad x \in \Omega\,; \quad \partial_n a = 0 \,, \qquad x \in \partial\Omega \,, \tag{3.25a}$$

$$\tau h_t = -h + \frac{\varepsilon^{-N}}{|\Omega|} \int_\Omega \frac{a^m}{h^s} \, dx \,. \tag{3.25b}$$

In (3.25b), $|\Omega|$ denotes the volume of Ω. In this section we will consider the case where the reaction-time constant τ in (3.25b) is zero. The possibility of Hopf bifurcations when $\tau > 0$ is discussed in the next section.

An interior one-spike equilibrium solution to (3.25) in $\mathbb{R}^N$ is given by

$$a = h^{q/(p-1)} w \left[\varepsilon^{-1}|x - x_0|\right] \,, \quad h = \left(\frac{\omega_N}{|\Omega|} \int_0^\infty w^m \rho^{N-1} \, d\rho \right)^{-1/\zeta} \,. \tag{3.26}$$

Here ζ is given in (1.2), ω_N is the surface area of the unit N-dimensional sphere, and $w(\rho)$ satisfies (1.7). The finite-domain nonlocal eigenvalue problem of the form (3.8), obtained by linearizing (3.25) around this equilibrium solution, is

$$M_\varepsilon = \epsilon^2 \Delta \phi_\varepsilon + (-1 + p w^{p-1}) \phi_\varepsilon - \frac{mq\epsilon^{-N} w^p}{s+1} \left(\frac{\int_\Omega w^{m-1} \phi \, dx}{\int_\Omega w^m \, dx} \right) = \lambda^\varepsilon \phi_\varepsilon \,, \tag{3.27}$$

where $\partial_n \phi_\varepsilon = 0$ on $\partial\Omega$. The corresponding infinite-line nonlocal eigenvalue problem is

$$M_0 \Phi \equiv L_0 \Phi - \frac{mq}{s+1} w^p \left(\frac{\int_{\mathbb{R}^N} w^{m-1} \Phi \, dy}{\int_{\mathbb{R}^N} w^m \, dy} \right) = \lambda \Phi \,, \tag{3.28}$$

with $\Phi \to 0$ as $|y| \to \infty$, and L_0 defined in (3.14b).

From (3.14), and the condition (1.2) on the exponents (p, q, m, s), we see that $\alpha = mq/(s+1) > p-1$. From the second statement of Proposition 3.4 we conclude that for any nonzero eigenvalue of (3.28), we have that $\text{Re}(\lambda) < 0$ when $m = 2$ and $1 < p \leq 1 + 4/N$, or when $m = p+1$ and $1 < p < p_c$, where p_c is the critical Sobolev exponent. Therefore, under these conditions, the nonlocal term has pushed the unstable eigenvalue of the local operator L_0 into the stable left half-plane. Since the discrete eigenvalues of (3.28) are exponentially close to corresponding eigenvalues of (3.27), we conclude from the discussion following (3.8) that an interior one-spike solution to the shadow GM model will be metastable.

Then, by using the projection method in a similar way as was done in Sec. 2, the derivation in Sec. 2.5 of [44] yields the estimate (2.58) for the exponentially small eigenvalues of (3.27). Therefore, to leading order, the contribution of the nonlocal term in (3.27) is subdominant to that of the boundary layer calculation given in Sec. 2.5. Then, by using the projection method for the time-dependent problem, the following result for the metastable motion of an interior one-spike solution for the shadow GM model was obtained in [44]:

Proposition 3.8. *Let $\varepsilon \to 0$, and assume that either $m = 2$ and $1 < p \leq 1 + 4/N$, or $m = p+1$ and $1 < p < p_c$, where p_c is the critical Sobolev exponent. Then, a one-spike solution for the shadow GM model (3.25) with $\tau = 0$ is given asymptotically by $a(x,t) \sim h^{q/(p-1)} w\left(\varepsilon^{-1}|x - x_0(t)|\right)$, where $x_0(t)$ satisfies the differential equation*

$$x_0' \sim \frac{\varepsilon N c^2}{\beta_N \omega_N} \int_{\partial\Omega} \hat{r}\, r^{1-N} e^{-2r/\varepsilon} (1 + \hat{r}\cdot\hat{n})\, \hat{r}\cdot\hat{n}\, d\partial\Omega \,, \quad \beta_N \equiv \int_0^\infty w'^2 \rho^{N-1}\, d\rho \,. \tag{3.29}$$

Here $r = |x - x_0|$, $\hat{r} = (x - x_0)/r$, $\hat{n}$ is the unit outward normal to $\partial\Omega$, and c is defined in (1.7b). Next, assume that there is a unique point x_m on $\partial\Omega$, where r is minimized. Then, the spike moves exponentially slowly in a straight line towards x_m and the distance $r_m(t) \equiv |x_m - x_0(t)|$ satisfies

$$r_m' \sim -\xi r_m \left(\frac{\varepsilon}{r_m}\right)^{(N+1)/2} H(r_m)\, e^{-2r_m/\varepsilon} \,, \quad \xi \equiv \frac{2Nc^2}{\omega_N \beta_N} \pi^{(N-1)/2} \,. \tag{3.30}$$

In terms of the principal radii of curvature R_i, $i = 1, \ldots, N-1$ of $\partial\Omega$ at x_m, the function $H(r_m)$ is defined by $H(r_m) \equiv \left(1 - \frac{r_m}{R_1}\right)^{-1/2} \cdots \left(1 - \frac{r_m}{R_{N-1}}\right)^{-1/2}$.

This result was first derived formally in Proposition 2 and Corollary 2 of [44] and was later proved rigorously in [14]. Metastability will also occur for other (p, q, m, s) whenever we can guarantee that for any nonzero eigenvalue of (3.28) we have $\text{Re}(\lambda) < 0$ (see Question 3.2 above).

This analysis shows that an interior one-spike solution to the shadow GM model with $\tau = 0$ is ultimately unstable, and the spike will drift exponentially slowly towards the closest point on the boundary. An open problem concerns how the spike attaches to the boundary of the domain.

Question 3.5. Analyze the time-dependent motion of a spike when $\text{dist}(x_0, \partial\Omega) = O(\varepsilon)$. How does a spike attach itself to the boundary?

We remark that if we were to change the boundary conditions from Neumann to the Robin condition

$$\varepsilon \partial_n a + \kappa a = 0\,, \quad x \in \partial\Omega\,, \tag{3.31}$$

where $\partial_n a$ is the outward normal derivative, then from Eq. (2.58) of Sec. 2.5 we would expect that the exponentially small eigenvalues of (3.27) will all be negative when $\kappa > 1$ (see Proposition 2.1 for the analogous formula in one-dimension). Therefore, when $\kappa > 1$ an interior one-spike equilibrium solution will be stable. This leads to the next question.

Question 3.6. Consider (3.25) (with $\tau = 0$) and with the Robin condition (3.31) for a with $k > 1$. Prove that an interior one-spike equilibrium solution is stable, and that a one-spike solution drifts exponentially slowly towards the point in $\partial\Omega$ that maximizes the distance to the boundary.

In [45] a formal asymptotic analysis was done for (3.25) when $\tau = 0$ to derive an equation of motion for a spike on the boundary $\partial\Omega$ of a domain. Since the spike is localized, to leading order we have a spike on the boundary of a half-space. In view of the Neumann boundary conditions, the stability of the spike profile on an $O(1)$ time-scale is again determined by the infinite-line nonlocal eigenvalue problem (3.28). The following result was given in Proposition 2.1 of [45]:

Proposition 3.9. *Let $\varepsilon \to 0$ and assume that either $m = 2$ and $1 < p \leq 3$, or $m = p+1$. Then, the motion of a spike for (3.25) that is confined to the* *smooth boundary of a two-dimensional evolves according to*

$$a \sim h^{q/(p-1)} w \left(\varepsilon^{-1} \left[(s - s_0(t))^2 + \eta^2 \right]^{1/2} \right) + O(\varepsilon) \,, \tag{3.32a}$$

$$s_0'(t) \sim \frac{4b}{3\pi} \varepsilon^3 \kappa'(s_0) \,, \quad b \equiv \frac{\int_0^\infty \rho^2 \left[w'(\rho) \right]^2 \, d\rho}{\int_0^\infty \rho \left[w'(\rho) \right]^2 \, d\rho} \,. \tag{3.32b}$$

Here $w(\rho)$ satisfies (1.7) when $N = 2$, η is the distance from $x \in \Omega$ to $\partial\Omega$, and s is the corresponding orthogonal coordinate, which measures arclength along $\partial\Omega$ when $\eta = 0$. In addition, κ is the curvature of $\partial\Omega$, taken with the sign convention that $\kappa > 0$ for a circle.

This result shows that the speed of the spike is $O(\varepsilon^3)$, and that stable equilibrium points correspond to points on the boundary where κ has local maxima. An analogous result for the spike motion on the boundary of a three-dimensional domain is given in Proposition 3.1 of [45]. For the full GM model (1.1) it was proved in [18) that there is an equilibrium boundary spike solution that concentrates at a local maximum of the curvature of $\partial\Omega$ whenever the inhibitor diffusivity D in (1.1b) is sufficiently large. Therefore, for equilibrium boundary spike solutions, the shadow GM model closely predicts behavior in the full GM model for D large.

The result (3.32b) predicts that a spike is stationary on a flat segment of the boundary where $\kappa' = 0$. In this case, as was shown in [45], the motion of a spike is exponentially slow and is determined by the local behavior of the boundary at the ends of the flat segment. Consider the two-dimensional case where $x = (x, y)$, and suppose that the spike is located on the straight-line boundary segment joining the points $(x_L, 0)$ and $(x_R, 0)$ as shown in Fig. 5. The spike is centered at $x_0 = (\xi, 0)$ where $x_L < \xi < x_R$. We decompose $\partial\Omega$ as $\partial\Omega = \partial\Omega_c \cup \partial\Omega_s$ where $\partial\Omega_s$ is the straight-line segment of the boundary and $\partial\Omega_c$ denotes the remaining curved part of $\partial\Omega$. The distance between the spike and $\partial\Omega_c$ is assumed to be a minimum at either of the two corners $(x_L, 0)$ or $(x_R, 0)$.

The local behavior of Ω_c near the corner points is critical to determining the motion. Near these corner points, we assume the local behavior

$$(x_L, 0) \,; \quad y = \psi_L(x), \quad \psi_L'(x) \sim -K_L (x_L - x)^{\alpha_L} \,, \quad \text{as } x \to x_L^- \,, \tag{3.33a}$$

$$(x_R, 0) \,; \quad y = \psi_R(x), \quad \psi_R'(x) \sim K_R (x - x_R)^{\alpha_R} \,, \quad \text{as } x \to x_R^+ \,, \tag{3.33b}$$

where $\alpha_L > 0$ and $\alpha_R > 0$. When $\alpha_L = \alpha_R = 1$, K_L and K_R are proportional to the curvature of $\partial\Omega_c$ at the left and right corners, respectively.

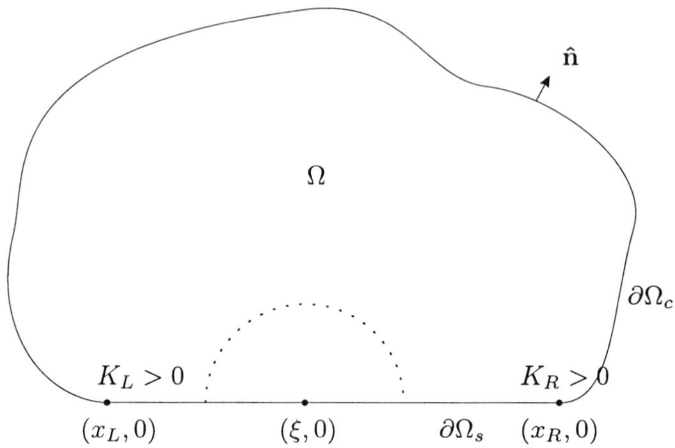

Figure 5. Plot of a two-dimensional domain Ω with a flat boundary segment. The spike is centered at $x = \xi$ on the flat segment. The dotted line indicates an approximate equipotential for a.

The following result characterizing the motion of a spike on the flat segment was derived in Proposition 5.2 of [45] using the projection method.

Proposition 3.10. *Let $\varepsilon \to 0$ and assume that either $m = 2$ and $1 < p \leq 3$, or $m = p + 1$. Assume that the distance between ξ and $\partial\Omega_c$ is a minimum at either of the two corners $(x_L, 0)$ or $(x_R, 0)$. Then, for $\varepsilon \to 0$, the x-coordinate, $\xi(t)$, of the center of a spike that is attached to the flat segment $\partial\Omega_s$ satisfies the asymptotic differential equation,*

$$\xi'(t) \sim \frac{2\varepsilon c^2}{\pi\beta}\left\{\frac{K_R}{x_R - \xi}\left(\frac{\varepsilon}{2}\right)^{\alpha_R+1}\Gamma(\alpha_R + 1)\, e^{-2(x_R-\xi)/\varepsilon} - \frac{K_L}{\xi - x_L}\left(\frac{\varepsilon}{2}\right)^{\alpha_L+1}\Gamma(\alpha_L + 1)\, e^{-2(\xi-x_L)/\varepsilon}\right\}. \tag{3.34}$$

Here c is given in (1.7b), $\beta \equiv \int_0^\infty \rho w'^2(\rho)\, d\rho$, and the constants K_L, K_R, α_L and α_R are defined in (3.33) in terms of the corner behavior of $\partial\Omega_c$.

A similar differential equation for the motion of a straight-line interface in a constant width neck region of a dumbbell-shaped domain was derived in [60]. The result (3.34) shows that there is a unique steady-state one-spike boundary solution on $x_L < \xi < x_R$ whenever K_L and K_R have the same sign. This solution is unstable when Ω is convex near $(x_L, 0)$ and $(x_R, 0)$ so that $K_R < 0$ and $K_L < 0$. It is stable when $K_L > 0$ and $K_R > 0$. If $K_L < 0$ and $K_R > 0$, then there is no steady-state and $\xi(t)$ will move exponentially slowly towards x_R. Similarly, $\xi(t)$ will move towards

x_L if $K_L > 0$ and $K_R < 0$. When the spike touches $(x_L, 0)$ or $(x_R, 0)$, its subsequent evolution is determined by (3.32b).

Question 3.7. Give a rigorous proof of the characterization of time-dependent spike motion on the boundary of the domain as given in Proposition 3.9–3.10 above, and in Proposition 3.1 of [45]. Formulate a numerical method that is able to numerically compute interior and boundary spike solutions to (3.25) over, possibly, exponentially long time-scales.

3.2. *Hopf Bifurcations for the Shadow GM Model*

We now analyze (3.25) when $\tau > 0$. A equilibrium one-spike solution is again given by (3.26). For this shadow GM problem, it was shown numerically in [78] that a Hopf bifurcation in the spike amplitude can occur when τ is large enough. Since this instability occurs on an $O(1)$ time-scale, for this problem we are not concerned with the exponentially small eigenvalues. By computing the full numerical solution to (3.25) for $N = 2$, in Fig. 6 we plot the amplitude $a_m \equiv a(0, t)$ of a spike centered at $r = |x| = 0$ versus t for two values of τ. This figure suggests that there is a Hopf bifurcation for some τ in $0.53 < \tau < 0.58$.

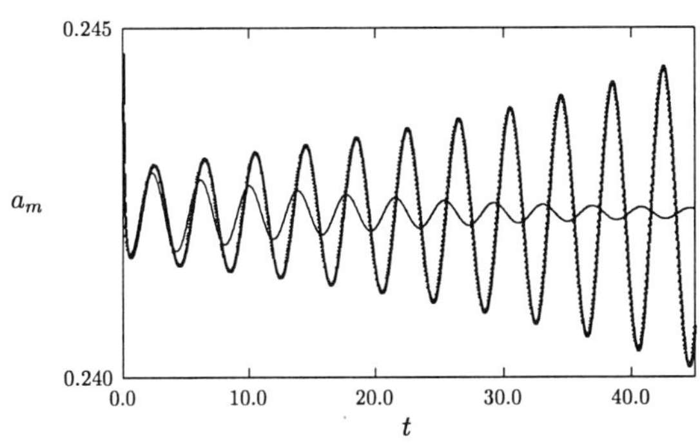

Figure 6. The Hopf bifurcation. We plot a at $x = 0$ versus t for $\tau = 0.53$ (solid curve) and $\tau = 0.58$ (heavy solid curve). The exponent set is $(p, q, m, s) = (2, 1, 2, 0)$, with $N = 2$, and $\varepsilon = 0.05$.

To analyze the Hopf bifurcation we must study the associated spectral problem. Upon linearizing (3.25) around the equilibrium solution in (3.26), we obtain the following nonlocal eigenvalue problem in place of (3.28):

$$M_0\Phi \equiv L_0\Phi - \frac{mq}{s+1+\tau\lambda} w^p \left(\frac{\int_{\mathbb{R}^N} w^{m-1}\Phi \, dy}{\int_{\mathbb{R}^N} w^m \, dy} \right) = \lambda\Phi \,, \tag{3.35}$$

where $\Phi \to 0$ as $|y| \to \infty$ and L_0 is defined in (3.14b). This eigenvalue problem is more complicated than (3.28) since the multiplier of the nonlocal term now depends on $\tau\lambda$.

The spectrum of (3.35) was studied in [106], where the following result was obtained.

Proposition 3.11. *Assume that $m = 2$ and $1 < p \leq 1 + 4/N$. Then, there exists a unique $\tau = \tau_0 > 0$ such that (3.35) has an eigenvalue $\lambda = i\lambda_I^0$, with $\lambda_I^0 > 0$. For all $\tau > \tau_0$ there are exactly two eigenvalues of (3.35) in the right half-plane Re$(\lambda) > 0$. In addition, there exists a τ_c with $\tau_c > \tau_0$ such that for any τ with $\tau > \tau_c$, these two eigenvalues are on the positive real axis, with one eigenvalue tending to ν_0 and the other eigenvalue tending to zero as $\tau \to \infty$. Here $\nu_0 > 0$ is the unstable eigenvalue of the local operator L_0. Alternatively, for any τ with $0 < \tau < \tau_0$ there are no eigenvalues of (3.35) in the right half-plane.*

The existence of a complex conjugate pair of eigenvalues for some $\tau = \tau_0$ was proved in Theorem 2.3 of [106]. The remainder of the proposition was proved in Lemma 3.1 and in Sec. 3 of [106]. Since there are no eigenvalues in the left half-plane when $0 < \tau < \tau_0$ and exactly two when $\tau > \tau_0$, this result shows a transversal crossing condition for τ near τ_0 and proves the existence of a Hopf bifurcation point. In Lemma 2.5 and Eq. (2.36) of [106], the following bounds on the frequency λ_I^0 were derived:

Proposition 3.12. *Assume that $m = 2$ and $1 < p \leq 1 + 4/N$. Let ζ and $C_{N,p}$ be defined by*

$$C_{N,p} \equiv \left[\frac{N + (p+1)(1 - N/2)}{p+1} \right] , \quad \zeta \equiv \frac{2q}{p-1} - (s+1) > 0 \,. \tag{3.36}$$

Then, the Hopf bifurcation values τ_0 and $\lambda_I^0 > 0$ satisfy

$$\lambda_I^0 \geq \frac{\zeta(p-1)^2}{qC_{N,p}} \,, \quad \lambda_I^0 \leq \frac{2q}{(s+1)} \left(\frac{\int_{\mathbb{R}^N} w^{2p} \, dy}{\int_{\mathbb{R}^N} w^2 \, dy} \right)^{1/2} \,. \tag{3.37}$$

For the exponent set $(p, q, m, s) = (2, 1, 2, 0)$, in Fig. 7 we illustrate Proposition (3.11) by plotting the path of $\lambda = \lambda_R + i\lambda_I$ as τ is increased. This type of path in the spectrum is very similar to what was shown for

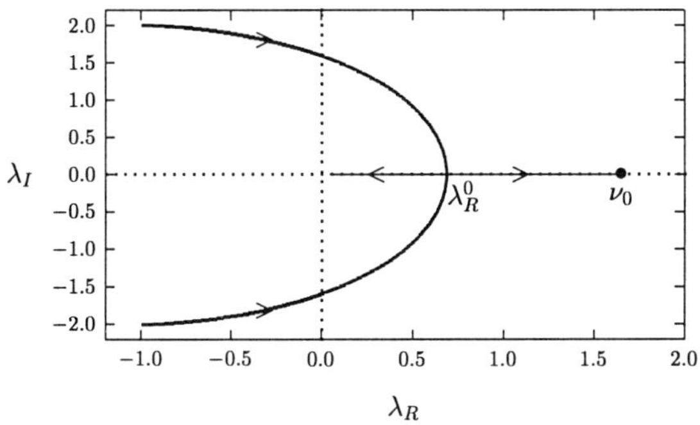

Figure 7. Plot of the path of $\lambda = \lambda_R \pm i\lambda_I$ for $(p, q, m, s) = (2, 1, 2, 0)$ with $N = 2$ (arrows indicate direction of increasing τ). There is a pair of pure imaginary eigenvalues when $\tau = \tau_0$. The complex conjugate pair merge onto the real axis at λ_R^0 when $\tau = \tau_c$. For $\tau > \tau_c$, one eigenvalue then tends to the eigenvalue $\nu_0 > 0$ of L_0 as $\tau \to \infty$.

the GS model in one spatial dimension in [25] and [26]. The existence of two real positive eigenvalues for $\tau \gg 1$ was proved in [16].

In Table 1 we give some numerical results for τ_0 and λ_I^0, obtained from full numerical solutions of (3.35) for different exponent sets (p, q, m, s) and dimension N. The lower and upper bounds on λ_I^0 predicted in (3.37) are also shown. Notice that the lower bound is actually quite close to the numerically computed value.

Table 1. Numerical values for τ_0, λ_I^0, and τ_c for different exponent sets (p, q, m, s) and dimension N for which $m = 2$ and $1 < p \leq 1 + 4/N$. The lower and upper bounds for λ_I^0 given in (3.37) are given in the fifth and sixth columns.

(p, q, m, s)	N	τ_0	λ_I^0	(lower bound)	(upper bound)	τ_c
$(2, 1, 2, 0)$	1	0.771	1.238	1.200	2.484	4.560
$(2, 1, 2, 0)$	2	0.561	1.593	1.500	3.230	3.338
$(2, 1, 2, 0)$	3	0.373	2.174	2.000	4.519	2.189
$(3, 2, 2, 0)$	1	0.304	2.859	2.667	5.842	1.800
$(3, 2, 2, 0)$	2	0.150	4.477	4.000	9.849	0.843
$(4, 2, 2, 0)$	1	0.149	2.525	2.143	6.628	0.471

We now discuss a few of the ideas that are needed for the proof of Proposition 3.11. As in the derivation of (3.18), the eigenvalues of (3.35) are the roots of the function $g(\lambda) = 0$, where

$$g(\lambda) \equiv \frac{s+1+\tau\lambda}{qm} - f(\lambda), \qquad f(\lambda) \equiv \frac{\int_{\mathbb{R}^N} w^{m-1} \left(L_0 - \lambda\right)^{-1} w^p \, dy}{\int_{\mathbb{R}^N} w^m \, dy}. \tag{3.38}$$

The function $g(\lambda)$ is analytic in $\text{Re}(\lambda) > 0$ except at the simple pole $\lambda = \nu_0$. In place of (3.19), the number M of zeroes of $g(\lambda)$ in $\text{Re}(\lambda) > 0$ is

$$M = \frac{3}{2} + \frac{1}{\pi} \left[\arg g\right]_{\Gamma_I}. \tag{3.39}$$

Here $[\arg g]_{\Gamma_I}$ denotes the change in the argument of $g(\lambda)$ along the semi-infinite imaginary axis $\Gamma_I = i\lambda_I$, $0 \leq \lambda_I < \infty$, traversed in the downwards direction. By letting $\lambda = i\lambda_I$, we find, in place of (3.20), that the eigenvalues of (3.35) along the positive imaginary axis are the roots of the coupled system $\tilde{g}_R = \tilde{g}_I = 0$, given by

$$\tilde{g}_R(\lambda_I) \equiv \frac{s+1}{qm} - \tilde{f}_R\left(\lambda_I\right), \qquad \tilde{g}_I(\lambda_I) \equiv \frac{\tau\lambda}{qm} - \tilde{f}_I\left(\lambda_I\right), \tag{3.40}$$

where $\tilde{f}_R$ and $\tilde{f}_I$ are defined in (3.20c).

Since $(s+1)/qm < (p-1)^{-1}$ from (1.2), we have that $\tilde{g}_R(0) < 0$ and $\tilde{g}_R \to (s+1)/qm > 0$ as $\lambda_I \to +\infty$. For $m = 2$ and $1 < p \leq 1 + 4/N$, we have from Proposition 3.6 that $\tilde{f}_R^{\prime} < 0$ for $\lambda_I > 0$, and hence there is a unique root to $\tilde{g}_R = 0$. If τ is sufficiently large, then $\tilde{g}_I > 0$ at the unique root of $\tilde{g}_R = 0$, and so we calculate that $[\arg g]_{\Gamma_I} = \pi/2$. Then, from (3.39) we get that $M = 2$ for τ sufficiently large. Alternatively, for τ small enough we have that $\tilde{g}_I < 0$ at the unique root of $\tilde{g}_R = 0$, so that $[\arg g]_{\Gamma_I} = -3\pi/2$, which yields $M = 0$. To show that there are two eigenvalues in the right half-plane along the positive real axis when τ is sufficiently large, we look for roots of $g(\lambda) = 0$ for $\lambda = \lambda_R > 0$. From (3.38), we get

$$g_R(\lambda_R) \equiv \frac{s+1+\tau\lambda_R}{qm} - f_R(\lambda_R), \quad f_R(\lambda_R) \equiv \frac{\int_{\mathbb{R}^N} w^{m-1} \left(L_0 - \lambda_R\right)^{-1} w^p \, dy}{\int_{\mathbb{R}^N} w^m \, dy}. \tag{3.41}$$

A simple calculation shows that $f_R(0) = 1/(p-1)$, so that $g_R(0) < 0$. In addition, we have that $f_R(\lambda_R) \to +\infty$ as $\lambda_R \to \nu_0^-$, where $\nu_0 > 0$ is the unique positive eigenvalue of L_0 (see Proposition 3.1). Hence, $g_R \to -\infty$ for $\lambda_R \to \nu_0^-$. Therefore, it is clear from (3.41) that for $\tau \to \infty$, there are two roots to (3.41). For $\tau \to \infty$, one root tends to ν_0^- while the other root tends to zero. These are the main ideas of the proof of Proposition 3.11.

Question 3.8. Can one determine more general conditions on the exponents (p, q, m, s) and on N for the existence of a unique Hopf bifurcation point for $\tau > 0$?

Question 3.9. Analyze the large-scale oscillatory motion for (3.25) for values of τ well-beyond the Hopf bifurcation point. Using a weakly nonlinear analysis determine whether the Hopf bifurcation is subcritical or supercritical. In the presence of the fast oscillation, derive an ODE for the slow motion of the center of the spike. Is the drift still exponentially slow?

3.3. *A Microwave Heating Model*

Another problem where nonlocal singularly perturbed PDE's arise is in the study of hot-spot formation in the microwave heating of ceramics (cf. [62], [63], [9], [10], and [11]). As shown in [62], when a thin cylindrical ceramic sample is placed in a resonant single-mode microwave cavity in such a way that the intensity of the electric field is constant along the axis of the cylinder, a localized region of elevated temperature, known as a hot-spot, can arise. The mechanism for the formation of a stable hot-spot is the detuning of the cavity that occurs as a result of a large increase in the electrical conductivity of the sample for high temperatures. Depending on the parameters, this detuning shifts the resonant point of the cavity, which reduces the strength of the electric field, and thereby stabilizes the temperature profile. In the small Biot number limit, and for a thin sample, the modeling of this detuning effect leads to a nonlocal reaction-diffusion equation for the dimensionless temperature $u(x, t)$ along the axis of the sample of the form (cf. [62]):

$$u_t = \varepsilon^2 u_{xx} - 2\left(u + b[(u+1)^4 - 1]\right) + \frac{p_c f(u)}{1 + \chi^2 \left[\int_{-1}^{1} f(u)\, dx\right]^2}, \qquad |x| \leq 1, \tag{3.42}$$

with $u_x = 0$ at $x = \pm 1$. Here $\varepsilon \ll 1$, $\chi > 0$, $p_c > 0$, and $b \ll 1$, and $f(u)$ is electrical conductivity of the sample, which is modeled as $f(u) = 1 + u^2$.

An equilibrium hot-spot solution has $u \to \infty$ as $\varepsilon \to 0$. The appropriate rescaling is $u = \varepsilon^{-2/3} U$ with $U = O(1)$ as derived in [9]. Using this rescaling in (3.42), and with $b \ll O(\varepsilon^2)$, we obtain the following rescaled problem

$$U_t = \varepsilon^2 U_{xx} - 2U + \frac{p_c U^2}{\chi^2 I_0^2}, \quad |x| \leq 1; \qquad U_x(\pm 1, t) = 0, \tag{3.43}$$

where $I_0 \equiv \varepsilon^{-1} \int_{-1}^{1} U^2\, dx$. This problem has the form in (3.7). The equilibrium hot-spot solution for (3.43) is given by

$$U \sim U_\varepsilon[x;x_0] \equiv \gamma w\left[\frac{\sqrt{2}(x-x_0)}{\varepsilon}\right], \quad \gamma^3 \equiv \left(\frac{p_c}{36\chi^2}\right), \quad I_0 = 3\sqrt{2}\left(\frac{p_c}{36\chi^2}\right)^{\frac{2}{3}}. \tag{3.44}$$

Here $w(y)$ is the solution (1.8) to Carrier's problem with $p=2$.

Linearizing (3.43) around this equilibrium solution, we obtain the following finite-domain nonlocal eigenvalue problem:

$$M_\varepsilon\phi_\varepsilon \equiv \varepsilon^2\phi_{\varepsilon xx} + \left(-2+\frac{2p_cU_\varepsilon}{\chi^2 I_0^2}\right)\phi_\varepsilon - \frac{4p_cU_\varepsilon^2}{\varepsilon\chi^2 I_0^3}\int_{-1}^{1} U_\varepsilon\phi_\varepsilon\, dx = \lambda^\varepsilon\phi_\varepsilon, \quad |x|\le 1, \tag{3.45}$$

with $\phi_{\varepsilon x}(\pm 1)=0$. Letting, $\phi_\varepsilon(x)=\Phi(y)$, where $y=\sqrt{2}\varepsilon^{-1}(x-x_0)$, the discrete spectrum of (3.45) can be approximated by the spectrum of the corresponding infinite-line nonlocal eigenvalue problem

$$\Phi_{yy} - \Phi + 2w\Phi - \alpha w^2\left(\frac{\int_{-\infty}^{\infty} w\Phi\, dy}{\int_{-\infty}^{\infty} w^2\, dy}\right) = \frac{\lambda}{2}\Phi, \quad -\infty < y < \infty, \tag{3.46a}$$

where $\Phi \to 0$ as $|y|\to\infty$. The constant α in (3.46a) is

$$\alpha \equiv \frac{6\sqrt{2}\,p_c\,\gamma^3}{\chi^2 I_0^3}, \tag{3.46b}$$

where γ and I_0 are defined in (3.44).

This infinite-line problem has the form (3.14) with $p=m=2$. Therefore, the second statement of Proposition 3.3 proves that any nonzero eigenvalue of (3.46a) satisfies $\text{Re}(\lambda)<0$ when $\alpha>1$. A simple calculation using the expressions for γ and I_0 in (3.44) yields that $\alpha=4$ for all $\chi>0$ and $p_c>0$. Therefore, the principal eigenvalue of (3.45) is exponentially small, and a hot-spot will exhibit metastable behavior. In this way, the following main result was derived in [50]:

Proposition 3.13. *For $\varepsilon \ll 1$, and for any $\chi>0$ and $p_c>0$, the exponentially small eigenvalue of (3.45) has the asymptotic estimate*

$$\lambda_0 \sim 120\left(e^{-2\sqrt{2}(1-x_0)/\varepsilon} + e^{-2\sqrt{2}(1+x_0)/\varepsilon}\right). \tag{3.47}$$

A metastable hot-spot solution to (3.43) is given by

$$u(x,t) \sim \frac{3\gamma}{2}\,sech^2\left(\frac{\varepsilon^{-1}[x-x_0(t)]}{\sqrt{2}}\right), \tag{3.48a}$$

where the hot-spot location $x_0(t)$ *satisfies the differential equation,*

$$\frac{dx_0}{dt} \sim 60\varepsilon\sqrt{2}\left(e^{-2\sqrt{2}(1-x_0)/\varepsilon} - e^{-2\sqrt{2}(1+x_0)/\varepsilon}\right). \tag{3.48b}$$

This shows that the equilibrium hot-spot solution centered at $x_0 = 0$ is unstable, and that the hot-spot tends to the closest of the two boundaries. As a remark, a similar analysis can be done for the generalized polynomial conductivity model $f(u) = 1 + c_1 u^p$. In place of (3.46a), the corresponding finite-domain nonlocal eigenvalue problem is

$$\Phi_{yy} - \Phi + 2w^{p-1}\Phi - 2pw^p \left(\frac{\int_{-\infty}^{\infty} w^{p-1}\Phi\, dy}{\int_{-\infty}^{\infty} w^p\, dy}\right) = \frac{\lambda}{2}\Phi\,, \quad -\infty < y < \infty\,. \tag{3.49}$$

The only result for the spectrum of this problem is given in the third statement of Proposition 3.3.

Question 3.10. Can a similar analysis be done for (3.49) and for the multi-dimensional version of (3.42)?

3.4. *A Flame-Front Evolution Model*

This problem concerns the evolution of a flame-front in a vertical channel. For a channel with a constant cross-section Ω in the $x \equiv (x_1, x_2)$ plane, and under certain physical conditions, the flame-front interface $z = Z(x_1, z_2, t)$ satisfies the nonlocal PDE (cf. [7], [92])

$$Z_t - \frac{1}{2}|\nabla Z|^2 = \varepsilon\Delta Z + Z - \frac{1}{|\Omega|}\int_\Omega Z\, dx\,, \quad x \in \Omega\,; \quad \partial_n Z = 0\,, \quad x \in \partial\Omega\,. \tag{3.50}$$

Here $\varepsilon \ll 1$, $|\Omega|$ is the area of the cross-section and ∂_n is the outward normal derivative. It is observed in physical experiments and numerical computations that the flame-front interface assumes a roughly paraboloidal shape and that the tip of the paraboloid moves very slowly.

The simpler one-dimensional problem in the slab geometry $\Omega = [0, 1]$ was studied in [7] and [92] by first converting (3.50) into a local problem by using the transformation $v = -Z_x$. This yields a convection-diffusion problem known as the Burgers-Sivashinsky equation

$$v_t + vv_x - v = \varepsilon v_{xx}\,, \qquad v(0,t) = v(1,t) = 0\,. \tag{3.51}$$

By analyzing (3.51), it was shown in [92] that the tip $x_0 = x_0(t)$ of a parabolic flame-front interface drifts exponentially slowly towards the closest point on the channel wall (i.e. either $x = 0$ or $x = 1$) according to the ODE

$$x_0' \sim \sqrt{\frac{2}{\pi\varepsilon}} \left[\left((1 - x_0)^2 + \frac{\pi^2 \varepsilon}{3} \right) e^{-(1-x_0)^2/2\varepsilon} - \left((x_0^2 + \frac{\pi^2\varepsilon}{3}) \right) e^{-x_0^2/2\varepsilon} \right] . \tag{3.52}$$

The simple substitution $v = -Z_x$ to eliminate the nonlocal term for the one-dimensional case has no counterpart for the two-dimensional problem (3.50). However, the change of variables

$$Z = 2\varepsilon \log u \,, \tag{3.53}$$

readily reduces (3.50) to the nonlocal problem

$$u_t = \varepsilon \triangle u + u \left(\log u - \frac{1}{|\Omega|} \int_\Omega \log u \, dx \right) , \quad x \in \Omega ; \quad \partial_n u = 0 \,, \quad x \in \partial\Omega \,, \tag{3.54}$$

which is a special case of (3.7). The corresponding local steady-state problem is $\varepsilon \triangle u + u \log u = 0$, which has a degenerate nonlinearity of the type studied in [21]. This leads to the following question:

Question 3.11. Can one analyze (3.54) and the spectrum of the linearized problem? In one-dimension, can one use (3.54) to give a rigorous proof of (3.52). Can the two-dimensional problem for a paraboloidal-shaped interface be analyzed? Investigate spike equilibria for Carrier's problem $\varepsilon^2 u'' + u \log u = 0$, and its multi-dimensional counterpart.

4. Dynamics and Equilibria of a Spike in an R-D System

We begin this section by giving some results for the dynamics and equilibrium locations for an interior one-spike solution to the GM model (1.1) with $\tau = 0$ in a bounded domain $\Omega \in \mathbb{R}^2$. The results in this section cover the range $D \gg O(\varepsilon^2)$, where D is not exponentially large as $\varepsilon \to 0$. The near-shadow limit where $D = O(e^{c/\varepsilon})$ for some $c > 0$ is discussed in §5.1.

Even in the simple case where $\tau = 0$, there are different dynamical laws of spike motion that hold for different ranges of D. For the shadow problem where $D = \infty$, we found above in Proposition 3.8 that a one-spike solution is metastable and that the spike drifts exponentially slowly towards the closest point on $\partial\Omega$. Alternatively, for D small with $\varepsilon^2 \ll D \ll 1$, we find below that the motion of an interior one-spike solution is again metastable, but now the spike tends to a point in Ω that maximizes the distance function. In contrast, for $D \gg 1$, the motion of a one-spike solution is proportional to the gradient of the regular part of the Neumann Green's function for the Laplacian. Finally, for $D = O(1)$, the motion of a spike is proportional to the gradient of the regular part of the reduced wave Green's function. This second Green's function depends on D.

For $D \gg 1$ and $D = O(1)$, the equilibrium location of a one-spike solution is given by the zeroes of the gradient of the regular part of the Neumann Green's function and the reduced wave Green's function, respectively. In this way, we examine how both the shape of the domain and the inhibitor diffusivity D determine the possible equilibrium locations for a one-spike solution. For D small, we find that stable equilibrium spike locations tend to the centers of the disks of largest radii that can fit within the domain. Hence, for D small, there are two stable equilibrium locations for a dumbbell-shaped domain. In contrast, from the Neumann Green's function, we predict that for a family of dumbbell-shaped domains there is only one possible equilibrium location when D is sufficiently large. This change in the multiplicity of an equilibrium spike-layer location is explained from a certain bifurcation behavior of the zeroes of the gradient of the regular part of the reduced wave Green's function that occurs at a certain critical value of D.

The results below have been derived in [54] and [55]. There have been very few analytical studies of the dynamics of spikes for the GM model (1.1) or for other reaction-diffusion systems in $\mathbb{R}^2$. In [15] and [103] the motion of a one-spike solution for the GM model (1.1) was studied for $D \gg 1$. In [34] the metastable motion of a two-spike solution in $\mathbb{R}^2$ was studied in the weak interaction limit where $D = O(\varepsilon^2)$. This problem is equivalent to a one-spike solution in a half-space with a Neumann boundary condition. For this case, the spike interaction is repulsive, as was found in [34].

In the results below, there are two Green's functions that play a prominent role. The reduced wave Green's function $G(x; x_0)$ satisfies

$$\triangle G - \frac{1}{D} G = -\delta(x - x_0), \quad x \in \Omega; \quad \partial_n G = 0, \quad x \in \partial\Omega. \tag{4.1}$$

The regular part $R(x; x_0)$ of this function is defined by

$$R(x, x_0) = G(x, x_0) + \frac{1}{2\pi} \log |x - x_0|. \tag{4.2}$$

Alternatively, the Neumann Green's function $G_m(x; x_0)$ satisfies

$$\triangle G_m = \frac{1}{|\Omega|} - \delta(x - x_0)\,, \quad x \in \Omega\,; \quad \partial_n G_m = 0, \quad x \in \partial\Omega\,; \quad \int_\Omega G_m \, dx = 0\,. \tag{4.3}$$

The regular part R_m of this Green's function is defined by

$$R_m(x, x_0) = \frac{1}{2\pi} \log |x - x_0| + G_m(x, x_0)\,. \tag{4.4}$$

In terms of R and R_m, we define $R_0 \equiv R(x_0, x_0)$ and

$$\nabla R_0 \equiv \nabla_x R(x, x_0)|_{x=x_0}\,, \qquad \nabla R_{m0} \equiv \nabla_x R_m(x, x_0)|_{x=x_0}\,. \tag{4.5}$$

For $D \gg 1$, it is easy to see that these two Green's functions are related by

$$R(x, x_0) \sim \frac{D}{|\Omega|} + R_m(x, x_0) + O\left(D^{-1}\right)\,. \tag{4.6}$$

We begin with the first result that holds for $D \gg -\log \varepsilon$ as derived in [15] and in Proposition 3.2 of [103]:

Proposition 4.1. *Let $\varepsilon \to 0$ and $D \gg -\log \varepsilon$, and assume that the spike profile is stable on an $O(1)$ time-scale. Then, the motion of a one-spike solution for (1.1) in $\Omega \in \mathbb{R}^2$ is characterized by*

$$a(x, t) \sim H^{q/(p-1)} w\left[\varepsilon^{-1}|x - x_0|\right]\,, \quad H \sim \left(\frac{D\nu}{b}\right)^{1/\zeta}\,, \tag{4.7}$$

where $\nu \equiv -1/\log \varepsilon$, $b \equiv \int_0^\infty w^m(\rho)\rho \, d\rho$, and $\zeta > 0$ is given in (1.2). The spike location $x_0 = x_0(t)$ satisfies

$$\frac{dx_0}{dt} \sim -\left(\frac{2q|\Omega|}{p-1}\right) \frac{\varepsilon^2}{D} \nabla R_{m0}\,. \tag{4.8}$$

Here ∇R_{m0} is defined in (4.5).

The following result for $D \gg O(\varepsilon^2)$ is derived in Proposition 2.1 of [54]:

Proposition 4.2. *Let $\varepsilon \to 0$ and $D \gg O(\varepsilon^2)$, and assume that the spike profile is stable on an $O(1)$ time-scale. Then, the trajectory of an interior one-spike solution for (1.1) in $\Omega \in \mathbb{R}^2$ is given by*

$$\frac{dx_0}{dt} \sim -\left(\frac{4\pi q}{p-1}\right) \frac{\varepsilon^2}{-\log \varepsilon + 2\pi R_0} \nabla R_0\,, \tag{4.9}$$

where R_0 and its gradient are defined in (4.5).

For $D \gg 1$, we can use (4.6) in (4.9), to show that (4.9) reduces to (4.8). For $D \ll 1$, we can approximate R_0 by the distance function as outlined in Proposition 3.2 of [54]. More specifically, let $r(x) = \text{dist}(\partial\Omega, x)$. Suppose that x_0 is a local minimum of $R(x_0, x_0)$ for $D \ll 1$. Then, for $D \ll 1$, x_0 is a local maximum of $r(x)$. A more detailed result as given in Proposition 3.1 of [54] determines the following formula for the speed x_0':

Proposition 4.3. *Let $\varepsilon \to 0$ and $O(\varepsilon^2) \ll D \ll O(1)$, and assume that the spike profile is stable on an $O(1)$ time-scale. Assume that at some time t, there is a unique point $x_m \in \partial\Omega$ that is closest to x_0. Then, at that time t, the speed of a spike for (1.1) in $\Omega \in \mathbb{R}^2$ is given by*

$$\frac{dx_0}{dt} \sim -\frac{\sqrt{\pi\lambda}\, q}{p-1}\left(\frac{\varepsilon^2}{\log 2 - \gamma - \log[\varepsilon\lambda]}\right) r_m^{-1/2} e^{-2\lambda r_m}\left(1 - \frac{r_m}{\kappa_m}\right)^{-\frac{1}{2}} \hat{r}_m \,. \tag{4.10}$$

Here $\lambda \equiv D^{-1/2}$, γ is Euler's constant, $r_m = |x_m - x_0|$, $\hat{r}_m = (x_m - x_0)/r_m$, and κ_m is the curvature of $\partial\Omega$ at x_m. Since $dr_m/dt = -dx_0/dt \cdot \hat{r}_m > 0$ from (4.10), the spike moves away from the closest point on the boundary.

In Theorem 4.1 of [54], a complex variable method was used to determine an explicit formula for ∇R_{m0} for a certain class of mappings of the unit disk. From this formula, and from detailed boundary integral numerical computations, the following conjecture was made in [54] regarding the number of zeroes of ∇R_{m0}.

Conjecture 4.1. *Let Ω be any simply-connected bounded domain in $\mathbb{R}^2$, not necessarily convex, and assume that $\partial\Omega$ is smooth. Then, ∇R_{m0} has a unique root inside Ω. Thus, it is conjectured that for $D \gg 1$ there is a unique equilibrium location of an interior one-spike solution of (1.1), and that this location is stable.*

To illustrate the novelty of this conjecture consider a similar problem, but now for the Dirichlet Green's function G_d, which satisfies

$$\triangle G_d = -\delta(x - x_0) \quad x \in \Omega\,; \quad G_d = 0\,, \quad x \in \partial\Omega\,. \tag{4.11}$$

The regular part of G_d and its gradient are defined by

$$R_d(x, x_0) = G_d(x, x_0) + \frac{1}{2\pi}\log|x - x_0|\,, \quad \nabla R_{d0} \equiv \nabla\, R_d(x, x_0)\big|_{x=x_0}\,. \tag{4.12}$$

Many properties of R_d are given in the survey article [2]. The uniqueness of a root to $\nabla R_{d0} = 0$ in a *convex* domain was established in [43] and

[12]. Consider a class of possibly non-convex domains Ω, generated by the following map $f(z)$ of the unit disk $|z| = 1$, parameterized by b,

$$f(z) = \frac{(1-b^2)z}{z^2 - b^2}, \quad b > 1. \tag{4.13}$$

In Fig. 8 we plot Ω for several values of b. When $b \to \infty$, Ω is a perturbation of the unit circle. However, when $b \to 1^+$, Ω becomes the union of two disjoint circles each of radius $1/2$. It is easy to verify that Ω is non-convex when $1 < b < 1 + \sqrt{2}$. By using this map, it was proved in [43] that ∇R_{d0} can have multiple roots in a non-convex domain.

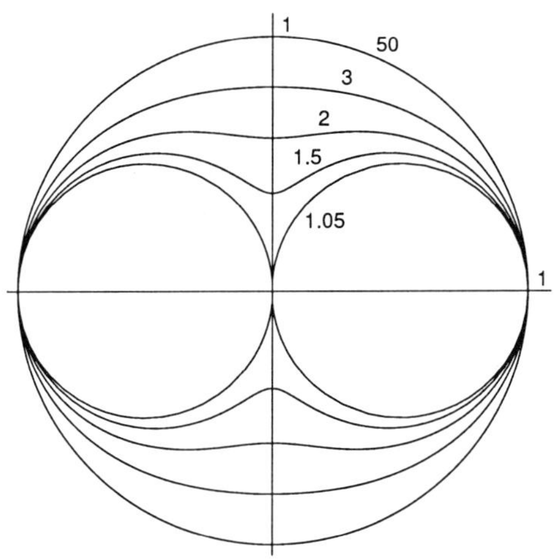

Figure 8. The boundary of $\Omega = f(B)$, with $f(z)$ as given in (4.13), where B is the unit disk. The values of b are as indicated.

From Conjecture 4.1 it follows that for D large enough, there is exactly one possible location for an interior one-spike equilibrium solution, and this location is stable. However, for a dumbbell-shaped domain such as shown in Fig. 8 for $1 < b < 1 + \sqrt{2}$, we know that when D is small enough, the only possible minima of the regular part R_0 of the reduced wave Green's function are near the centers of the lobes of the two dumbbells. In addition, R_0 has a saddle-point at the origin. For D small enough, the equilibria in the lobes of the dumbbell are stable, while the equilibrium at the origin is unstable. Hence, this suggests that a pitchfork bifurcation occurs for the zeroes of ∇R_0 as D is increased past some critical value D_c. As D approaches D_c from below, the two equilibria in the lobes of the dumbbell should simultaneously merge into the origin. To investigate this effect, a boundary integral method was used in [55] to compute the zeroes of ∇R_0 as a function of $\lambda = D^{-1/2}$ for different values of the shape-parameter b in the class of mappings (4.13). The bifurcation diagram, as shown in Fig. 9, does indeed show a pitchfork bifurcation as predicted, with a subcritical bifurcation if b is small and a supercritical bifurcation if b is large.

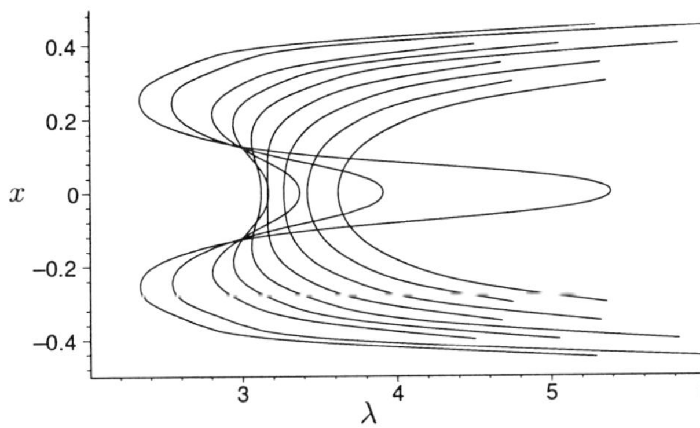

Figure 9. Plot of the bifurcation diagram for the spike equilibria versus $\lambda = D^{-1/2}$ for various values of the dumbbell shape-parameter b. The leftmost curve is $b = 1.15$. Successive curves from left to right correspond to increments in b of 0.05.

This numerical study, and Conjecture 4.1, shows that further work is required to understand the properties of the zeroes of ∇R_0 and ∇R_{m0}. The pitchfork bifurcation behavior for the zeroes of ∇R_0 in nonconvex domains having two axes of symmetry should be a generic feature.

Question 4.1. Can one determine general properties of the zeroes of the gradients of the regular parts of the Neumann Green's function and the reduced wave Green's function? Does Conjecture 4.1 hold?

Question 4.2. Do the zeroes of ∇R_0 have a subcritical bifurcation with respect to $\lambda = D^{-1/2}$ in a dumbbell-shaped domain, whenever the neck of the dumbbell is sufficiently thin, and a supercritical bifurcation in λ when a dumbbell-shaped domain is sufficiently close to a circular domain?

Question 4.3. Can one study the bifurcation behavior for an interior two-spike equilibrium solution as a function of D and the domain topology?

4.1. *The Near-Shadow System*

Next, we consider a one-spike solution for the near-shadow limit of the GM model (1.1) where D is exponentially large as $\varepsilon \to 0$. For the shadow problem where $D = \infty$, and under certain conditions on the exponents (p, q, m, s), it was shown in Sec. 3.1 that a one-spike profile is stable, but that the spike will drift towards the boundary with an asymptotically exponentially small speed as $\varepsilon \to 0$. The asymptotic equilibrium location of this metastable spike is determined by critical points of the distance function. However, when D is exponentially large, the spike location is determined by a balance between the Neumann Green's function R_m defined in (4.4) and the exponentially weak interaction between the far-field behavior of the spike and the boundary $\partial\Omega$ of the domain. In the one-dimensional case this balance leads to the following result (cf. [55]):

Proposition 4.4. *Let $\varepsilon \ll 1$, $\Omega = [-1, 1]$, and $x_0 \in (-1, 1)$. Assume that the spike profile is stable on an $O(1)$ time-scale. Then, when D is exponentially large as $\varepsilon \to 0$, the spike location x_0 satisfies the ODE*

$$\frac{dx_0}{dt} \sim \frac{4\varepsilon c^2}{\beta} \left(\frac{\varepsilon \sinh(2x_0/\varepsilon)}{2x_0} - \frac{D_c}{D} \right) e^{-2/\varepsilon} x_0 \,, \tag{4.14}$$

where D_c and β are defined by

$$D_c = \frac{\varepsilon^2 q \beta}{2c^2(p-1)} e^{2/\varepsilon} \,, \quad \beta \equiv \int_0^\infty \left[w'(y) \right]^2 \, dy \,. \tag{4.15}$$

This result shows that the spike at $x_0 = 0$ is stable when $D < D_c$, and is unstable when $D > D_c$. When $D < D_c$, there are two unstable equilibria, which are the nonzero roots of $2y/\sinh(2y) = D/D_c$ with $y = \varepsilon^{-1} x_0$.

Similar bifurcation behavior occurs for dumbbell-shaped domains. For these domains, an unstable spike for the shadow problem with $D = \infty$ can be located either in the neck or in one of the two lobes of the dumbbell. However, when D is exponentially large, the next result, as derived in Proposition 4.1 of [55], shows that the bifurcation behavior is such that the spike in the neck of the dumbbell becomes stable through a pitchfork bifurcation. A further result in [55] shows that the spikes in the lobes of the dumbbell tend to the boundary when D is exponentially large. Therefore, this suggests how Conjecture 4.1 arises from the near-shadow limit.

Proposition 4.5. *Let $\varepsilon \ll 1$ and consider the class of dumbbell-shaped domains generated by the mapping $f(z)$ given in (4.13). Assume that the*

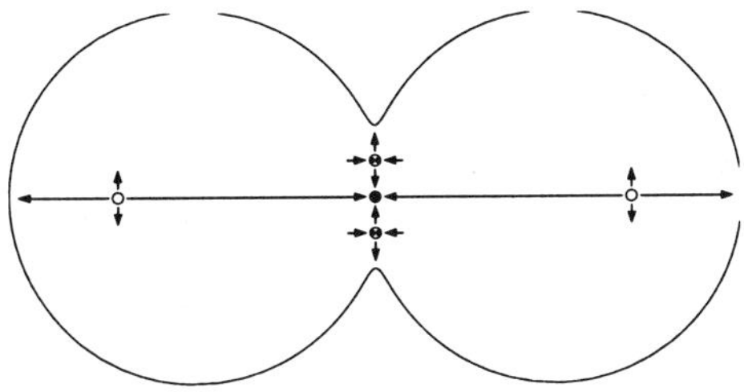

Figure 10. Plot of a dumbbell-shaped domain and the bifurcation behavior of spikes in the neck of the dumbbell when $D = O(e^{c/\varepsilon})$.

spike location $(0, y_0)$ *on the* y*-axis satisfies* $y_0 = O(\varepsilon)$. *Then, the local trajectory* $y_0(t)$ *satisfies*

$$\frac{dy_0}{dt} \sim \varepsilon^{1/2}\,\mu_0\, e^{-2y_m/\varepsilon}\left[\frac{\sinh(2y_0/\varepsilon)}{(2y_0/\varepsilon)} - \frac{D_c}{D}\right] y_0\,, \tag{4.16a}$$

where D_c *and* μ_0 *satisfy*

$$D_c \equiv \frac{\varepsilon^2 \beta}{4c^2}\left(\frac{q}{p-1}\right)\left(\frac{\pi}{\varepsilon}\right)^{1/2}\left[y_m\left(1-\kappa_m y_m\right)\right]^{1/2} G_b e^{2y_m/\varepsilon}\,, \tag{4.16b}$$

$$\mu_0 \equiv \frac{8c^2}{\sqrt{\pi}\beta}\left[y_m\left(1-\kappa_m y_m\right)\right]^{-1/2}, \qquad G_b \equiv \frac{(b^2-1)}{2(b^4-1)^2}\left[2b^6+3b^4+2b^2-1\right]. \tag{4.16c}$$

Here κ_m *is the curvature of the boundary at the point* $(0, y_m)$*, where*

$$\kappa_m \equiv \left(\frac{b^2+1}{b^2-1}\right)^3\left[1 - \frac{8b^2}{\left(b^2+1\right)^2}\right], \qquad y_m \equiv \left(\frac{b^2-1}{b^2+1}\right). \tag{4.16d}$$

Therefore, when $D > D_c$, $y_0 = 0$ is the unique, and unstable, equilibrium solution for (4.16a). For $D < D_c$, $y_0 = 0$ is stable, and there are two unstable equilibria with $|y_0| = O(\varepsilon)$, satisfying $2\zeta/\sinh(2\zeta) = D/D_c$ for $\zeta = y_0/\varepsilon$. Therefore, the local bifurcation is subcritical in D/D_c. In Fig. 10 we show the geometry and the bifurcation behavior indicated by Proposition 4.5.

Question 4.4. Provide a rigorous proof of Propositions 4.4 and 4.5. What is the behavior of equilibrium spikes that are $O(\varepsilon)$ close to the boundary?

Question 4.5. What is the relationship between k-spike equilibria of the shadow system (3.25) where $D = \infty$ and k-spike equilibria to the near-shadow system (1.1) where D is exponentially large.

4.2. *Pinning of a Spike*

For the GM model (1.1) with $\tau = 0$, we now examine the effect of spatial variations in the coefficients of the differential operators. We will consider two such problems. The first problem, in one space dimension, is

$$a_t = \varepsilon^2 a_{xx} - [1 + V(x)]\, a + \frac{a^p}{h^q}\,, \quad -1 < x < 1\,; \quad a_x(\pm 1, t) = 0\,, \tag{4.17a}$$

$$0 = D h_{xx} - \mu(x) h + \varepsilon^{-1} \frac{a^m}{h^s}\,, \quad -1 < x < 1\,; \quad h_x(\pm 1, t) = 0\,. \tag{4.17b}$$

Here $\mu(x) > 0$ and $V(x) > 0$ are precursor gradients (cf. [103]), and (p, q, m, s) satisfy (1.2). The second problem is in $\mathbb{R}^2$, and is given by

$$a_t = \varepsilon^2 \Delta a - [1 + V(x)]\, a + \frac{a^p}{h^q}\,, \quad x \in \Omega\,; \quad \partial_n a = 0 \quad x \in \partial\Omega\,, \tag{4.18a}$$

$$0 = D \Delta h - \mu(x) h + \varepsilon^{-2} \frac{a^m}{h^s}\,, \quad x \in \Omega\,; \quad \partial_n h = 0 \quad x \in \partial\Omega\,. \tag{4.18b}$$

In Corollary 2.2 of [103], the effect of $V(x)$ for a spatially independent μ was obtained for (4.17).

Proposition 4.6. *Let $\varepsilon \ll 1$ and suppose that the spike profile is stable on an $O(1)$ time-scale. Let $\mu(x) = \mu$ be a positive constant. Then, the motion of a one-spike solution for (4.17) is characterized by*

$$a(x,t) \sim H^{q/(p-1)} \left[1 + V(x_0)\right]^{1/(p-1)} w \left[\sqrt{1 + V(x_0)} \frac{(x - x_0)}{\varepsilon} \right], \tag{4.19}$$

where $h(x_0) = H$ and H is given in Eq. (2.14b) of [103]. Moreover, $w(y)$ satisfies (1.8), $\theta \equiv \sqrt{\mu/D}$, and the spike location $x_0(t)$ satisfies

$$\frac{dx_0}{dt} \sim -\frac{\varepsilon^2 q\theta}{p-1} \left(\tanh\left[\theta(1 + x_0)\right] - \tanh\left[\theta(1 - x_0)\right]\right) - \frac{\varepsilon^2}{2} \left(\frac{p+3}{p-1}\right) \frac{V'(x_0)}{1 + V(x_0)}\,. \tag{4.20}$$

It is easy to see from (4.20) that if V is convex with a minimum at some point in $[-1, 1]$ then there exists a unique equilibrium solution to (4.20) and this equilibrium solution is stable. When $V(x)$ is not convex the situation is more complex. For instance suppose that $V(x)$ is a double-well potential of the form $V(x) = b(1 - x^2)^2$ with $b > 0$. Define ω by $\omega \equiv \frac{q\theta^2 \operatorname{sech}^2\theta}{p+3}$. Then, it readily follows that when $b/(1 + b) > \omega$, we have that $x_0 = 0$ is an unstable equilibrium solution to (4.20), and that there exists a stable equilibrium solution on each of the subintervals $-1 < x < 0$ and $0 < x < 1$.

Alternatively, if $\frac{b}{1+b} < \omega$, then $x_0 = 0$ is the only equilibrium solution to (4.20) and it is stable. This is a classic pitchfork bifurcation scenario.

The next result, given in Corollary 2.4 of [103], shows the effect of $\mu(x)$:

Proposition 4.7. *Let $\varepsilon \ll 1$ and suppose that the spike profile is stable on an $O(1)$ time-scale. Let $V(x) = 0$ and assume that $D \gg 1$. Then, the spike location for (4.17) satisfies*

$$\frac{dx_0}{dt} \sim -\frac{2\varepsilon^2 q}{D(p-1)} \left(\int_{-1}^{x_0} \mu(y)\, dy - \mu_a \right), \qquad \mu_a \equiv \frac{1}{2} \int_{-1}^{1} \mu(x)\, dx\,. \quad (4.21)$$

The results (4.20) and (4.21) show that the spike dynamics depends only on pointwise properties of $V(x)$, but on global properties of $\mu(x)$. The final result, given in Proposition 3.1 of [103], is for (4.18).

Proposition 4.8. *Let $\varepsilon \ll 1$ and suppose that the spike profile is stable on an $O(1)$ time-scale. Suppose that $D \gg -\log \varepsilon$ as $\varepsilon \to 0$. Then, the motion of a one-spike solution for (4.18) is characterized by*

$$a(x,t) \sim H^{q/(p-1)} \left[1 + V(x_0)\right]^{1/(p-1)} w \left[\sqrt{1+V(x_0)} \frac{|x - x_0|}{\varepsilon} \right], \quad (4.22)$$

where $h(x_0) = H$ and H is given in Eq. (3.9) of [103], and $w(y)$ satisfies (1.7) in $\mathbb{R}^2$. The spike location $x_0(t)$ satisfies the following gradient flow

$$\frac{dx_0}{dt} \sim -\frac{2\varepsilon^2}{(p-1)} \nabla W(x_0)\,, \quad \textit{where} \quad W(x_0) \equiv \log\left[1 + V(x_0)\right]. \quad (4.23)$$

The spike motion is orthogonal to level curves of $W(x)$ and $dW(x_0)/dt < 0$ except at critical points of W.

When $D \gg -\log \varepsilon$, we observe from (4.23) that stable equilibria for the location of the spike occur at points where the potential $V(x)$ has a local minimum. Notice the qualitative similarity between this result and the result mentioned in Sec. 2.4 for Eq. (2.45).

Question 4.6. Can one derive an equation of motion for (4.18) when $D = O(1)$, that involves both ∇R_0 and ∇V, where R_0 is the reduced wave Green's function of (4.5)? If so, then a delicate competition will determine the possible equilibrium locations for an interior one-spike solution.

4.3. *Sudden Oscillatory Instabilities*

We now illustrate a dynamical instability that can occur for a one-spike solution to the GM model (1.1) in one space dimension when τ is sufficiently large. For a one-spike solution on the $\Omega = [-1,1]$, it is known from (4.20) with $V(x) \equiv 0$, that when $\tau = 0$ the motion of a spike satisfies

$$\frac{dx_0}{dt} \sim -\frac{\varepsilon^2 q\theta}{p-1} \left(\tanh\left[\theta(1+x_0)\right] - \tanh\left[\theta(1-x_0)\right]\right) . \qquad (4.24)$$

A simple analysis shows that this ODE is still asymptotically valid for $\tau > 0$ provided that the spike profile is stable. For certain ranges of τ and D, the spike profile may be stable for x_0 sufficiently close to the right endpoint $x = 1$, but may be unstable on an $O(1)$ time-scale due to a Hopf bifurcation in the spike amplitude that occurs at some critical value of x_0 near $x_0 = 0$. Since a spike moves slowly towards the origin under (4.24), this suggests that the spike will enter a zone of instability at some time t with $t = O(\varepsilon^{-2})$. The instability will initiate a rapid oscillation in the spike amplitude relative to a possible continued slow drift of the spike towards the origin. We call this phenomena a sudden oscillatory instability. This phenomenon was first observed in [95], and also occurs for the SC model (1.6) (cf. [95].

To study this instability, we construct a quasi-equilibrium one-spike solution to (1.1) in $\Omega = [-1,1]$. Then, we derive a nonlocal eigenvalue problem for the stability of the spike profile. A simple calculation shows that the quasi-equilibrium solution a_c, h_c is given for $\varepsilon \ll 1$ by

$$a_c \sim H^{q/(p-1)} w\left[\varepsilon^{-1}(x - x_0(t))\right] , \quad H = \left[G_0(x_0; x_0) \int_{-\infty}^{\infty} w^m \, dy\right]^{-1/\zeta} , \qquad (4.25a)$$

$$h_c \sim G_0\left[x; x_0\right] / G_0\left[x_0; x_0\right] . \qquad (4.25b)$$

Here $w(y)$ is the homoclinic solution in (1.8), ζ is given in (1.2), and $G_0(x; x_0)$ is the Green's function satisfying

$$DG_{0xx} - G_0 = -\delta(x - x_0) , \quad -1 < x < 1 ; \quad G_{0x}(\pm 1; x_0) = 0 . \qquad (4.25c)$$

In Sec. 4.1 of [95] the following nonlocal eigenvalue problem was derived to study the stability of a_c and h_c on an $O(1)$ time-scale:

Proposition 4.9. *Let* $0 < \varepsilon \ll 1$, $\tau \geq 0$, *and* $x_0 \in (-1,1)$. *Then, the stability of the quasi-equilibrium profile (4.25) for the GM model (1.1) on* $\Omega = [-1,1]$ *is determined by the spectrum of the nonlocal eigenvalue problem*

$$L_0 \Phi - \chi_m w^p \left(\frac{\int_{-\infty}^{\infty} w^{m-1} \Phi \, dy}{\int_{-\infty}^{\infty} w^m \, dy} \right) = \lambda \Phi , \quad -\infty < y < \infty , \qquad (4.26a)$$

with $\Phi \to 0$ *as* $|y| \to \infty$. *Here* L_0 *is the local operator defined in (3.16), and the multiplier* χ_m *is defined by*

$$\chi_m = \chi_m(z; x_0) \equiv qm \left[s + \sqrt{1+z} \left(\frac{\beta(\theta_\lambda; x_0)}{\beta(\theta_0; x_0)} \right) \right]^{-1}, \tag{4.26b}$$

where the function $\beta(\xi; x_0)$ *is defined by*

$$\beta(\xi; x_0) \equiv \tanh\left[\xi(1+x_0)\right] + \tanh\left[\xi(1-x_0)\right]. \tag{4.26c}$$

Here z, θ_λ, *and* θ_0, *are given by*

$$z \equiv \tau\lambda, \qquad \theta_\lambda \equiv \theta_0\sqrt{1+z}, \qquad \theta_0 \equiv D^{-1/2}. \tag{4.26d}$$

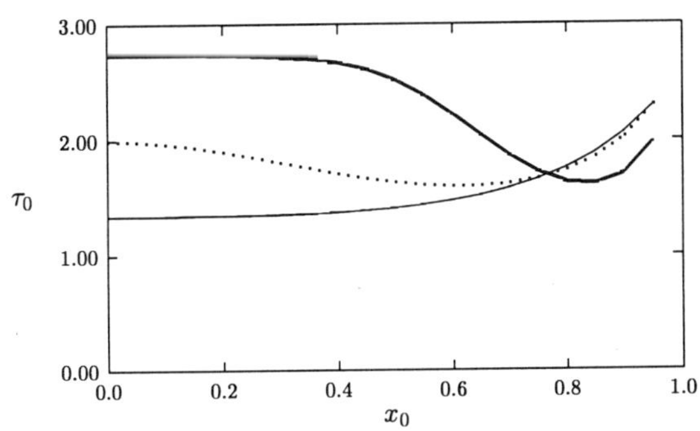

Figure 11. Plots of τ_0 versus x_0 for the GM model with $(p, q, m, s) = (2, 1, 2, 0)$. Here we have $D = 1.0$ (solid curve), $D = 0.5$ (dashed curve), and $D = 0.1$ (heavy solid curve).

This eigenvalue problem is very similar to (3.14) except that the constant multiplier α of the nonlocal term in (3.14) is replaced by the multiplier χ_m, which depends on x_0, on D, and on the product $\tau\lambda$. Consider the exponent set $(p, q, m, s) = (2, 1, 2, 0)$. Then, since $\chi_m > 1$ when $\tau = 0$, we conclude from the second statement of Proposition 3.3 that the spike profile is stable for $\tau = 0$. However, for each fixed D, the spike profile centered at x_0 will become destabilized due to a Hopf bifurcation when τ exceeds some critical value $\tau_0(x_0)$. By symmetry τ_0 is an even function of x_0 and so we need only consider $x_0 \in (0, 1)$. The results for $\tau_0(x_0)$ obtained from a numerical computation for different values of D are shown in Fig. 11. Since τ_0 is a monotone increasing function of x_0 for $x_0 > 0$ when $D = 1$, by the discussion in the beginning of this subsection, a sudden oscillatory instability will occur for certain parameter values.

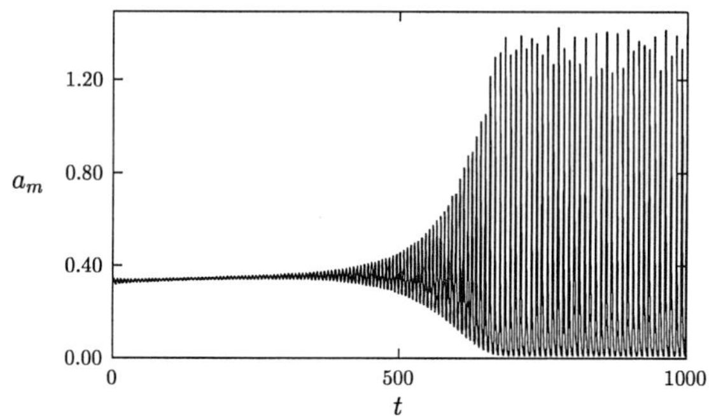

Figure 12. Plots of the spike amplitude $a_m = a(x_0, t)$ versus t for $D = 1.0$, $\varepsilon = 0.03$, $\tau = 1.35$, $x_0(0) = 0.6$, and $(p, q, m, s) = (2, 1, 2, 0)$.

To illustrate this, we take the parameter values $D = 1.0$, $\varepsilon = 0.03$, $\tau = 1.35$, $x_0(0) = 0.6$, and $(p, q, m, s) = (2, 1, 2, 0)$. Important values for $\tau_0(x_0)$ are $\tau_0(0.6) = 1.477$, $\tau_0(0) = 1.343$ and $\tau_0(0.35) = 1.36$. Since $\tau < \tau_0(0.6)$, the spike is initially stable. However, since $\tau_0(0) < \tau$, the slowly drifting spike will experience a sudden instability before it reaches the origin. To verify this prediction, full numerical solutions to (1.1) were computed. The theory is indeed confirmed in Fig. 12 where we plot the spike amplitude a_m defined by $a_m(t) \equiv a[x_0(t); t]$ versus t. The corresponding spike location $x_0(t)$ is plotted in Fig. 13. Before the onset of the instability, the spike motion is given asymptotically by (4.24).

Question 4.7. Can one analyze the large-scale spike oscillations for a slowly drifting spike beyond the Hopf bifurcation point? What is the ODE for spike motion after the onset of the oscillation? Can a similar phenomenon occur in a two-dimensional domain when τ is large enough?

5. Multi-Spike Solutions in Reaction-Diffusion Systems

In this section we begin by constructing multi-spike equilibrium solutions in one spatial dimension to the Gray-Scott (GS) model in the low feed-rate regime (1.5), the Gierer-Meinhardt model (GM) (1.1), and the Schnakenburg (SC) model (1.6). The analysis assumes semi-strong spike interactions

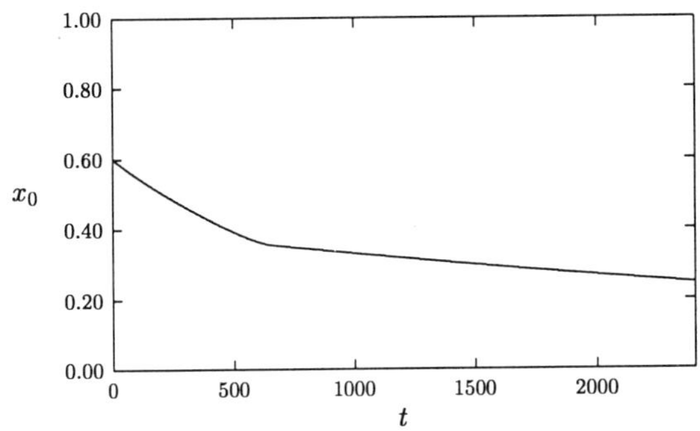

Figure 13. Plots of the spike location $x_0(t)$ versus t for $D = 1.0$, $\varepsilon = 0.03$, $\tau = 1.35$, $x_0(0) = 0.6$, and $(p, q, m, s) = (2, 1, 2, 0)$.

and is done for a one dimensional spatial domain. Two different types of spike patterns are constructed: symmetric patterns where the spikes have equal height, and asymmetric patterns where the spikes have different heights. The first study on the construction of k-spike symmetric patterns for (1.1) in one space dimension was done in [96]. For each of the three reaction-diffusion models, we show that each asymmetric pattern is characterized by two different spikes, B (big) and S (small). For a k-spike asymmetric equilibria, there are $k_1 > 0$ small spikes S and $k_2 = k - k_1 > 0$ large spikes B arranged in any order $SBBS \ldots SBB$ across the interval. Neglecting the orientation of large and small spikes in a spike sequence, we show that there are $k - 1$ asymmetric k-spike equilibrium patterns that bifurcate from a symmetric k-spike solution branch at some critical value of the parameters.

We begin by constructing asymmetric multi-spike equilibria for the GM model (1.1) as done in [104]. A different approach was used in [28]. We first construct a symmetric one-spike equilibrium solution centered at the origin for (1.1) posed on a finite domain of length $2l$, where $l > 0$ is a parameter. Therefore, we look for an even solution to

$$\varepsilon^2 a_{xx} - a + \frac{a^p}{h^q} = 0\,, \quad -l < x < l\,; \quad a_x(\pm l) = 0\,, \tag{5.1a}$$

$$Dh_{xx} - h + \varepsilon^{-1} \frac{a^m}{h^s} = 0\,, \quad -l < x < l\,; \quad h_x(\pm l) = 0\,. \tag{5.1b}$$

The basic idea is that we seek all values of l, labeled by $l_1, \ldots, l_n$, such that $h(l_1) =, \ldots, = h(l_n)$. For a certain range of the parameters, it is found that there are exactly two such values of l. These "local" solutions are then used to obtain a global asymmetric pattern for (1.1) on $[-1,1]$.

The asymptotic solution to (5.1) is

$$a(x) \sim [h(0)]^{q/(p-1)} \, w\left(x/\varepsilon\right) , \tag{5.2}$$

where w is given in (1.8) of Sec. 1. Since a is localized near $x = 0$, the term $\varepsilon^{-1} a^m / h^s$ in (5.1b) can be asymptotically approximated as a Dirac mass. Therefore, on $-l < x < l$, with $h^{'}(\pm l) = 0$, $h(x)$ satisfies

$$Dh^{''} - h = -\left[h(0)\right]^{\zeta+1} \left(\int_{-\infty}^{\infty} [w(y)]^m \, dy \right) \, \delta(x) \, , \quad -l < x < l \, , \tag{5.3}$$

where ζ is given in (1.2). The solution is

$$h(x) = [h(0)]^{\zeta+1} \left(\int_{-\infty}^{\infty} [w(y)]^m \, dy \right) \, G_l(x;0) \, , \tag{5.4}$$

where $G_l(x;0)$ is the Green's function satisfying

$$DG_{lxx} - G_l = -\delta(x) \, , \quad -l < x < l \, ; \quad G_{lx}(\pm l; 0) = 0 \, . \tag{5.5}$$

A simple calculation gives

$$G_l(x;0) = \frac{\cosh\left[(l - |x|) D^{-1/2}\right]}{2\sqrt{D} \sinh\left(l D^{-1/2}\right)} \, . \tag{5.6}$$

Setting $x = 0$ in (5.4), we obtain

$$[h(0)]^{\zeta} = \left[G_l(0;0) \int_{-\infty}^{\infty} [w(y)]^m \, dy \right]^{-1} \, . \tag{5.7}$$

We can then write $h(l)$ as $h(l) = h(0) G_l(l;0)/G_l(0;0)$. From (5.6) and (5.7), we then determine $h(l)$ in terms of some constant C as

$$h(l) = C \beta_{gm} \left(l D^{-1/2} \right) \, . \tag{5.8}$$

The function, $\beta_{gm}(z)$, for $z > 0$, is given by

$$\beta_{gm}(z) \equiv \frac{(\tanh z)^{1/\zeta}}{\cosh z} \, , \qquad \zeta \equiv \frac{qm}{p-1} - (s+1) > 0 \, . \tag{5.9}$$

We will analyze (5.8) below.

Symmetric and asymmetric spike patterns also occur in the SC model (1.6). The existence and stability of symmetric patterns in the SC model was studied in [49]. In [107] asymmetric spike patterns were constructed.

To construct these patterns, we first must calculate a symmetric one-spike solution centered at the origin for

$$\varepsilon^2 u_{xx} - u + vu^2 = 0\,, \quad -l < x < l\,; \quad u_x(\pm l) = 0\,, \tag{5.10a}$$

$$Dv_{xx} + \frac{1}{2} - \frac{b}{\varepsilon} vu^2 = 0\,, \quad -l < x < l\,; \quad v_x(\pm l) = 0\,. \tag{5.10b}$$

The asymptotic solution to (5.10a) is

$$u(x) \sim \frac{1}{v(0)} w\left(x/\varepsilon\right)\,, \tag{5.11}$$

where $w(y) = \frac{3}{2}\text{sech}^2(y/2)$. Since u is localized near $x = 0$, the term $\varepsilon^{-1}bvu^2$ in (5.10b) is a multiple of a Dirac mass. Thus, for $\varepsilon \ll 1$, $v(x)$ satisfies

$$Dv'' + \frac{1}{2} - \frac{6b}{v(0)}\delta(x) = 0\,, \quad -l < x < l\,; \quad v'(\pm l) = 0\,. \tag{5.12}$$

By integrating (5.12) we get $v(0) = 6b/l$. The solution to (5.12) is

$$v(x) = lG_n(x;0)\,, \tag{5.13}$$

where $G_n(x;0)$ is the Neumann Green's function satisfying

$$DG_{nxx} + \frac{1}{2l} - \delta(x) = 0\,, \quad -l < x < l\,; \quad G_{nx}(\pm l;0) = 0\,, \tag{5.14}$$

with $G_n(0;0) = 6bl^{-2}$. A simple calculation gives

$$G_n(x;0) = -\frac{x^2}{4lD} + \frac{l}{2D} + \frac{6b}{l^2} - \frac{1}{2D}\left(l - |x|\right)\,, \quad -l \le x \le l\,. \tag{5.15}$$

In terms of this solution, we can calculate $v(l) = lG_n(l;0)$ explicitly as

$$v(l) = \frac{r^2}{4D}\left[\beta_{sc}\left(l/r\right)\right]^{-1}\,, \qquad r \equiv \left[24bD\right]^{1/3}\,, \tag{5.16}$$

where the function $\beta_{sc}(z)$ is defined by

$$\beta_{sc}(z) \equiv \frac{z}{z^3+1}\,. \tag{5.17}$$

We now analyze some properties of β_{gm} and β_{sc} in (5.9) and (5.17), respectively, and show how these properties lead to the existence of asymmetric spike patterns.

The function $\beta_{gm}(z) > 0$ in (5.9) has a unique global maximum point at $z = z_{gm}$, where

$$z_{gm} = \log\left(\sqrt{\zeta^{-1}} + \sqrt{\zeta^{-1}+1}\right)\,. \tag{5.18}$$

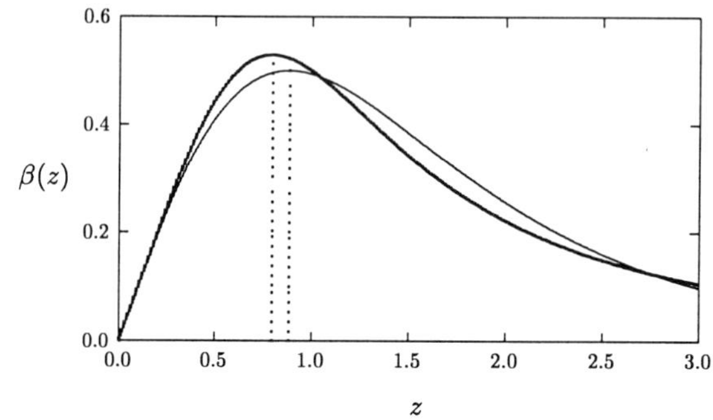

Figure 14. Plot of β_{gm} (heavy solid curve) versus z for the GM model with $\zeta = 1$. The solid curve is β_{sc} for the SC model. The maxima of β_{gm} and β_{sc} occur at z_{gm} and z_{sc} (dotted lines), respectively.

In addition, it satisfies $\beta'_{gm}(z) > 0$ on $[0, z_{gm})$ and $\beta'_{gm}(z) < 0$ on (z_{gm}, ∞). Therefore, given any $z \in (0, z_{gm})$, there exists a unique point $\tilde{z} = f_{gm}(z)$, with $\tilde{z} > z_{gm}$, such that $\beta_{gm}(z) = \beta_{gm}(\tilde{z})$.

Identical qualitative properties hold for the SC model. More specifically, the function $\beta_{sc}(z) > 0$ in (5.17) has a unique global maximum point at $z = z_{sc}$, where

$$z_{sc} = 2^{-1/3} . \tag{5.19}$$

In addition, $\beta'_{sc}(z) > 0$ on $[0, z_{sc})$ and $\beta'_{sc}(z) < 0$ on (z_{sc}, ∞). Therefore, given any $z \in (0, z_{sc})$, there exists a unique point $\tilde{z} = f_{sc}(z)$, with $\tilde{z} > z_{sc}$ such that $\beta_{sc}(z) = \beta_{sc}(\tilde{z})$. In Fig. 14 we plot β_{gm} and β_{sc}. One implication of these properties of β_{gm} and β_{sc} is the following:

Proposition 5.1. *Let $\xi > 0$ for the GM model. Given any l with $lD^{-1/2} < z_{gm}$, there exists a unique $\tilde{l}$, with $\tilde{l}D^{-1/2} \equiv \tilde{z} > z_{gm}$, such that $h(l) = h(\tilde{l})$. Similarly, for the SC model, given any l, with $l/r < z_{sc}$, there exists a unique $\tilde{l}$, with $\tilde{l}/r \equiv \tilde{z} > z_{sc}$, such that $v(l) = v(\tilde{l})$.*

We refer to solutions of length l and $\tilde{l}$ as S-type (small) and B-type (big) spikes, respectively. For both the GM and the SC models, we now show how to construct k-spike asymmetric equilibria on $-1 < x < 1$ with $k_1 > 0$ spikes of type S and $k_2 = k - k_1 > 0$ spikes of type B arranged in any particular order from left to right across the interval as

$$\text{SBSSB...B}\,, \qquad \text{with} \quad k_1 - \text{S's} \quad \text{and} \quad k_2 - \text{B's}\,. \tag{5.20}$$

For the GM model, we use translation invariance and the fact that $h(l) = h(\tilde{l})$ to glue S and B type spikes together to satisfy C^1 continuity for the inhibitor concentration h defined on $[-1,1]$. A similar construction holds for the SC model. We use translation invariance and $v(l) = v(\tilde{l})$ to glue S and B type spikes together to satisfy a C^1 continuity for the global function v defined on $[-1,1]$.

Since the support of an S-spike and a B-spike is $2l$ and $2\tilde{l}$, respectively, we get the length constraint $2k_1 l + 2k_2 \tilde{l} = 2$. For the GM model, we have that $z = lD^{-1/2}$, $\tilde{z} = \tilde{l}D^{-1/2}$, and that $\beta_{gm}(z) = \beta_{gm}(\tilde{z})$. Therefore, for the GM model z and $\tilde{z}$ satisfy the coupled algebraic system

$$k_1 z + k_2 \tilde{z} = D^{-1/2}\,, \quad \beta_{gm}(z) = \beta_{gm}(\tilde{z})\,. \tag{5.21}$$

Using the inverse function $\tilde{z} = f_{gm}(z)$ defined for $0 < z < z_{gm}$, we can reduce (5.21) to the study of the roots of the equation

$$f_{gm}(z) = \frac{D^{-1/2}}{k_2} - \frac{k_1 z}{k_2}\,, \tag{5.22a}$$

on $0 < z < z_{gm}$. In terms of the solution to (5.22a), l and $\tilde{l}$ are given by

$$l = zD^{1/2}\,, \quad \tilde{l} = f_{gm}(z)D^{1/2}\,. \tag{5.22b}$$

An identical construction can be done for the SC model. In place of (5.22), we obtain that

$$f_{sc}(z) = \frac{1}{rk_2} - \frac{k_1 z}{k_2}\,, \tag{5.23a}$$

on $0 < z < z_{sc}$, where l and $\tilde{l}$ are given by

$$l = zr\,, \quad \tilde{l} = f_{sc}(z)r\,. \tag{5.23b}$$

Here r is defined in (5.16)

Therefore, the problem of constructing asymmetric patterns for the GM and SC models has been reduced to the study of a single transcendental equation (5.22a) or (5.23a). To study these equations, we first use (5.9) and (5.17) to derive some key properties of the inverse functions $f_{gm}(z)$ and $f_{sc}(z)$. We summarize the result as follows:

Proposition 5.2. *For any $\zeta > 0$, $f_{gm}(z)$ is convex on $(0, z_{gm})$, $f'_{gm}(z) < -1$ on $(0, z_{gm})$, and $f'_{gm}(z_{gm}) = -1$. For the special case $\zeta = 1$, corresponding to the exponent set $(p,q,m,s) = (2,1,2,0)$ for the GM model, we calculate explicitly that*

$$f_{gm}(z) \equiv \log\left[cschz + \coth z\right]\,, \qquad \textit{for} \quad z \in [0, z_{gm})\,. \tag{5.24}$$

Similar properties hold for the SC model. In particular, $f_{sc}(z)$ *is convex on* $(0, z_{sc})$, $f'_{sc}(z) < -1$ *on* $(0, z_{sc})$, *and* $f'_{sc}(z_{sc}) = -1$. *We calculate,*

$$f_{sc}(z) = -\frac{z}{2} + \frac{\sqrt{z^2 + 4/z}}{2} . \tag{5.25}$$

We now determine conditions for which (5.22a) and (5.23a) have solutions on the interval $0 < z \leq z_{gm}$ and $0 < z \leq z_{sc}$, respectively. For the GM model, we must find the intersection points of $\tilde{z} = f_{gm}(z)$ with the straight line $\tilde{z} = -k_1 z/k_2 + k_2^{-1} D^{-1/2}$. A similar graphical analysis of (5.23a) holds for the SC model. The properties of $f_{gm}(z)$ and $f_{sc}(z)$ given in Proposition 5.2 show that there are two cases to consider:

$$\text{Case (i)} \quad k_1/k_2 \leq 1 ; \qquad \text{Case (ii)} \quad k_1/k_2 > 1 . \tag{5.26}$$

We readily obtain the first result for $k_1/k_2 \leq 1$.

Proposition 5.3. *Let* $k_1/k_2 \leq 1$. *Then, for the GM model, there exists a unique solution to (5.22a) on* $z \in (0, z_{gm})$ *when* $D < D_{gm}(k)$, *where*

$$D_{gm}(k) = \frac{1}{k^2 z_{gm}^2} , \qquad k = k_1 + k_2 , \tag{5.27}$$

and z_{gm} *satisfies (5.18). When* $D = D_{gm}(k)$, *then* $z = \tilde{z} = z_{gm}$ *and hence* $l = \tilde{l} = 1/k$. *In this case, we get a symmetric* k*-spike solution with spikes of equal height. A similar result holds for the SC model. When* $k_1/k_2 \leq 1$, *there exists a unique solution to (5.23a) on* $z \in (0, z_{sc})$ *when* $D < D_{sc}(k)$, *where*

$$D_{sc}(k) = \left[12bk^3\right]^{-1} , \qquad k = k_1 + k_2 . \tag{5.28}$$

When $D = D_{sc}(k)$, *we get* $z = \tilde{z} = z_{sc}$ *and* $l = \tilde{l} = 1/k$, *which yields a symmetric* k*-spike solution with spikes of equal height.*

Since $f'_{gm}(z) < -1$ for $0 < z < z_{gm}$ and $f'_{sc}(z) < -1$ for $0 < z < z_{sc}$, with f_{gm} and f_{sc} convex, there can be solution multiplicity to (5.22a) and (5.23a) when $k_1/k_2 > 1$. In particular, as D is decreased from a large value, the straight line of slope $-k_1/k_2$ first intersects the curve $\tilde{z} = f_{gm}(z)$ at a point of tangency at the critical value $D = D^*_{gm}$. As D is decreased slightly below D^*_{gm}, there are exactly two roots to (5.22a) until D is decreased to the value $D_{gm}(k)$. For D below $D_{gm}(k)$ there is only one solution to (5.22a). For the GM model with $\zeta = 1$, this result is illustrated in Fig. 15. We summarize our result as follows:

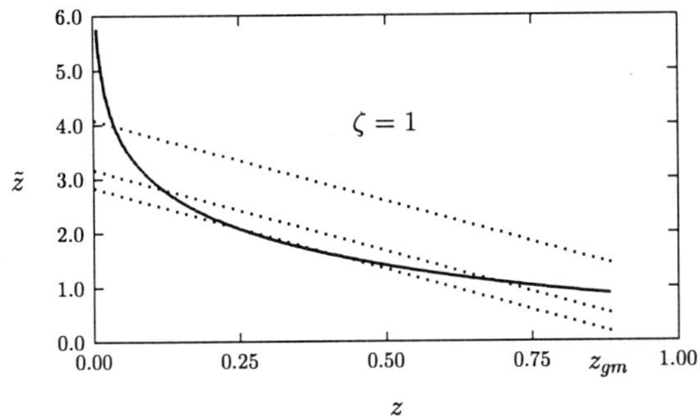

Figure 15. Plot of the graphical solution to (5.22a) when $\zeta = 1$, $k_1 = 3$, and $k_2 = 1$. Depending on D, $\tilde{z} = f_{gm}(z)$ (solid) will intersect the straight line in (5.22a) (dotted lines) at either zero, one, or two points. When $D = D^*_{gm}$ the straight line is tangent to $f_{gm}(z)$. When $D = D_{gm}(k)$ the straight line intersects $f_{gm}(z)$ at $z = z_{gm}$.

Proposition 5.4. *Consider the GM model with $k_1/k_2 > 1$. Then, there exists a critical value $D^*_{gm} > D_{gm}(k)$ such that the solution multiplicity for (5.22a) on $z \in (0, z_{gm})$ is as follows:*

$$\begin{aligned} &\textit{If}\ \ D > D^*_{gm} \qquad \text{there are no solutions}\,, \\ &\textit{If}\ \ D_{gm}(k) < D < D^*_{gm} \qquad \text{there are exactly two solutions}\,, \\ &\textit{If}\ \ D < D_{gm}(k) \qquad \text{there is exactly one solution}\,. \end{aligned} \tag{5.29}$$

The critical value D^*_{gm} is the value of D where there is a double root to (5.22a). For a GM model with $\zeta = 1$, we calculate explicitly that

$$D^*_{gm} = \left[k_1 \sinh^{-1}\left(\frac{k_2}{k_1}\right) + k_2 \sinh^{-1}\left(\frac{k_1}{k_2}\right)\right]^{-2}, \quad \textit{for} \quad k_1/k_2 > 1\,. \tag{5.30}$$

An identical solution multiplicity result holds for the SC model except that z_{gm}, $D_{gm}(k)$, and D^*_{gm}, are to be replaced with z_{sc}, $D_{sc}(k)$, and D^*_{sc}, respectively, where

$$D^*_{sc} = \frac{1}{24b}\left[k_2 f_{sc}(z_*) + k_1 z_*\right]^{-3}. \tag{5.31a}$$

Here z_* is defined by

$$z_* = \left[\frac{2(1+\gamma^2) - 2\sqrt{(1+\gamma^2)^2 - (1-\gamma^2)}}{(1-\gamma^2)}\right]^{1/3}, \quad \gamma \equiv 1 - \frac{2k_1}{k_2}\,. \tag{5.31b}$$

In summary, consider a particular fixed ordering of k_1 spikes of type S and k_2 spikes of type B across the interval. For $k_1 \leq k_2$, and for the GM or the SC models, there is only one asymmetric spike pattern with the particular ordered sequence when D is below $D_{gm}(k)$. However, when there are more S (small) than B (big) spikes in the sequence, then for some range of D there are exactly two such patterns that have the same ordering. For the GM model we obtain the following result for asymmetric patterns (see Proposition 2.1 of [104]):

Proposition 5.5. *For the range of D in Proposition 5.4, let l and $\tilde{l}$ be determined from (5.22). Then, for $\varepsilon \to 0$, the activator concentration a for an asymmetric equilibrium k-spike pattern for (1.1) with k_1 spikes of type S and $k_2 = k - k_1$ spikes of type B is given by*

$$a(x) \sim \sum_{j=1}^{k} \left[h_{l_j}\right]^{q/(p-1)} w\left[\varepsilon^{-1}(x - x_j)\right], \quad h_{l_j} \equiv \left(\frac{2\sqrt{D}\, \tanh\left(l_j D^{-1/2}\right)}{\int_{-\infty}^{\infty} [w(y)]^m \, dy} \right)^{1/\zeta}. \tag{5.32}$$

Here for each j, $l_j = l$ or $l_j = \tilde{l}$. The value $l_j = l$ must occur $k_1 > 0$ times, while $l_j = \tilde{l}$ must occur $k_2 = k - k_1 > 0$ times. The small and large spikes can be arranged in any sequence. The spike locations x_j are given by

$$x_1 = l_1 - 1, \quad x_k = 1 - l_k, \quad x_{j+1} = x_j + l_{j+1} + l_j, \quad j = 1, \ldots, k-2. \tag{5.33}$$

The symmetric k-spike solution occurs when $l_j = 1/k$ for $j = 1, \ldots, k$.

In Fig. 16 we illustrate this result for the GM model by plotting a 5-spike asymmetric pattern of the form SBBSB for a specific parameter set.

The result in Proposition 5.3 shows that asymmetric spike patterns bifurcate off of the symmetric patterns at the value $D = D_{gm}(k)$ for $k \geq 2$. To illustrate this result graphically, in Fig. 17 we plot a bifurcation diagram of an L_1 type- norm of a, defined by $|a|_1 \equiv \sum_{j=1}^{k} h_{l_j}^{\gamma}$, versus D for both symmetric and asymmetric k-spike patterns. The portions of the branches that are unstable to $O(1)$ instabilities are indicated by the dashed lines in this figure. Some stability results are given below.

A similar result follows for the construction of asymmetric spike patterns in the SC model (see Proposition 1 of [107]):

Proposition 5.6. *For the range of D in Proposition 5.4, let l and $\tilde{l}$ be determined from (5.23). Then, for $\varepsilon \to 0$, the u component for an asymmetric equilibrium k-spike pattern for (1.6) with k_1 spikes of type S and*

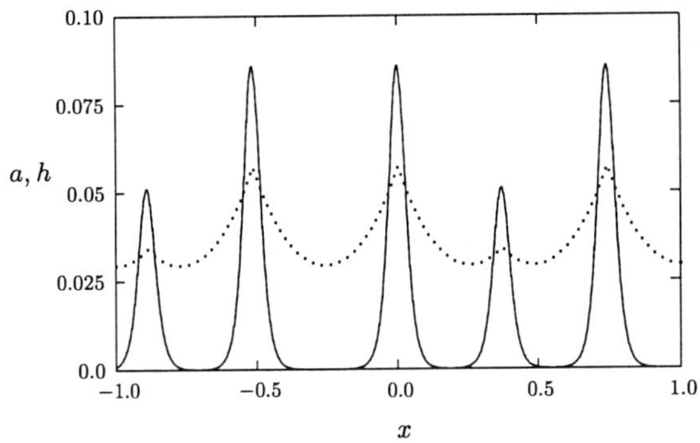

Figure 16. Plot of a (solid curve) and h (dotted curve) for a 5-spike asymmetric pattern of the type SBBSB for the GM model (1.1). The parameter values are $\varepsilon = .02$, $D = .04$, and $(p, q, m, s) = (2, 1, 2, 0)$.

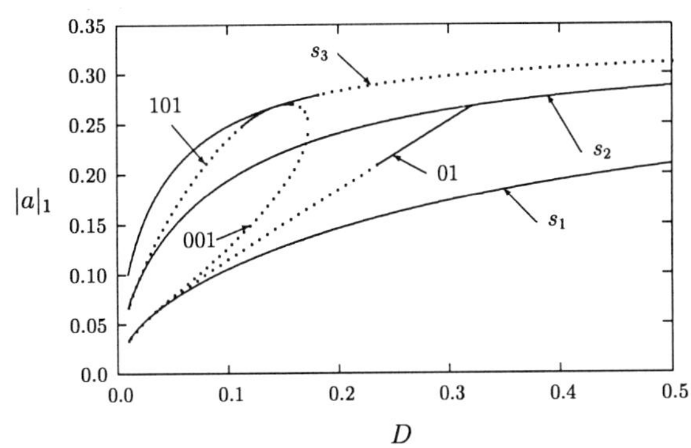

Figure 17. Plot of $|a|_1 \equiv \sum_{j=1}^{k} h_{l_j}^{\gamma}$ versus D for the GM model with $(p, q, m, s) = (2, 1, 2, 0)$ and for solutions with $k = 1, 2, 3$ spikes. The k-spike symmetric branch is s_k. The asymmetric patterns SB, BSB, and SSB are labeled by 01, 101, and 001, respectively. The dashed portions of these branches are unstable with on an $O(1)$ time-scale.

$k_2 = k - k_1$ spikes of type B is given by

$$u(x) \sim \sum_{j=1}^{k} v_j^{-1} w\left[\varepsilon^{-1}(x - x_j)\right], \quad v_j = \frac{6b}{l_j}. \tag{5.34}$$

Here l_j and x_j for $j = 1, \ldots, k$ are given in Proposition 5.5. The symmetric k-spike solution is found by setting $l_j = 1/k$ for $j = 1, \ldots, k$ in (5.34).

5.1. *Gray-Scott Model (Low Feed-Rate): Equilibrium*

The asymptotic construction of symmetric and asymmetric k-spike equilibria to the GS model in the low feed-rate regime (1.5) has many similarities with that for the GM model, except that now the equilibria can exhibit a saddle-node bifurcation structure.

We follow [56] by first constructing symmetric k-spike equilibria to (1.5) for $\varepsilon \ll 1$. The spike locations x_j, for $j = 1, \ldots, k$, for such a pattern satisfy

$$x_j = -1 + \frac{(2j-1)}{k}, \qquad j = 1, \ldots, k. \tag{5.35}$$

Since the spikes have equal height, then $u(x_j) = U$, for some constant U.

In the inner region near the jth spike, we let $y = \varepsilon^{-1}(x - x_j)$. In each inner region, we get that $u \sim U + O(\varepsilon)$. Therefore, from (1.5a), the leading-order inner solution for ν is $\nu \sim [\tilde{A}U]^{-1} w(y)$, where $w(y)$ is given in (1.8) with $p = 2$.

Approximating the term $\varepsilon^{-1} u \nu^2$ in (1.5b) as a multiple of a Dirac mass, and using $\int_{-\infty}^{\infty} w^2 dy = 6$, we get that the outer solution for u satisfies

$$Du_{xx} + (1-u) - \frac{6}{\tilde{A}^2 U} \sum_{j=1}^{k} \delta(x - x_j) = 0, \quad u_x(\pm 1) = 0. \tag{5.36}$$

The solution to (5.36) is

$$u(x) = 1 - \frac{6}{\tilde{A}^2 U} \sum_{j=1}^{k} G(x; x_j), \tag{5.37}$$

where $G(x; x_j)$ is the Green's function, satisfying

$$DG_{xx} - G = -\delta(x - x_j), \qquad -1 < x < 1; \qquad G_x(\pm 1; x_j) = 0. \tag{5.38}$$

We define $a_g \equiv \sum_{i=1}^{k} G(x_j; x_i)$, where the x_j satisfy (5.35). By solving (5.38) explicitly, the find that a_g is independent of j and is given by

$$a_g \equiv \sum_{i=1}^{k} G(x_j; x_i) = \left[2\sqrt{D} \tanh\left(\theta_0/k\right)\right]^{-1}. \tag{5.39}$$

Setting $u(x_j) = U$ in (5.37), we obtain a quadratic equation for U

$$U(U-1) = -\frac{6a_g}{\tilde{A}^2}. \tag{5.40}$$

By solving this equation, the following result was derived in [56]:

Proposition 5.7. *Let* $\varepsilon \to 0$, *with* $\tilde{A} = O(1)$ *and* $D = O(1)$ *in (1.5). Then, when* $\tilde{A} > \tilde{A}_{ke}$, *there are two symmetric* k*-spike equilibrium solutions to (1.5) given asymptotically by*

$$\nu_{\pm}(x) \sim \frac{1}{\tilde{A}U_{\pm}} \sum_{j=1}^{k} w\left[\varepsilon^{-1}(x - x_j)\right] , \tag{5.41}$$

where $w(y) = \frac{3}{2}\text{sech}^2(y/2)$. *We label* ν_+ *and* ν_- *as the large and small solution, respectively. Here* $U_{\pm}$, *and the saddle-node bifurcation value* $\tilde{A}_{ke} = \sqrt{24a_g}$ *are given by*

$$U_{\pm} \equiv \frac{1}{2}\left[1 \pm \sqrt{1 - \frac{\tilde{A}_{ke}^2}{\tilde{A}^2}}\right] , \quad \tilde{A}_{ke} \equiv \sqrt{\frac{12\theta_0}{\tanh(\theta_0/k)}} , \quad \theta_0 \equiv D^{-1/2} . \tag{5.42}$$

We now construct asymmetric k-spike equilibria, where the spikes can have different heights. The analysis, given in Sec. 2.1 of [56], is very similar to that for the GM and SC models given above. The idea is to construct a symmetric one-spike equilibrium to (1.5) centered at $x = 0$ on a domain $-l < x < l$ with $u_x(\pm l) = 0$. The goal is to find all different values of l, labeled by $l_1, \ldots, l_n$, such that $u(l_1) =, \ldots, = u(l_n)$. Once again we find that there are only two types of spikes S and B.

A straightforward calculation shows that $u(l)$ is given by

$$u(l) = E_{\pm}(z) \equiv 1 + \frac{1}{2}\text{sech}(z)\left[-1 \pm \sqrt{1 - \frac{\coth(z)}{\tilde{A}_0^2}}\right] , \quad \text{for } z > z_0 , \tag{5.43a}$$

where

$$z \equiv \theta_0 l , \quad \theta_0 \equiv D^{-1/2} , \quad z_0 \equiv \frac{1}{2}\log\left[\frac{\tilde{A}_0^2 + 1}{\tilde{A}_0^2 - 1}\right] , \quad \tilde{A} \equiv \sqrt{12\theta_0}\tilde{A}_0 . \tag{5.43b}$$

Clearly $E_{\pm}(z) > 0$ for $z > z_0$. For $z > z_0$, it is easy to see that $E_+(z)$ is monotonically increasing. Therefore, asymmetric patterns cannot be obtained with $E_+(z)$.

A simple calculation as given in Proposition 2.2 of [56] shows that $E_-(z_0) < 1$ with $E_-'(z) < 0$ for $z_0 < z < z_{gs}$, and $E_-'(z) > 0$ for $z > z_{gs}$. Moreover, $E_-(z) \to 1$ as $z \to \infty$. The point z_{gs} where $E_-(z)$ has its minimum value is the unique root of

$$\tilde{A}_0 = [\tanh z]^{-1/2}\,[\tanh(2z)]^{-1} . \tag{5.44}$$

In Fig. 18 we plot E_- versus z for $z \geq z_0$ when $\tilde{A}_0 = 3$.

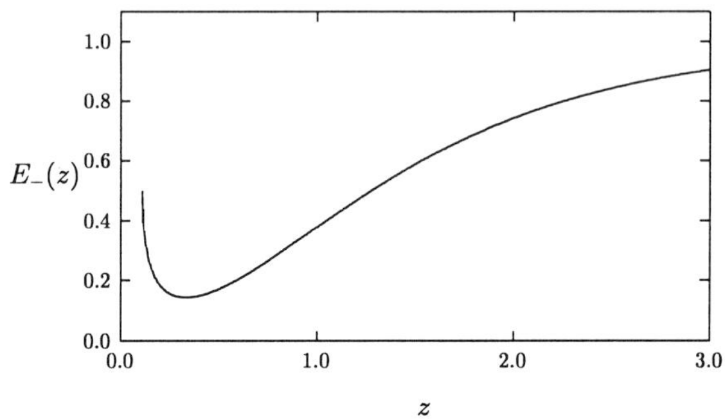

Figure 18. Plot of E_- versus z for $z \geq z_0$ when $\tilde{A}_0 = 3$.

Therefore, when $\tilde{A}_0 > 1$, we conclude that for any z in $z_0 < z < z_{gs}$, there exists a unique $\tilde{z} = f_{gs}(z)$, with $\tilde{z} > z_{gs}$, such that $E_-(z) = E_-(\tilde{z})$. Since $z = \theta_0 l$ and $\tilde{z} = \theta_0 \tilde{l}$, the implication of this result is that for any l with $z_0 < l\theta_0 < z_{gs}$, there exists a unique $\tilde{l}$, with $\tilde{l}\theta_0 > z_{gs}$, such that $u(l) = u(\tilde{l})$. This implies that in any asymmetric pattern of this form using $E_-(z)$ there are again only two types of spikes S and B. Therefore, in place of (5.23), we obtain that

$$f_{gs}(z) = \frac{\theta_0}{k_2} - \frac{k_1 z}{k_2}, \tag{5.45a}$$

on $z_0 < z < z_{gs}$. In terms of the solution to (5.45a), we get

$$l = z\theta_0^{-1}, \quad \tilde{l} = f_{gs}(z)\theta_0^{-1}. \tag{5.45b}$$

To recover the symmetric k-spike equilibrium solutions, we set $z = f_{gs}(z) = \theta_0/k$. This has a solution only when $z = z_{gs}$. Therefore, setting $z = \theta_0/k$ in (5.44), we we obtain the critical value of $\tilde{A}_0$ for the emergence of the asymmetric branch. This leads to the following bifurcation result:

Proposition 5.8. *Let l and $\tilde{l}$ be found from (5.45) for a given $\tilde{A} > \sqrt{12\theta_0}$ and $\theta_0 = D^{-1/2}$. Then, for $\varepsilon \to 0$, the ν component of an asymmetric equilibrium k-spike pattern for (1.5) with k_1 spikes of type S and $k_2 = k - k_1$*

spikes of type B is given by

$$\nu_-(x) \sim \sum_{j=1}^{k} \frac{1}{\tilde{A}U(l_j)} w\left[\frac{(x-x_j)}{\varepsilon}\right], \quad U(l_j) \equiv \frac{1}{2}\left(1-\sqrt{1-\frac{12\theta_0\coth(\theta_0 l_j)}{\tilde{A}^2}}\right). \tag{5.46}$$

Here for each j, $l_j = l$ or $l_j = \tilde{l}$. The value $l_j = l$ must occur $k_1 > 0$ times, while $l_j = \tilde{l}$ must occur $k_2 = k - k_1 > 0$ times. The spike locations x_j are given by (5.33). Finally, the asymmetric k-spike equilibrium solutions bifurcate from the k-spike symmetric small equilibrium solution branch of Proposition 5.7 at the value $\tilde{A} = \tilde{A}_{ka}$, where

$$\tilde{A}_{ka} \equiv \tilde{A}_{ke}\left[\tanh\left(\frac{2\theta_0}{k}\right)\right]^{-1}. \tag{5.47}$$

Here $\tilde{A}_{ke}$ is the saddle-node bifurcation value given in (5.42).

To display our results graphically in a bifurcation diagram it is convenient to define a norm of ν by $|\nu|_2 \equiv \left(\varepsilon^{-1}\int_{-1}^{1}\nu^2\,dx\right)^{1/2}$. For symmetric and asymmetric spike patterns, a simple calculation using (5.46) shows that

$$|\nu|_2 \sim \frac{\sqrt{6k}}{\tilde{A}U_\pm} \ (\text{symm}), \quad |\nu|_2 \sim \frac{1}{\tilde{A}}\left(\frac{6k_1}{[U(l)]^2} + \frac{6k_2}{\left[U(\tilde{l})\right]^2}\right) \ (\text{asymm}). \tag{5.48}$$

In Fig. 19 we plot the symmetric and asymmetric solution branches for $k = 1, \ldots, 4$ when $D = 0.75$. It is easy to see that an asymmetric branch with k_1 small spikes asymptotes to the symmetric branch with $k - k_1$ spikes as $\tilde{A} \to \infty$. There are several questions suggested by these calculations.

Question 5.1. What general properties of a reaction-diffusion system will, necessarily, lead to asymmetric patterns involving ony two types of spikes in the semi-strong spike interaction regime?

Question 5.2. Can one find singularly perturbed reaction-diffusion systems having similar asymmetric patterns, but that now involve three or more types of spikes?

5.2. *Gray-Scott Model (High Feed-Rate): Equilibrium*

We now construct k-spike symmetric spike patterns for the GS model in the high feed-rate regime (1.3). In this regime, where $A = O(1)$, the scalings

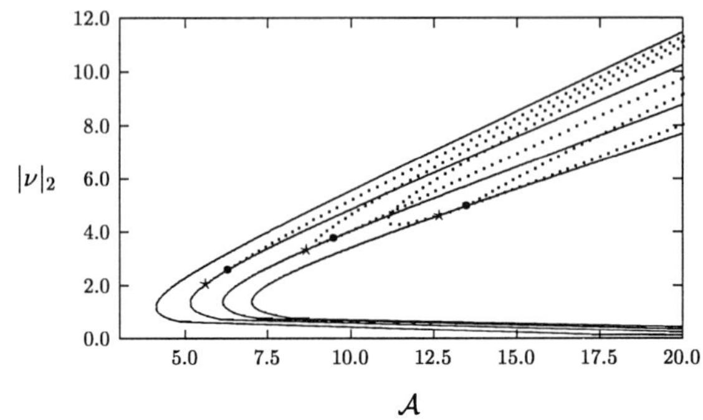

Figure 19. Bifurcation diagram of symmetric (solid curves) and asymmetric (dotted curves) spike patterns in the low feed-rate regime $A = \tilde{A}\varepsilon^{1/2}$ for $D = 0.75$ and $k = 1, 2, 3, 4$. The saddle-node values $\tilde{A}_{ke}$ increase with k.

for equilibria are $v = O(\varepsilon^{-1})$ and $u = O(\varepsilon)$ (cf. [57]). Although we will only give results for spike patterns on the interval $[-1, 1]$, other essentially one-dimensional patterns can be constructed. Specifically, in [72] and [58], the existence and stability of radially symmetric solutions, known as ring solutions, was investigated.

In [57] it was shown that k-spike equilibria can be constructed in terms of the solutions $V(y) > 0$ and $U(y) > 0$ to a *core problem* defined by

$$V'' - V + V^2 U = 0\,, \qquad U'' = UV^2\,, \qquad 0 < y < \infty\,, \tag{5.49a}$$

$$V'(0) = U'(0) = 0\,; \quad V \to 0\,, \quad U \sim By\,, \quad \text{as} \quad y \to \infty\,, \tag{5.49b}$$

where B is related to A by

$$B \equiv A \tanh(\theta_0/k)\,, \quad \theta_0 = D^{-1/2}\,. \tag{5.49c}$$

The precise result, as given in Proposition 3.1 of [57], is as follows:

Proposition 5.9. *Let $\varepsilon \to 0$, $A = O(1)$, $\varepsilon A/\sqrt{D} \ll 1$, and suppose that (5.49) has a solution. Then, the v-component for a k-spike equilibrium solution to (1.3) is given by*

$$v \sim \frac{\sqrt{D}}{\varepsilon} \sum_{j=1}^{k} \left(V\left[\varepsilon^{-1}(x - x_j)\right] + O\left(\frac{\varepsilon A}{\sqrt{D}}\right)\right)\,. \tag{5.50}$$

In the jth inner region, defined by $|x - x_j| = O(\varepsilon)$, we also have that $u \sim \frac{\varepsilon}{A\sqrt{D}} U\left[\varepsilon^{-1}(x - x_j)\right]$.

The core problem (5.49) was studied qualitatively and numerically in [57] in terms of B. We first determine some analytical properties of (5.49). By integrating (5.49a) from $0 < y < \infty$, and using (5.49b), we get

$$\int_0^\infty V^2 U \, dy = \int_0^\infty V \, dy = B \,. \tag{5.51}$$

To determine a good parameterization of the solution branch for (5.49), we define $U_0 \equiv U(0)$ and $V_0 \equiv V(0)$. Then, we can readily derive that

$$\frac{V_0^2}{2} + \int_0^\infty \frac{U}{3} \left[V^3\right]' \, dy = \frac{V_0^2}{2} - \frac{U_0 V_0^3}{3} - \int_0^\infty \frac{1}{3} V^3 U' \, dy = 0 \,. \tag{5.52}$$

This gives,

$$\int_0^\infty V^3 U' \, dy = V_0^2 \left[\frac{3}{2} - U_0 V_0\right] \,. \tag{5.53}$$

Since $U' > 0$ on $0 < y < \infty$, we get the key inequality that

$$0 < \gamma < \frac{3}{2} \,, \qquad \gamma \equiv U_0 V_0 \,. \tag{5.54}$$

In [57] numerical solutions to (5.49) were computed for which V has a single maximum at $y = 0$ as $\gamma \to 3/2$ from below. By using Euler continuation a resulting curve $B = B(\gamma)$ was computed. This limiting solution for V asymptotically matches onto the solution constructed in the low feed-rate regime in Sec. 5.1. It was found in [57] that the curve $B = B(\gamma)$ is double-valued with $B \to 0$ as $\gamma \to 0$ and as $\gamma \to 3/2$, and it has a saddle-node bifurcation point at the maximum value B_c of B given by $B_c = 1.347$, where $\gamma = \gamma_c = 1.02$. This critical value of B was also computed in [74]. We refer to the range $\gamma_c < \gamma < 3/2$ and $0 < \gamma < \gamma_c$ as the *primary* and *secondary* branches of the $B = B(\gamma)$ bifurcation diagram. Using (5.49c), these results show that equilibrium k-spike solutions in the high feed-rate regime exist only when A is small enough. The equilibrium result, derived in Proposition 3.2 of [57], is summarized as follows:

Proposition 5.10. *Let $\varepsilon \ll 1$, $A = O(1)$, and $\varepsilon A/\sqrt{D} \ll 1$. Then, there will be no k-spike equilibrium solution to (1.3) that merges onto the low feed-rate regime solution when*

$$A > A_{pk} \equiv 1.347 \coth\left(\frac{1}{k\sqrt{D}}\right) \,. \tag{5.55}$$

In Fig. 20 we plot the norm $|v|_2$, defined by $|v|_2^2 \equiv \int_{-1}^{1} v^2\, dx$ versus A when $D = 0.1$. This norm can be evaluated asymptotically using (5.50) as

$$|v|_2^2 \sim \varepsilon^{-1} k D \int_{-\infty}^{\infty} V^2\, dy\,. \tag{5.56}$$

Notice that the saddle-node bifurcation values A_{pk} occur at approximately the same value when $D \ll 1$, since for k fixed we have that $\tanh\left(1/k\sqrt{D}\right) \approx 1$ for $D \ll 1$. This is called the *lining-up property* of saddle-node equilibria.

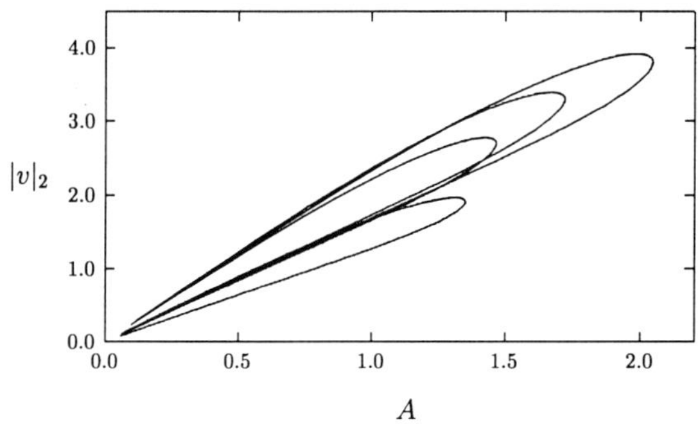

Figure 20. Bifurcation diagram of $|v|_2$ versus A when $A = O(1)$ for k-spike solutions to the GS model in the high feed-rate regime with $\varepsilon = 0.02$, $D = 0.1$, and $k = 1, \ldots, 4$. The saddle-node values A_{pk} increase with k.

In [57] the limiting behavior of (5.49) was studied asymptotically for $\gamma \to 3/2$ from below and for $\gamma \to 0$ from above. The main result of this analysis is that on the secondary branch of the $B = B(\gamma)$ bifurcation diagram a one-spike solution to (5.49) splits into a two-spike solution where the spikes are separated by $O(-\log B)$ as $B \to 0$. This multi-bump behavior is closely related to a phenomenon known as pulse-splitting, as described in [32], [82], [83], [81], [98], and [57].

For $\gamma \to 3/2$ from below, the following result obtained in Proposition 4.1 of [57] shows that the solution matches onto the low feed-rate regime solution:

Proposition 5.11. *Along the primary branch, the core problem (5.49) has a solution where $\gamma \to 3/2$ from below as $\delta \to 0$, where $\delta = 1/U(0) \ll 1$. This solution has the form*

$$V \sim \delta \mathring{w}(y) + O(\delta^2)\,, \quad U \sim \delta^{-1} + O(\delta)\,, \tag{5.57a}$$

where $w(y)$ satisfies (1.8) with $p = 2$. We have that $B \sim 3\delta$, and that

$$\gamma \equiv U_0 V_0 \sim \frac{3}{2} - \frac{11B^2}{48}, \quad \text{as} \quad B \to 0. \tag{5.57b}$$

Alternatively, as $\gamma \to 0$ from above, the solution V has two bumps. This behavior, as derived in Proposition 4.2 of [57] is characterized as follows:

Proposition 5.12. *Let $\delta = 1/U_0 \ll 1$. Then, along the secondary branch, the core problem (5.49) has a solution where $\gamma \equiv U_0 V_0 \to 0$ from above as $\delta \to 0$. This solution has the form*

$$V \sim \delta \left[w(y - y_1) + w(y + y_1) \right] , \quad U \sim \frac{1}{\delta} + O(\delta) , \tag{5.58a}$$

where $w(y)$ satisfies (1.8) when $p = 2$. We have that $B \sim 6\delta$ as $\delta \to 0$, and

$$\gamma \equiv U_0 V_0 \sim \sqrt{2} B / \sqrt{5} , \quad y_1 \sim -\log B + O(1) , \quad \text{as} \quad B \to 0 . \tag{5.58b}$$

The existence of multi-bump solutions to the core problem along the secondary branch is closely related to a similar multi-bump phenomena for the modified version (2.45) of Carrier's problem, as discussed in Sec. 2.4.

Question 5.3. Can one determine rigorous properties of multi-bump solutions to the core problem and of the $B = B(\gamma)$ bifurcation curve? In particular, can we prove the existence of a saddle-node bifurcation and determine analytical bounds for the saddle-node value?

5.3. *Stability of Multi-Spike Equilibria: One Dimension*

In this subsection we give some results for the stability of symmetric k-spike patterns for the GS model (1.5) in the low feed-rate regime and the GM model (1.1). For each of these models, there are two types of eigenvalues in the spectrum of the linearization. There are eigenvalues that are $O(1)$ as $\varepsilon \to 0$, referred to as the *large* eigenvalues, and eigenvalues of order $O(\varepsilon^2)$, referred to as the *small* eigenvalues. The large eigenvalues are associated with the initiation of profile instabilities, whereby the spike amplitudes will either oscillate, typically with a common frequency and phase, or else undergo a competition instability leading to the monotonic annihilation of spikes in a spike sequence. Alternatively, the small eigenvalues of order

$O(\varepsilon^2)$, determine the stability with respect to translations of the spike profile, and are associated with the bifurcation of asymmetric branches of spike equilibria off of the symmetric branches.

We first consider the stability of symmetric k-spike patterns for the GS model (1.5) with respect to the large eigenvalues. To analyze the stability of these solutions we let $u(x,t) = u_{\pm}(x) + e^{\lambda t}\eta(x)$, and $\nu(x,t) = \nu_{\pm}(x) + e^{\lambda t}\phi(x)$, where $\nu_{\pm}$ and $u_{\pm}$ are defined in Proposition 5.8 with $l_j = 1/k$. Here ϕ is a localized eigenfunction of the form

$$\phi(x) \sim \sum_{j=1}^{k} c_j \Phi\left[\varepsilon^{-1}(x - x_j)\right], \tag{5.59}$$

for some coefficients c_j to be found, and $\int_{-\infty}^{\infty} w(y)\Phi(y)\,dy \neq 0$. A lengthy derivation leading to Proposition 3.2 of [56], yields the following spectral problem for $\Phi(y)$:

Proposition 5.13. *Let $0 < \varepsilon \ll 1$. Then, with $\Phi = \Phi(y)$, the $O(1)$ eigenvalues determining the stability of k-spike symmetric equilibria in the GS model (1.5) satisfy the nonlocal eigenvalue problem*

$$L_0\Phi - \chi_{gs} w^2 \left(\frac{\int_{-\infty}^{\infty} w\Phi\,dy}{\int_{-\infty}^{\infty} w\,dy} \right) = \lambda\Phi\,, \quad -\infty < y < \infty\,; \quad \Phi \to 0\,, \quad \text{as } |y| \to \infty\,. \tag{5.60}$$

Here $w(y)$ is the homoclinic solution (1.8) with $p = 2$, and L_0 is the local operator $L_0\Phi \equiv \Phi'' - \Phi + 2w\Phi$. The multiplier $\chi_{gs} = \chi_{gs}(z;j)$ is defined by

$$\chi_{gs} \equiv 2s_g \left(s_g + \frac{\sqrt{1+z}}{\tanh(\theta_0/k)} \left[\tanh(\theta_\lambda/k) + \frac{(1 - \cos[\pi(j-1)/k])}{\sinh(2\theta_\lambda/k)} \right] \right)^{-1}, \tag{5.61}$$

where $z \equiv \tau\lambda$, $\theta_\lambda \equiv \theta_0\sqrt{1+z}$, and $\theta_0 \equiv D^{-1/2}$. Here s_g is defined in terms of $U_{\pm}$, given in (5.42), by

$$s_g = \frac{1 - U_{\pm}}{U_{\pm}}\,, \qquad 0 < s_g < \infty\,. \tag{5.62}$$

The coefficients $\mathbf{c}^t = (c_1, \ldots, c_k)$ in (5.59) are given by

$$\mathbf{c}_1^t = \frac{1}{\sqrt{k}}(1, \ldots, 1)\,; \quad c_{l,j} = \sqrt{\frac{2}{k}} \cos\left(\frac{\pi(j-1)}{k}(l - 1/2) \right), \quad j = 2, \ldots, k\,. \tag{5.63}$$

The method for deriving (5.60) is similar to the approach used in [80] for studying the stability of pulses in the Fitzhugh-Nagumo model. Notice that the small solution ν_- corresponds to the range $1 < s_g < \infty$, while the large solution corresponds to $0 < s_g < 1$. The saddle-node bifurcation value $\tilde{A}_{ke}$ corresponds to $s_g = 1$. Using (5.42), we can write $\tilde{A}$ as

$$\frac{\tilde{A}}{\tilde{A}_{ke}} = \frac{(1+s_g)}{2\sqrt{s_g}} \,. \tag{5.64}$$

As in the derivation (3.16)–(3.18), there is an equivalent formulation of (5.60) which states that the eigenvalues of (5.60) with $\int_{-\infty}^{\infty} w\Phi \, dy \neq 0$ are the union of the zeros of the functions $g_j(\lambda) = 0$ for $j = 1,..,k$, where

$$g_j(\lambda) \equiv C_j(\lambda) - f(\lambda) \,, \quad f(\lambda) \equiv \frac{\int_{-\infty}^{\infty} w \left(L_0 - \lambda\right)^{-1} w^2 \, dy}{\int_{-\infty}^{\infty} w^2 \, dy} \,, \quad C_j(\lambda) \equiv \frac{1}{\chi_{go}} \,. \tag{5.65}$$

Notice that this reformulation is very similar to (3.18). By analyzing the zeroes of $g_j(\lambda)$ in the right half-plane $\text{Re}(\lambda) \geq 0$ in a similar way as in Sec. 3, the main stability result for multi-spike solutions is as follows (cf. Proposition 3.10 and 3.13 of [56]:

Proposition 5.14. *The large solution u_+, ν_+ is unstable for any $0 < s_g < 1$, $k \geq 1$, and $D > 0$. Next, let $k \geq 1$, and consider the k-spike small solution u_-, ν_-, where $s_g > 1$. For $\tilde{A} > \tilde{A}_{kL}$, the solution is stable on an $O(1)$ time-scale when $0 < \tau < \tau_{hL}$. Alternatively, for $\tilde{A}_{ke} < \tilde{A} < \tilde{A}_{kL}$, the small solution is unstable for any $\tau > 0$. The threshold $\tilde{A}_{kL}$ is given analytically by*

$$\tilde{A}_{kL} = \frac{\tilde{A}_{ke}\left((\gamma_k/2) + 2\sinh^2\left(\theta_0/k\right)\right)}{\left(\left[(\gamma_k/2) + 2\sinh^2\left(\theta_0/k\right)\right]^2 - \left(\gamma_k/2\right)^2\right)^{1/2}} \,, \quad \gamma_k \equiv 1 + \cos\left(\frac{\pi}{k}\right) \,. \tag{5.66}$$

Notice that $\tilde{A}_{kL} = \tilde{A}_{ke}$ when $k = 1$.

Let $\tilde{A}$ satisfy $\tilde{A} > \tilde{A}_{kL}$. Then, as τ increases beyond τ_{hL}, a Hopf bifurcation in the spike amplitudes was computed numerically in [56]. The threshold τ_{hL} is given by the *minimum value* of the set τ_j, $j = 1,..,k$, for which $g_j(\lambda) = 0$, $j = 1,..,k$, has complex conjugate roots on the imaginary axis. Let $\lambda = \pm i\lambda_h$ be the corresponding value of λ. Then, as was shown in [56], the unstable eigenfunction generically has the form of a *synchronous oscillatory instability* with

$$\nu = \nu_- + \delta e^{i\lambda_h t}\phi + \text{c.c}\,, \quad \phi(x) = \sum_{n=1}^{k} \Phi\left[\varepsilon^{-1}(x - x_n)\right]\,. \tag{5.67}$$

Here c.c denotes complex conjugate and $\delta \ll 1$. In other words, the instability threshold is set by the $j = 1$ mode in (5.61), for which $c_n = 1$ for $n = 1, \ldots, k$. This type of instability is illustrated in Fig. 21.

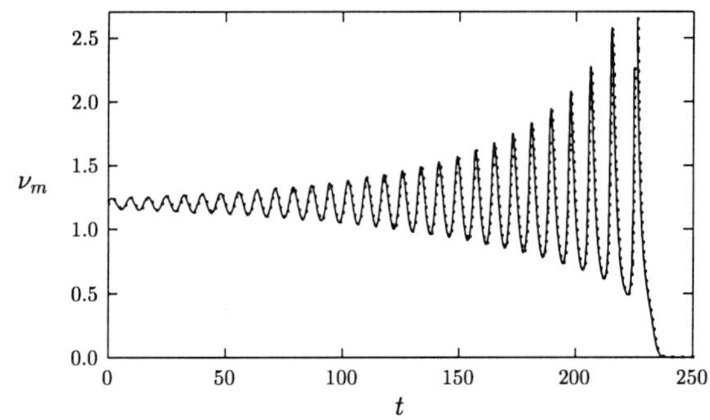

Figure 21. Synchronous Oscillatory Instability: Plot of the spike amplitudes for $k = 3$, $D = 0.75$, $\tilde{A} = 8.86$, $\tau = 7.6$ and $\varepsilon = 0.01$. The amplitudes of the spikes trace out identical trajectories after a short initial transient.

Alternatively, suppose that $\tilde{A}_{ke} < \tilde{A} < \tilde{A}_{kL}$. Then, for any $\tau > 0$, the dominant initial instability was shown in [56] to correspond to the $j = k$ mode in (5.61). This instability has the form

$$v = v_- + \delta e^{\lambda_{Rk} t}\phi\,, \quad \phi(x) = \sum_{n=1}^{k} c_n \Phi\left[\varepsilon^{-1}(x - x_n)\right]\,, \tag{5.68a}$$

where

$$c_n = \cos\left(\frac{\pi(k-1)}{k}(n - 1/2)\right)\,, \quad n = 1, .., k\,. \tag{5.68b}$$

Here $\delta \ll 1$, and $\lambda_{Rk} > 0$ is the unique root of $g_k(\lambda_R) = 0$. Since $\sum_{n=1}^{k} c_n = 0$, this instability locally conserves the sum of the heights of the spikes. Hence, it is referred to as a *competition instability*. The numerical experiments in Sec. 3.3 of [56] show that this instability leads to a spike competition process whereby certain spikes in a spike sequence are ultimately annihilated. This type of instability is illustrated in Fig. 22.

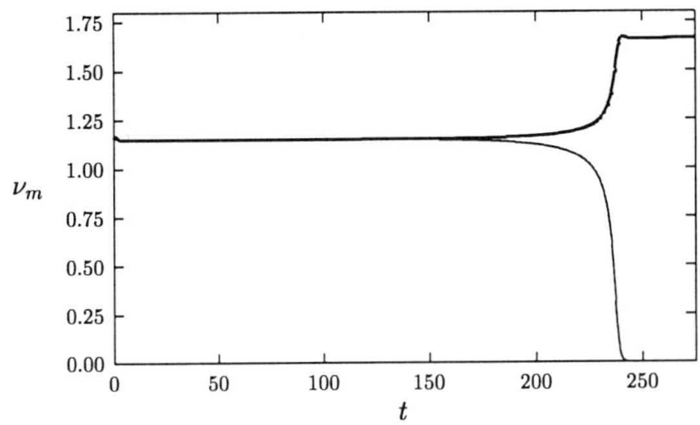

Figure 22. Competition Instability: Plot of th spike amplitudes $\nu_m \equiv \nu(x_j, t)$ versus t for $k = 3$, $D = 0.75$, $\tilde{A} = 8.6$, $\varepsilon = 0.01$, and $\tau = 2.0$. The middle spike is annihilated, and the other two spikes have a common amplitude.

An interesting limiting case of (5.60) is to take the limit where $\tilde{A} \gg 1$, but $\tilde{A} \ll O(\varepsilon^{-1/2})$. This represents the intermediate limit between the GS model in the low feed-rate regime (1.5) and the high feed-rate regime (1.3). This intermediate regime was first studied in the pioneering works of [25]–[27], and [29]. In this intermediate regime there are no longer any competition instabilities, and there is a scaling law for the Hopf bifurcation. The main result as given in [56] is as follows:

Proposition 5.15. *Let $\varepsilon \ll 1$, $D = O(1)$, and $O(1) \ll \tilde{A} \ll O(\varepsilon^{-1/2})$. Then, the k-spike equilibrium small solution u_-, ν_-, is stable with respect to the large eigenvalues when $\tau < \tau_H$, where*

$$\tau_H \sim \frac{\tilde{A}^4 D}{9} \tanh^4(\theta_0/k)\, \tau_{0h} \left(1 - \frac{6\theta_0}{\tilde{A}^2 \tanh(\theta_0/k)}\right)^2 + o(1)\,. \tag{5.69}$$

Here $\tau_{0h} = 1.748$ is the minimum value of τ_0 for which the nonlocal eigenvalue problem

$$\Phi'' - \Phi + 2w\Phi - \frac{2w^2}{1+\sqrt{\tau_0 \lambda}} \left(\frac{\int_{-\infty}^{\infty} w\Phi\, dy}{\int_{-\infty}^{\infty} w\, dy} \right) = \lambda\Phi\,, \quad -\infty < y < \infty\,, \tag{5.70}$$

with $\Phi \to 0$ as $|y| \to \infty$, has a pair of pure imaginary eigenvalues.

This limiting eigenvalue problem was studied [25], [26], and [73]. As for the small eigenvalues, the following result was obtained in Proposition 4.4 of [56]:

Proposition 5.16. *Let $\varepsilon \ll 1$, and $\tau = O(1)$. For $k = 1$, the small solution u_- and ν_- is always stable with respect to the small eigenvalue. For $k > 1$, and for the small solution u_-, ν_-, one small eigenvalue satisfies $\lambda_k < 0$, and so is stable. The other small eigenvalues λ_j, for $j = 1, \ldots, k-1$ are negative at a fixed value of D if and only if $\tilde{A}$ satisfies*

$$\tilde{A} > \tilde{A}_{kS}\,, \qquad \tilde{A}_{kS} \equiv \tilde{A}_{ke} \left[\tanh\left(\frac{2\theta_0}{k} \right) \right]^{-1} . \tag{5.71}$$

Here $\tilde{A}_{ke}$ are the existence thresholds of (5.42).

By comparing (5.71) with (5.47) of Proposition 5.8 it follows that $k-1$ branches of asymmetric equilibria bifurcate from the symmetric k-spike branch when $k-1$ small eigenvalues simultaneously cross through zero.

Competition and synchronous oscillatory instabilities also occur for the GM model (cf. [105]). By linearizing around a symmetric k-spike pattern, the following nonlocal eigenvalue problem was found to determine the stability of the solution on an $O(1)$ time-scale (see Proposition 2.3 of [105]):

Proposition 5.17. *Let $0 < \varepsilon \ll 1$. Then, the $O(1)$ eigenvalues determining the stability of k-spike symmetric equilibria in the GM model (1.1) satisfy the nonlocal eigenvalue problem*

$$L_0\Phi - \chi_{gm} w^p \left(\frac{\int_{-\infty}^{\infty} w^{m-1}\Phi\, dy}{\int_{-\infty}^{\infty} w^m\, dy} \right) = \lambda\Phi\,, \qquad -\infty < y < \infty\,, \tag{5.72}$$

with $\Phi \to 0$ as $|y| \to \infty$. Here $w(y)$ is the homoclinic solution in (1.8), and L_0 is the local operator $L_0\Phi \equiv \Phi'' - \Phi + pw^{p-1}\Phi$. The multiplier χ_{gm} is

$$\chi_{gm} = qm \left(s + \frac{\sqrt{1+z}}{\tanh(\theta_0/k)} \left[\tanh(\theta_\lambda/k) + \frac{(1-\cos[\pi(j-1)/k])}{\sinh(2\theta_\lambda/k)} \right] \right)^{-1}, \tag{5.73}$$

where $z \equiv \tau\lambda$, $\theta_\lambda \equiv \theta_0\sqrt{1+z}$, and $\theta_0 \equiv D^{-1/2}$.

By comparing the eigenvalue problems in Propositions 5.13 and 5.17, an interesting *spectral equivalence principle* is found to hold. Namely, the GS model has the same nonlocal eigenvalue problem as a GM model with exponent set $(2, s_g, 2, s_g)$.

As described above for the GS model, there are two types of instabilities of k-spike patterns with $k > 1$: synchronous oscillatory instabilities and competition instabilities. Competition instabilities, due to real positive eigenvalues in the spectrum of (5.72), will only occur when

$$D > D_k \equiv \frac{4}{k^2}\left(\log\left[a_k + \sqrt{a_k^2 - 1}\right]\right)^{-2} \qquad k = 2, 3, \ldots \,, \tag{5.74}$$

where $a_k \equiv 1 + \left[1 + \cos\left(\frac{\pi}{k}\right)\right]\zeta^{-1}$, and ζ is defined in (1.2). This result was given in [47] and Proposition 2.6 of [105]. Finally, it was shown in Proposition 11 of [47] that there are $k-1$ eigenvalues that simultaneously cross through zero when $D = D_{gm}$, where D_{gm} was the threshold given in Proposition 5.3 for the bifurcation of asymmetric k-spike branches from a symmetric k-spike branch. Many further results on the spectrum of (5.72) are given in [105]. Similar results have been found in [49] for the SC model (1.6).

There are several key open problems related to the stability of k-spike patterns in one spatial dimension. Although, the existence of a Hopf bifurcation value is guaranteed, it appears to be difficult to establish a transver sal crossing condition as was done for the shadow GM model in Sec. 3.2. This leads to the first question.

Question 5.4. Can one give a rigorous proof that there is a unique value of τ where the GM model (1.1) and the GS model (1.5) have a Hopf bifurcation?

The theory described above is based solely on the linearization around a symmetric k-spike patterns. The stability of asymmetric patterns is largely open.

Question 5.5. What are the stability properties of asymmetric spike equilibria for the GM model (1.1) and the GS model (1.5) as a function of $\tau > 0$?.

Question 5.6. Is the Hopf bifurcation subcritical or supercritical? Can one perform a weakly nonlinear theory to determine the local behavior of the bifurcating solutions?

Finally, there are many questions relating to the dynamics of quasi-equilibrium spike patterns for the GM and the GS models. The dynamics of spikes for $\tau = 0$ was studied for the GM model in [48], while a two-spike evolution for the GS model (1.5) was analyzed in [23] and [24]. This leads to the next question.

Question 5.7. Determine the dynamical behavior of k-spike quasi-equilibria for the GM and GS models and the coarsening behavior of these solutions whereby spikes are annihilated through synchronous oscillations or competition instabilities.

In addition to some of the one-dimensional results given here, there have been a few key studies of the existence and stability of k-spike patterns in the two-dimensional GM and GS models (cf. [19], [113], [114], [115], and [116]). However, there are still many open problems in this direction, including studying the dynamics of multi-spike patterns in multi-dimensional domains.

6. Concluding Remarks

This survey has focussed only on some questions and results related to scalar quasilinear problems and to reaction-diffusion systems with semi-strong spike interactions. For each of our problems involving semi-strong interactions, one of the components of the reaction can be approximated by the solution to either (2.2) or (2.47), while the other component varies on the length scale of the domain.

Alternatively, there have been several studies of spike behavior for systems having weak spike interactions. These systems are characterized by a localization of both components of the reaction. For the Gray-Scott model (1.3) with weak spike interactions, where $D = O(\varepsilon^2)$, spike-replication and chaotic behavior was analyzed in the pioneering studies of [82], [83], [98], and [81]. The initial numerical study of [86] was in the weak interaction regime. For the GM model (1.1) with $D = O(\varepsilon^2)$, the spikes also exhibit self-replication behavior (cf. [30], [95]), and multi-bump phenomena (cf. [20]). Exponentially weak interactions of localized structures were studied for general classes of systems in [32], [33], and [90].

Acknowledgments

I am very grateful to Prof. Chen Hua of Wuhan University and Prof. Roderick Wong of the City University of Hong Kong for inviting me to participate in the meeting in Wuhan, China, in October 2003, and for inviting me to write this survey. It is my great pleasure to acknowledge the contributions of my collaborator Juncheng Wei and my former graduate students David Iron, Theodore Kolokolnikov, and Xiaodi Sun to most of the work described herein. This work was supported by NSERC grant 81541.

References

1. S. Ai, J . Math. Anal. Appl., **277**, No. 2, (2003), pp. 405–422.

2. C. Bandle, M. Flucher, SIAM Rev., **38**, No. 2, (1996), pp. 191–238.
3. P. Bates, N. Fusco, J. Differential Equations, **160**, No. 2, (2000), pp. 283–356.
4. P. Bates, J. Shi, J. Funct. Anal., **196**, No. 2, (2002), pp. 211–264.
5. G. Bellettini, G. Fusco, Asymptot. Anal. **26**, No. 3-4, (2001), pp. 307–357.
6. G. Bellettini, G. Fusco, Asymptot. Anal., **33**, No. 1, (2003), pp. 9–50.
7. H. Berestycki, S. Kamin, G. Sivashinsky, Interfaces Free Bound. **3**, No. 4, (2001), pp. 361–392.
8. H. Berestycki, J. Wei, Ann. Sc. Norm. Super. Pisa Cl. Sci (5), **2**, No. 1, (2003), pp. 199–230.
9. A. Bose, G. Kriegsmann, Methods Appl. Anal., **5**, No. 4, (1998), pp. 351–366.
10. A. Bose, G. Kriegsmann, Methods Appl. Anal., **7**, No. 2, (2000), pp. 295–311.
11. A. Bose, SIAM J. Math. Anal., **31**, No. 2, (2000), pp. 431–454.
12. L. A. Caffarelli, A. Friedman, Duke Math. J., **52**, No. 2, (1985), pp. 431–456.
13. G. Carrier, C. Pearson, *Ordinary Differential Equations*, Blaisdell Publishing Co., Waltham, MA, (1968). (Reprinted in SIAM's *Classics in Applied Mathematics*, series, Vol. **6**, SIAM, Philadelphia, (1991)).
14. X. Chen, M. Kowalczyk, Adv. Differential Equations, **6**, No. 7, (2001), pp. 847–872.
15. X. Chen, M. Kowalczyk, SIAM. J. Math. Anal., **33**, No. 1, (2001), pp. 172–193.
16. E. N. Dancer, Methods Appl. Anal., **8**, No. 2, (2001), pp. 245–256.
17. P. De Groen, G. Karadzhov, Electron. J. Differential Equations, No. 50, (2002), pp. 1–22.
18. M. Del Pino, P. Felmer, M. Kowalczyk, Commun. Pure Appl. Anal., **1**, No. 4, (2002), pp. 437–456.
19. M. Del Pino, M. Kowalczyk, J. Wei, Ann. Inst. H. Poincare, Anal. Non Lineaire, **20**, No. 1, (2003), pp. 53–85.
20. M. Del Pino, M. Kowalczyk, X. Chen, Commun. Contemp. Math. **3**, No. 3, (2001), pp. 419–439.
21. M. Del Pino, P. Felmer, Indiana Univ. Math. J., **48**, No. 3, (1999), pp. 883–898.
22. M. Del Pino, P. Felmer, K. Tanaka, Nonlinearity, **15**, No. 5, (2002), pp. 1653–1671.
23. A. Doelman, W. Eckhaus, T. J. Kaper, SIAM J. Appl. Math., **61**, No. 3, (2000), pp. 1080–1102.
24. A. Doelman, W. Eckhaus, T. J. Kaper, SIAM J. Appl. Math., **61**, No. 6, (2000), pp. 2036–2061.
25. A. Doelman, R. A. Gardner, T. J. Kaper, Phys. D, **122**, No. 1-4, (1998), pp. 1–36.
26. A. Doelman, R. A. Gardner, T. Kaper, Mem. Amer. Math. Soc., **155**, No. 737, (2002), xii+64 pp.
27. A. Doelman, R. A. Gardner, T. Kaper, Indiana U. Math. Journ., **50**, No. 1, (2001), pp. 443-507.
28. A. Doelman, T. Kaper, H. van der Ploeg, Methods Appl. Anal., **8**, No. 3, (2001), pp. 387–414.
29. A. Doelman, T. J. Kaper, P. Zegeling, Nonlinearity, **10**, No. 2, (1997),

pp. 523–563.
30. A. Doelman, H. van der Ploeg, SIAM J. Appl. Dyn. Syst., **1**, No. 1, (2002), pp. 65–104.
31. G. Domokos, P. Holmes, R. Soc. Lond. Proc. Ser. A. Math. Phys. Eng. Sci., **459**, No. 2034, (2003), pp. 1535–1561.
32. S. Ei, Y. Nishiura, K. Ueda, Japan J. Indust. Appl. Math., **18**, No. 2, (2001), pp. 181–205.
33. S. Ei, J. Dynam. Differential Equations, **14**, No. 1, (2002), pp. 85–137.
34. S. Ei, J. Wei, Japan J. Indust. Appl. Math., **19**, No. 2, (2002), pp. 181–226.
35. P. Freitas, Proc. Roy. Soc. Edinburgh Sect. A, **124**, No. 1, (1994), pp. 169–188.
36. A. Gierer, H. Meinhardt, Kybernetik, **12**, (1972), pp. 30–39.
37. P. Gray, S. K. Scott, Chem. Eng. Sci. **39**, (1984), pp. 1087–1097.
38. C. Gui, Commun. in Part. Diff. Eq., **21**, No. 5-6, (1996), pp. 787–820.
39. C. Gui, Duke Math. J., **84**, No. 3, (1996), pp. 739–769.
40. C. Gui, J. Wei, J. Differential Equations, **158**, No. 1, (1999), pp. 1–27.
41. C. Gui, J. Wei, Canad. J. Math., **52**, No. 3, (2000), pp. 522–538.
42. C. Gui, J. Wei, M. Winter, Ann. Inst. H. Poincare Anal. Non Lineaire, **17**, No. 1, (2000), pp. 47–82.
43. B. Gustafsson, Duke Math. J., **60**, No. 2, (1990), pp. 303–311.
44. D. Iron, M. J. Ward, SIAM J. Appl. Math., **60**, No. 3, (2000), pp. 778–802.
45. D. Iron, M. J. Ward, European. J. Appl. Math., **11**, No. 5, (2000), pp. 491–514.
46. D. Iron, M. J. Ward, Math. Comput. Simulation, **55**, No. 4-6, (2001), pp. 419–431.
47. D. Iron, M. J. Ward, J. Wei, Phys. D, **150**, No. 1-2, (2001), pp. 25–62.
48. D. Iron, M. J. Ward, SIAM J. Appl. Math., **62**, No. 6, (2002), pp. 1924–1951.
49. D. Iron, J. Wei, M. Winter, to appear, J. Math. Biology, (2004).
50. D. Iron, M. J. Ward, Analysis and Applications, **2**, No. 1, (2004), pp. 1–50.
51. W. Kath, C. Knessl, B. Matkowsky, Stud. Appl. Math., **77**, No. 1, (1987), pp. 61–88.
52. W. Kelley, J. Math. Anal. Appl., **255**, No. 2, (2001), pp. 678–697.
53. W. Kelley, IMA J. Appl. Math., **60**, No. 1, (1998), pp. 91–108.
54. T. Kolokolnikov, M. J. Ward, European J. Appl. Math., **14**, No. 5, (2003), pp. 513–545.
55. T. Kolokolnikov, M. J. Ward, to appear, Discrete Contin. Dyn. Syst. Ser. B., (2004).
56. T. Kolokolnikov, M. Ward, J. Wei, submitted, Mem. Amer. Math. Soc., (2003).
57. T. Kolokolnikov, M. Ward, J. Wei, submitted, Phys. D, (2003).
58. T. Kolokolnikov, J. Wei, submitted, European J. Appl. Math., (2003).
59. M. Kowalczyk, Duke Math J., **98**, No. 1, (1999), pp. 59–111.
60. M. Kowalczyk, J. Differential Equations, **138**, No. 1, (1997), pp. 55–85.
61. E. Knobloch, Outstanding Problems in the Theory of Pattern Formation, book chapter in *Nonlinear Dynamics and Chaos. Where Do We Go From Here?*, edited by S. J. Hogan et al., Institute of Physics Publishing, Bristol,

U.K. (2003).
62. G. Kriegsmann, IMA J. Appl. Math. **59**, No. 2, (1997), pp. 123–148.
63. G. Kriegsmann, IMA J. Appl. Math., **66**, No. 1, (2001), pp. 1–32.
64. M. Kuwamura, S. Ei, M. Mimura, Japan J. Indust. Appl. Math., **9**, No. 1, (1992), pp. 35–77.
65. M. Kwong, Arch. Rational Mech. Anal., **105**, No. 3, (1989), pp. 243–266.
66. C. Lange, Stud. Appl. Math., **68**, No. 3, (1983), pp. 227–257.
67. J. Y. Lee, M. J. Ward, Stud. Appl. Math., **94**, No. 3, (1995), pp. 271–326.
68. C. S. Lin, W. M. Ni, I. Takagi, J. Diff. Eq., **72**, (1988), pp. 1-27.
69. A. D. MacGillivray, R. Braun, G. Tanouglu, Stud. Appl. Math., **104**, No. 4, (2000), pp. 293–311.
70. H. Meinhardt, *Models of Biological Pattern Formation*, Academic Press, London, (1982).
71. H. Meinhardt, *The Algorithmic Beauty of Sea Shells*, Springer-Verlag, Berlin, (1995).
72. D. Morgan, T. Kaper, submitted, Phys. D, (2003).
73. C. Muratov, V. V. Osipov, SIAM J. Appl. Math., **62**, No. 5, (2002), pp. 1463–1487.
74. C. Muratov, V. V. Osipov, J. Phys. A: Math Gen. **33**, (2000), pp. 8893–8916.
75. K. Nagasaki, T. Suzuki, Asymptotic Anal., **3**, No. 2, (1990), pp. 173–188.
76. W. M. Ni, I. Takagi, Comm. Pure Appl. Math., **44**, No. 7, (1991), pp. 819–851.
77. W. M. Ni, I. Takagi, Duke Math. J., **70**, No. 2, (1993), pp. 247–281.
78. W. M. Ni, Notices Amer. Math. Soc., **45**, No. 1, (1998), pp. 9–18.
79. W. M. Ni, J. Wei, Comm. Pure. Appl. Math., **48**, No. 7, (1995), pp. 731–768.
80. Y. Nishiura, in Dynamics Reported: Expositions in Dynamical Systems Volume 3 (editors: C. K. R. T. Jones, U. Kirchgraber), Springer-Verlag, New York, (1995).
81. Y. Nishiura, Methods Appl. Anal., **8**, No. 2, (2001), pp. 321–332.
82. Y. Nishiura, D. Ueyama, Phys. D, **130**, No. 1-2, (1999), pp. 73–104.
83. Y. Nishiura, D. Ueyama, Phys. D, **150**, No. 3-4, (2001), pp. 137–162.
84. R. O'Malley Jr., J. Math. Anal. Appl., **54**, No. 2, (1976), pp. 449–466.
85. C.H. Ou, R. Wong, Stud. Appl. Math., **111**, No. 4, (2003), pp. 377–408.
86. J. E. Pearson, Science, **216**, (1993), pp. 189–192.
87. L. Reyna, M. J Ward, European J. Appl. Math., **5**, No. 4, (1994), pp. 495-523.
88. S. Rosenbalt, R. Szeto, Stud. Appl. Math., **63**, No. 2, (1980), pp. 99–117.
89. B. Sandstede, Trans. Amer. Math. Soc., **350**, No. 2, (1998), pp. 429–472.
90. J. Schnakenberg, J. Theoret. Biology, **81**, (1979), pp. 389–400.
91. B. Sleeman, M. J. Ward, J. Wei, submitted, SIAM J. Appl. Math., (2004).
92. X. Sun, M. J. Ward, European J. Appl. Math., **10**, No. 1, (1999), pp. 27–53.
93. X. Sun, M. J. Ward, Methods Appl. Anal., **8**, No. 2, (2001), pp. 333–347.
94. X. Sun, IMA J. Numer. Anal., **21**, No. 4, (2001), pp. 817–842.
95. W. Sun, T. Tang, M. J. Ward, J. Wei, Stud. Appl. Math., **111**, No. 1, (2003), pp. 41–84.
96. I. Takagi, J. Differential Equations, **61**, No. 2, (1986), pp. 208–249.
97. A. Turing, Phil. Trans. Roy. Soc. B, **237**, (1952), pp. 37–72.

98. D. Ueyama, Hokkaido Math J., **28**, No. 1, (1999), pp. 175–210.
99. Z. Wang, J. Differential Equations, **159**, No. 1, (1999), pp. 102-137.
100. M. J. Ward, Stud. Appl. Math., **87**, No. 2, (1992), pp. 95–135.
101. M. J. Ward, Stud. Appl. Math., **97**, No. 2, (1996), pp. 103-126.
102. M. J. Ward, book chapter in *Analyzing Multiscale Phenomena Using Singular Perturbation Methods*, (J. Cronin, R. O'Malley editors), Proceedings of Symposia in Applied Mathematics, Vol. 56, AMS Short Course (1998), pp. 151–184.
103. M. J. Ward, D. McInerney, P. Houston, D. Gavaghan, P. Maini, SIAM J. Appl. Math., **62**, No. 4, (2002), pp. 1297–1328.
104. M. J. Ward, J. Wei, European J. Appl. Math., **13**, No. 3, (2002), pp. 283–320.
105. M. J. Ward, J. Wei, J. Nonlinear Science, **13**, No. 2, (2003), pp. 209–264.
106. M. J. Ward, J. Wei, to appear, European J. Appl. Math., (2004).
107. M. J. Ward, J. Wei, Stud. Appl. Math., **109**, No. 3, (2002), pp. 229–264.
108. J. Wei, European J. Appl. Math., **10**, No. 4, (1999), pp. 353–378.
109. J. Wei, J. Differential Equations, **134**, No. 1, (1997), pp. 104–133.
110. J. Wei, Tohoku Math. J., **50**, No. 2, (1998), pp. 159–178.
111. J. Wei, Proc. Roy. Soc. Edinburgh Sect. A, **131**, No. 6, (2001), pp. 1457–1480.
112. J. Wei, Phys. D. **148**, No. 1-2, (2001), pp. 20-48.
113. J. Wei, M. Winter, J. Differential Equations, **178**, No. 2, (2002), pp. 478–518.
114. J. Wei, M. Winter, J. Nonlinear Science, **11**, No. 6, (2001), pp. 415–458.
115. J. Wei, M. Winter, Internat. J. Bifur. Chaos Appl. Sci. Engrg., **13**, No. 6, (2003), pp. 1529-1543.
116. J. Wei, M. Winter, Stud. Appl. Math., **110**, No. 1, (2003), pp. 63-102.
117. J. Wei, M. Winter, Phys. D, **176**, No. 3-4, (2003), pp. 147–180.

FIVE LECTURES ON ASYMPTOTIC THEORY

RODERICK S C WONG

Faculty of Science and Engineering
City University of Hong Kong
83 Tat Chee Avenue, Kowloon, Hong Kong
E-mail:mawong@cityu.edu.hk

LECTURE I

STATIONARY PHASE APPROXIMATION

> Kelvin invented his method of stationary phase as a tool for approximating the solution of this particular problem, and it is indeed a beautiful and strikingly successful example of its usefulness.
>
> J. J. Stoker

1. Introduction

According to Watson [13, p.229-230], the principle of stationary phase was first introduced explicitly by Lord Kelvin in 1887, although the essence of this principle can be found in the earlier works of Cauchy, Stokes and Riemann. Kelvin's original problem was to find the asymptotic behavior of the integral

$$u(x) = \frac{1}{2\pi} \int_0^\infty \cos\{m[x - tf(m)]\} dm, \qquad \text{as } x \to \infty,$$

where all variables are real. However, it is now more common to state the principle in terms of the integral

$$I(\lambda) = \int_a^b g(t) e^{i\lambda f(t)} dt, \tag{1.1}$$

where a, b and the function $f(t)$ are real, and λ is a large real parameter.

The underlying principle of stationary phase is the assertion that the major asymptotic contribution to the integral (1.1) comes from points where the *phase function* $f(t)$ is *stationary*, i.e., where $f'(t)$ vanishes. The function $g(t)$ is known as the *amplitude function* in wave propagation theory. If $f(t)$ has only one stationary point, say c, in (a, b), then the principle states that asymptotically we have

$$I(\lambda) \sim g(c)\sqrt{\frac{2\pi}{\lambda|f''(c)|}}e^{i[\lambda f(c)+\sigma\pi/4]}, \tag{1.2}$$

as $\lambda \to +\infty$, where $\sigma = \operatorname{sgn} f''(c)$. A heuristic argument for (1.2) is given in Sec. 2, followed by a rigorous proof based on the Abel summability method introduced by Olver [8] in 1974.

In Sec. 3, we consider a case in which the phase function $f(x)$ involves an auxiliary parameter α, and present an asymptotic formula which holds uniformly when two stationary points of $f(x)$ coalesce as α tends to a critical value, say $\alpha = \alpha_0$.

The two-dimensional generalization of (1.1) is the double integral

$$I(\lambda) = \iint_D g(x, y)e^{i\lambda f(x,y)}dxdy, \tag{1.3}$$

where D is a bounded domain, $f(x, y)$ is a real-valued function, and λ is a large positive parameter. If $f(x, y)$ has only one stationary point, say (x_0, y_0), then the analogue of (1.2) is

$$\begin{aligned} I(\lambda) \sim & \frac{2\pi}{\lambda}|\det f''(x_0, y_0)|^{-\frac{1}{2}}g(x_0, y_0) \\ & \times \exp\left\{i\lambda f(x_0, y_0) + i\frac{\pi}{4}\sigma_f(x_0, y_0)\right\}, \end{aligned} \tag{1.4}$$

where

$$\det f''(x_0, y_0) = f_{xx}(x_0, y_0)f_{yy}(x_0, y_0) - f_{xy}^2(x_0, y_0) \tag{1.5}$$

and

$$\sigma_f(x_0, y_0) = \begin{cases} 2 & \text{if } \det f''(x_0, y_0) > 0 \text{ and } f_{xx}(x_0, y_0) > 0, \\ -2 & \text{if } \det f''(x_0, y_0) > 0 \text{ and } f_{xx}(x_0, y_0) < 0, \\ 0 & \text{if } \det f''(x_0, y_0) < 0. \end{cases} \tag{1.6}$$

We will again first give a heuristic derivation of (1.4) in Sec. 4, and then a rigorous demonstration based on the method of resolution of multiple integrals used by Jones and Kline [6].

The final section (Sec. 5) contains a two-dimensional generalization of the problem discussed in Sec. 3; that is, a problem in which the phase function $f(x, y)$ in the double integral (1.3) involves a parameter α, and has two simple stationary points in the integration domain D, which coalesce as α approaches a critical value α_0. An asymptotic formula is given, which holds uniformly in a neighborhood of α_0.

2. Stationary-Phase Approximation

Returning to the integral (1.1), we suppose that $f(c)$ is a simple minimum in the interval (a, b), i.e., $f'(c) = 0$ and $f''(c) > 0$. Then, according to Kelvin's principle, we have heuristically

$$\begin{aligned} I(\lambda) &\sim \int_{c-\varepsilon}^{c+\varepsilon} g(t) e^{i\lambda f(t)} dt \qquad (\varepsilon > 0) \\ &\sim g(c) \int_{c-\varepsilon}^{c+\varepsilon} e^{i\lambda[f(c)+f''(c)(t-c)^2/2]} dt \\ &\sim g(c) e^{i\lambda f(c)} \int_{-\infty}^{\infty} e^{i\lambda f''(c)(t-c)^2/2} dt \\ &\sim g(c) \sqrt{\frac{2\pi}{\lambda f''(c)}} e^{i[\lambda f(c)+\pi/4]}. \end{aligned} \tag{2.1}$$

Similarly, if $f(c)$ is a maximum, i.e., $f'(c) = 0$ and $f''(c) < 0$, then

$$I(\lambda) \sim g(c) \sqrt{\frac{2\pi}{-\lambda f''(c)}} e^{i[\lambda f(c)-\pi/4]}. \tag{2.2}$$

By coupling (2.1) and (2.2), one obtains the stationary-phase approximation (1.2)

To establish (2.1) rigorously, we subdivide the integration interval (a, b) at the stationary point $t = c$, so that

$$I(\lambda) = \left(\int_a^c + \int_c^b \right) g(t) e^{i\lambda f(t)} dt := I_1(\lambda) + I_2(\lambda). \tag{2.3}$$

For the moment, we consider only the second integral in which we introduce the variable u defined by

$$u := f(t) - f(c) = \sum_{s=0}^{\infty} f_s (t-c)^{s+2}, \tag{2.4}$$

and put

$$\rho = f(b) - f(c). \tag{2.5}$$

For convenience, we shall assume that ρ is finite. The treatment of the case in which ρ is infinite is actually simpler. In terms of the new variable, the integral $I_2(\lambda)$ becomes

$$I_2(\lambda) = e^{i\lambda f(c)} \int_0^{\rho} \varphi(u) e^{i\lambda u} du, \tag{2.6}$$

where

$$\varphi(u) = g(t)/f'(t). \tag{2.7}$$

For small $u, t-c$ and $\varphi(u)$ can be expanded in series of the form

$$t - c = \sum_{s=1}^{\infty} c_s u^{s/2}, \qquad \varphi(u) = \sum_{s=0}^{\infty} a_s u^{(s-1)/2}. \tag{2.8}$$

Since $f_0 = f''(c)/2$, it is readily verified that

$$a_0 = \frac{g(c)}{[2f''(c)]^{1/2}}. \tag{2.9}$$

For each $n \geq 0$, define $\varphi_n(u)$ by

$$\varphi(u) = \sum_{s=0}^{n-1} a_s u^{(s-1)/2} + \varphi_n(u). \tag{2.10}$$

Clearly,

$$\varphi_n(u) \sim a_n u^{(n-1)/2} + a_{n+1} u^{n/2} + \cdots, \qquad u \to 0^+, \tag{2.11}$$

and

$$\varphi_n^{(j)}(u) = \varphi^{(j)}(u) - \sum_{s=0}^{n-1} \frac{\Gamma[(s+1)/2]}{\Gamma[(s+1-2j)/2]} a_s u^{(s-1-2j)/2}, \tag{2.12}$$

$j = 0, 1, 2, \cdots$.

Now we quote a few results from the Abel summability theory. First, for any piecewise continuous function ψ in $(0, \infty)$, we have

$$\lim_{\varepsilon \to 0+} \int_0^{\infty} e^{-\varepsilon u} \psi(u) du = \int_0^{\infty} \psi(u) du, \tag{2.13}$$

if the integral on the right exists as an improper Riemann integral; see [12, p.26]. Second, for $\lambda > 0$ and Re $\alpha > 0$,

$$\lim_{\varepsilon \to 0+} \int_0^{\infty} e^{-\varepsilon u} e^{i\lambda u} u^{\alpha - 1} du = \frac{e^{\alpha\pi i/2} \Gamma(\alpha)}{\lambda^{\alpha}}. \tag{2.14}$$

Since the integral on the left-hand side of (2.14) does not exist when we set $\varepsilon = 0$, this result does not follow from (2.13). However, it can be proved directly, by expressing the integral in terms of the gamma function and then using the continuity of this function. Also, when $\lambda > 0$ and $\rho > 0$,

$$\lim_{\varepsilon\to 0+} \int_\rho^\infty e^{-\varepsilon u} e^{i\lambda u} u^{\alpha-1} du = \frac{e^{\alpha\pi i/2}}{\lambda^\alpha} \Gamma(\alpha, -i\lambda\rho), \tag{2.15}$$

where $\Gamma(\alpha, z)$ is the incomplete gamma function.

Substituting (2.10) into (2.6) and integrating term by term in the sense of Abel summability (2.13), we obtain from (2.13) and (2.14)

$$\int_0^\rho \varphi(u) e^{i\lambda u} du = \sum_{s=0}^{n-1} a_s \Gamma\left(\frac{s+1}{2}\right) \frac{e^{i(s+1)\pi/4}}{\lambda^{(s+1)/2}} + E_n(\lambda) - F_n(\lambda), \tag{2.16}$$

where

$$E_n(\lambda) = \lim_{\varepsilon\to 0+} \int_0^\rho e^{-(\varepsilon - i\lambda)u} \varphi_n(u) du \tag{2.17}$$

and

$$F_n(\lambda) = \sum_{s=0}^{n-1} a_s \lim_{\varepsilon\to 0} \int_\rho^\infty e^{-(\varepsilon - i\lambda)u} u^{(s-1)/2} du. \tag{2.18}$$

By repeated integration by parts, we have

$$\begin{aligned} E_n(\lambda) = \left(\frac{i}{\lambda}\right)^m \varphi_n^{(m-1)}(0) - e^{i\lambda\rho} \sum_{j=0}^{m-1} \varphi_n^{(j)}(\rho) \left(\frac{i}{\lambda}\right)^{j+1} \\ + \left(\frac{i}{\lambda}\right)^m \int_0^\rho e^{i\lambda u} \varphi_n^{(m)}(u) du, \end{aligned} \tag{2.19}$$

where $m = [(n+1)/2]$; i.e., $m = n/2$ if n is even, and $m = (n+1)/2$ if n is odd. Note that $\varphi_n^{(m-1)}(0) = 0$, when n is even.

To deal with $F_n(\lambda)$, we use (2.15) and obtain

$$F_n(\lambda) = \sum_{s=0}^{n-1} a_s \frac{e^{i(s+1)\pi/4}}{\lambda^{(s+1)/2}} \Gamma\left(\frac{s+1}{2}, -i\lambda\rho\right). \tag{2.20}$$

From the well known integral representation

$$\Gamma(\alpha, z) = \int_z^\infty e^{-t} t^{\alpha-1} dt, \tag{2.21}$$

we have by repeated integration by parts

$$\Gamma(\alpha, z) = e^{-z} z^{\alpha-1} \sum_{j=0}^{m-1} \frac{\Gamma(\alpha)}{\Gamma(\alpha - j)} \frac{1}{z^j} + \frac{\Gamma(\alpha)}{\Gamma(\alpha - m)} \Gamma(\alpha - m, z). \tag{2.22}$$

Hence

$$F_n(\lambda) = e^{i\lambda\rho}\sum_{s=0}^{n-1}\sum_{j=0}^{m-1} a_s\frac{\Gamma[(s+1)/2]}{\Gamma[(s+1-2j)/2]}\left(\frac{i}{\lambda}\right)^{j+1}\rho^{(s-1-2j)/2}$$
$$+\sum_{s=0}^{n-1} a_s\frac{e^{i(s+1)\pi/4}}{\lambda^{(s+1)/2}}\frac{\Gamma[(s+1)/2]}{\Gamma[(s+1-2m)/2]} \tag{2.23}$$
$$\times\Gamma\left(\frac{s+1-2m}{2}, -i\lambda\rho\right).$$

Subtracting (2.23) from (2.19), and making use of (2.12), we arrive at

$$E_n(\lambda) - F_n(\lambda) = -e^{i\lambda\rho}\sum_{j=0}^{m-1}\varphi^{(j)}(\rho)\left(\frac{i}{\lambda}\right)^{j+1} + \left(\frac{i}{\lambda}\right)^{m}\varphi_n^{(m-1)}(0)$$
$$+\{\delta_n(\lambda) - \varepsilon_n(\lambda)\}, \tag{2.24}$$

where

$$\delta_n(\lambda) = \left(\frac{i}{\lambda}\right)^{m}\int_0^{\rho} e^{i\lambda u}\varphi_n^{(m)}(u)du \tag{2.25}$$

and

$$\varepsilon_n(\lambda) = \sum_{s=0}^{n-1} e^{i(s+1)\pi/4}\frac{\Gamma[(s+1)/2]}{\Gamma[(s+1-2m)/2]}$$
$$\times\Gamma\left(\frac{s+1-2m}{2}, -i\lambda\rho\right)\frac{a_s}{\lambda^{(s+1)/2}}. \tag{2.26}$$

Recall that $\varphi_n^{(m-1)}(0) = 0$ when n is even. When n is odd, $m = (n+1)/2$ and we have $\varphi_n^{(m-1)}(0) = (m-1)!\, a_n$. Thus, by coupling (2.16) with (2.24), and with $m = [(n+1)/2]$, we can write

$$\int_0^{\rho}\varphi(u)e^{i\lambda u}du = \sum_{s=0}^{n-\nu} a_s\Gamma\left(\frac{s+1}{2}\right)\frac{e^{i(s+1)\pi/4}}{\lambda^{(s+1)/2}}$$
$$-e^{i\lambda\rho}\sum_{j=0}^{m-1}\varphi^{(j)}(\rho)\left(\frac{i}{\lambda}\right)^{j+1} + \{\delta_n(\lambda) - \varepsilon_n(\lambda)\}, \tag{2.27}$$

where $\nu = 0$ when n is odd (i.e., $2m = n+1$), and $\nu = 1$ when n is even (i.e., $2m = n$).

To estimate the error term in (2.27), we first have from (2.25)

$$|\delta_n(\lambda)| \le \frac{1}{\lambda^m}\int_0^{\rho}|\varphi_n^{(m)}(u)|du. \tag{2.28}$$

By deforming the integration path in (2.21) until it lies on the imaginary axis, it can be shown upon a partial integration that

$$|\Gamma(\alpha, \pm iy)| \leq 2y^{\alpha-1}, \qquad \alpha < 1, \qquad y > 0. \tag{2.29}$$

Hence we also have

$$|\varepsilon_n(\lambda)| \leq \frac{2}{\lambda^{m+1}} \sum_{s=0}^{n-1} |a_s| \frac{\Gamma[(s+1)/2]}{|\Gamma[(s+1-2m)/2]|} \rho^{-(2m-1-s)/2}. \tag{2.30}$$

The two estimates in (2.28) and (2.30) clearly establish the asymptotic nature of the expansion in (2.27). Taking just the first term, we have from (2.6) and (2.9)

$$I_2(\lambda) \sim g(c)\sqrt{\frac{\pi}{2f''(c)}}e^{i[\lambda f(c)+\pi/4]}. \tag{2.31}$$

The integral $I_1(\lambda)$ in (2.3) can be dealt with in exactly the same manner, after first changing the variable of integration from t to $-t$. The final result is that $I_1(\lambda)$ has the same asymptotic behavior as $I_2(\lambda)$ given in (2.31). The stationary-phase formula (2.1) now follows from (2.3).

3. Coalescing Stationary Points

We now consider a case in which the phase function $f(x)$ in the integral involves an auxiliary parameter α, say $f(t) = f(t, \alpha)$. We assume that $f(t, \alpha)$ has two local extrema in (a, b), say a maximum at $t = t_+(\alpha)$ and a minimum at $t = t_-(\alpha)$, and that these two extrema approach each other at $t = \bar{t} \in (a, b)$ when α tends to a critical value, say $\alpha = \alpha_0$. Thus,

$$f'(t_+, \alpha) = f'(t_-, \alpha) = 0, \qquad f''(t_\pm, \alpha) \neq 0 \tag{3.1}$$

for $\alpha \neq \alpha_0$, and

$$f'(\bar{t}, \alpha_0) = f''(\bar{t}, \alpha_0) = 0, \qquad f'''(\bar{t}, \alpha_0) \neq 0. \tag{3.2}$$

According to the approximation formulas in (2.1) and (2.2), the asymptotic behavior of $I(\lambda) = I(\lambda, \alpha)$ for large λ is given by

$$\begin{aligned} I(\lambda, \alpha) \sim\ & g(t_-)\sqrt{\frac{2\pi}{\lambda f''(t_-)}}e^{i[\lambda f(t_-)+\pi/4]} \\ & + g(t_+)\sqrt{\frac{2\pi}{-\lambda f''(t_+)}}e^{i[\lambda f(t_+)-\pi/4]}. \end{aligned} \tag{3.3}$$

In view of (3.2), the two terms on the right-hand side will blow up when α approaches α_0. Our aim here is to show that there is an asymptotic formula which holds uniformly for α in a neighborhood of α_0. Since the simplest function which exhibits the situation described in (3.1) – (3.2) is a cubic polynomial, Chester, Friedman and Ursell [3] introduced in their seminal paper the cubic transformation

$$f(t,\alpha) = \frac{1}{3}u^3 - \zeta u + \eta. \tag{3.4}$$

Furthermore, they showed that when $f(t,\alpha)$ is an analytic function of t and α, and when the coefficients ζ and η are chosen so that

$$\zeta^{3/2} = \frac{3}{4}[f(t_-,\alpha) - f(t_+,\alpha)], \tag{3.5}$$

$$\eta = \frac{1}{2}[f(t_-,\alpha) + f(t_+,\alpha)], \tag{3.6}$$

the cubic equation (3.4) has a real-valued solution $u(t,\alpha)$ such that $\partial u/\partial t \neq 0$ at $(\bar{t},\alpha_0)$. An extension of this result to C^∞ case can be found in the book of Hörmander [5, p. 204, Theorem 7.5.13].

From (2.27), it is readily seen that the contribution from an endpoint (in our case, it is $t = b$, i.e., $u = \rho$) is of order $O(\lambda^{-1})$, whereas the contribution from the stationary point (in our case, it is $t = c$, i.e., $u = 0$) is of order $O(\lambda^{-1/2})$. Thus, it is reasonable to expect that to the first approximation, the contribution from the end points can be neglected. We may therefore assume without loss of generality that the support of the amplitude function $g(t)$ in (1.1) is contained in the interval where the cubic transformation (3.4) holds and that it contains the two local extrema t_+ and t_-. Hence, upon changing the integration variable from t to u, we have

$$I(\lambda,\alpha) = \int_{-\infty}^{\infty} \varphi_0(u) e^{i\lambda(u^3/3-\zeta u+\eta)}\,du, \tag{3.7}$$

where

$$\varphi_0(u) = g(t)\frac{dt}{du} \tag{3.8}$$

vanishes smoothly and identically outside a finite u-interval. An asymptotic approximation of $I(\lambda,\alpha)$ can be obtained as follows. We write

$$\varphi_0(u) = a_0 + b_0 u + (u^2 - \zeta)\psi_0(u), \tag{3.9}$$

where

$$a_0 = \frac{1}{2}[\varphi_0(\zeta^{1/2}) + \varphi_0(-\zeta^{1/2})] \tag{3.10}$$

and

$$b_0 = \frac{1}{2\zeta^{1/2}}[\varphi_0(\zeta^{1/2}) - \varphi_0(-\zeta^{1/2})]; \tag{3.11}$$

see Wong [14, p. 368]. Inserting (3.9) in (3.7) yields

$$\begin{aligned} I(\lambda,\alpha) = 2\pi e^{i\lambda\eta}\Bigg[\mathrm{Ai}(-\lambda^{2/3}\zeta)\frac{a_0}{\lambda^{1/3}} \\ - i\mathrm{Ai}'(-\lambda^{2/3}\zeta)\frac{b_0}{\lambda^{2/3}} + \varepsilon_1(\lambda,\alpha)\Bigg], \end{aligned} \tag{3.12}$$

where

$$\varepsilon_1(\lambda,\alpha) = \frac{1}{2\pi}\int_{-\infty}^{\infty}(u^2-\zeta)\psi_0(u)e^{i\lambda(u^3/3-\zeta u)}du. \tag{3.13}$$

In (3.12), we have made use of the integral representation

$$\mathrm{Ai}(x) = \frac{1}{2\pi}\int_{-\infty}^{\infty}\exp\left\{i\left(\frac{1}{3}t^3 + xt\right)\right\}dt. \tag{3.14}$$

From (3.10) and (3.11), it is clear that

$$\lim_{\sqrt{\zeta}\to 0} a_0 = \varphi_0(0) \qquad \text{and} \qquad \lim_{\sqrt{\zeta}\to 0} b_0 = \varphi_0'(0). \tag{3.15}$$

With a_0 and b_0 so chosen, it is also easily seen that the function

$$\psi_0(u) = \frac{\varphi_0(u) - a_0 - b_0 u}{u^2 - \zeta} \tag{3.16}$$

is continuously differentiable in $(-\infty,\infty)$. To the integral in (3.13), we now apply an integration by parts and the result is

$$\varepsilon_1(\lambda,\alpha) = \frac{i}{2\pi\lambda}\int_{-\infty}^{\infty}\psi_0'(u)e^{i\lambda(u^3/3-\zeta u)}du. \tag{3.17}$$

Since $\varphi_0(u)$ vanishes identically outside a finite interval, it follows from (3.16) that $\psi_0(u) = O(u^{-1})$ and $\psi_0'(u) = O(u^{-2})$ as $u \to \pm\infty$. Therefore we have

$$|\varepsilon_1(\lambda,\alpha)| \le \frac{1}{2\pi\lambda}\int_{-\infty}^{\infty}|\psi_0'(u)|du = O(\lambda^{-1}), \tag{3.18}$$

where the O-term holds uniformly for α in a neighborhood of α_0. Since both $\sqrt{\zeta}$ and $-\sqrt{\zeta}$ are real stationary points in (3.4), ζ is always non-negative, a fact which also follows from (3.5). In view of the well known results

$$\mathrm{Ai}(-x) \sim \frac{1}{\pi^{1/2}x^{1/4}}\cos\left(\frac{2}{3}x^{3/2} - \frac{1}{4}\pi\right)$$

and

$$\mathrm{Ai}'(-x) \sim \frac{x^{1/4}}{\pi^{1/2}} \sin\left(\frac{2}{3}x^{3/2} - \frac{1}{4}\pi\right),$$

it is evident that the error term $\varepsilon_1(\lambda, \alpha)$ is indeed smaller than the first two terms in (3.12). Furthermore, the second term is of the same asymptotic order as the first term when $\zeta \neq 0$. Therefore, both terms must be kept in the first approximation to $I(\lambda, \alpha)$. Finally, we note that the integral in (3.17) is exactly of the same form as that in (3.7), except that $\varphi_0(u)$ is now replaced by $\varphi_1(u) := \psi_0'(u)$. Thus, the above procedure can be repeated to produce an infinite compound asymptotic expansion. The material in this section is taken from [7, Sec. 3].

4. Two-dimensional Approximation

A heuristic argument for (1.4) proceeds as follows. Since (x_0, y_0) is a stationary point, the Taylor expansion of $f(x, y)$ at (x_0, y_0) is given by

$$\begin{aligned} f(x,y) = f(x_0,y_0) + \frac{1}{2}\alpha(x-x_0)^2 + \frac{1}{2}\beta(y-y_0)^2 \\ + \gamma(x-x_0)(y-y_0) + \cdots, \end{aligned} \tag{4.1}$$

where $\alpha = f_{xx}(x_0, y_0), \beta = f_{yy}(x_0, y_0)$ and $\gamma = f_{xy}(x_0, y_0)$. Let $\xi = (x - x_0)$ and $\eta = (y - y_0)$. Then approximately, we have

$$f(x,y) \approx f(x_0,y_0) + \frac{1}{2}\alpha\xi^2 + \frac{1}{2}\beta\eta^2 + \gamma\xi\eta. \tag{4.2}$$

Substituting (4.2) in (1.3), and replacing $g(x, y)$ by $g(x_0, y_0)$, we obtain

$$\begin{aligned} I(\lambda) &\approx g(x_0,y_0)e^{i\lambda f(x_0,y_0)} \\ &\quad \times \int_{-\infty}^{\infty}\int_{-\infty}^{\infty} e^{\frac{i}{2}\lambda(\alpha\xi^2+2\gamma\xi\eta+\beta\eta^2)}d\xi d\eta. \end{aligned} \tag{4.3}$$

The double integral on the right-hand side can be evaluated in closed form, and this gives

$$I(\lambda) \approx \frac{2\pi i\sigma}{\lambda\sqrt{|\alpha\beta - \gamma^2|}} g(x_0,y_0)e^{i\lambda f(x_0,y_0)}, \tag{4.4}$$

where $\sigma = 1$ if $\alpha\beta > \gamma^2$ and $\alpha > 0; \sigma = -1$ if $\alpha\beta > \gamma^2$ and $\alpha < 0$; and $\sigma = -i$ if $\alpha\beta < \gamma^2$, thus proving (1.4). The quantity $\kappa = |\alpha\beta - \gamma^2|$ is the Gaussian curvature of the surface $z = f(x, y)$ at (x_0, y_0).

To give a rigorous derivation of (4.4), we first assume that the stationary point (x_0, y_0) occurs at the origin $(0, 0)$, and that the origin is an interior point of integration domain D in (1.3). Next, we point out that the cross-product term $(x - x_0)(y - y_0)$ in (4.1) can be eliminated by a linear transformation. Finally, we note that since the constant term $f(x_0, y_0)$ in (4.1) only contributes a factor $\exp\{i\lambda f(x_0, y_0)\}$ to the double integral $I(\lambda)$ in (1.3), we can take $f(x_0, y_0) = 0$ without loss of generality. The Taylor expansion (4.1) now reduces to the form

$$f(x, y) = f_{20}\, x^2 + f_{02}\, y^2 + \cdots, \tag{4.5}$$

which can be written as

$$f(x, y) = f_{20}\, x^2[1 + P(x, y)] + f_{02}\, y^2[1 + Q(x, y)], \tag{4.6}$$

where P and Q are power series in x and y satisfying $P(0, 0) = Q(0, 0) = 0$. In the integral in (1.3), we make the change of variables

$$u = x[1 + P(x, y)]^{1/2}, \qquad v = y[1 + Q(x, y)]^{1/2} \tag{4.7}$$

so that $f(x, y) = f_{20}\, u^2 + f_{02}\, v^2$. Let D' denote the image of D under this change of variables, and put

$$G(u, v) = g(x, y)\frac{\partial(x, y)}{\partial(u, v)} \qquad \text{and} \qquad F(u, v) = f_{20}\, u^2 + f_{02}\, v^2. \tag{4.8}$$

The double integral $I(\lambda)$ then becomes

$$I(\lambda) = \iint_{D'} G(u, v) e^{i\lambda F(u,v)} du dv. \tag{4.9}$$

The new amplitude function $G(u, v)$ has the Maclaurin expansion

$$G(u, v) = \sum G_{ij} u^i v^j, \qquad G_{00} = g(0, 0), \tag{4.10}$$

which is obtained by inverting the variables in (4.7). The coefficients G_{ij} in (4.10) can be expressed in terms of the derivatives of f and g at $(0, 0)$. Let m and M denote the infimum and supremum of F in D' (or equivalently, of f in D), respectively. By the method of resolution of multiple integrals [4, pp. 445-455], the double integral $I(\lambda)$ in (4.9) can be written as

$$I(\lambda) = \int_m^M h(t) e^{i\lambda t} dt, \tag{4.11}$$

where

$$h(t) = \int_{F(u,v)=t} \frac{G(u, v)}{\sqrt{F_u^2 + F_v^2}} d\sigma, \tag{4.12}$$

σ being the arc length of the curve $F(u, v) = t$.

In this lecture, we consider only the case when both f_{20} and f_{02} are positive, and refer the reader to [14, Chapter VIII] for a detailed discussion of other cases. Thus, in the present case, $(0,0)$ is a local minimum and $m=0$ in (4.11). Since the contribution from an end-point is of lower order than that from a stationary point, as remarked earlier in Sec. 3, we may assume that the support of $h(t)$ in (4.11) lies in a small interval containing the origin $t=0$. This can be achieved by making the support of $g(x,y)$, and hence that of $G(u,v)$, sufficiently small. From (4.6), it is clear that

$$\left.\frac{\partial(x,y)}{\partial(u,v)}\right|_{(0,0)} = 1.$$

We assume that $\partial(x,y)/\partial(u,v)$ is positive in the support of G. To derive an asymptotic expansion for $h(t)$ as $t \to 0^+$, we use (4.12). The second equation in (4.8) shows that the level curves $F(u,v)=t$ are ellipses. Let

$$u = \left(\frac{\xi}{f_{20}}\right)^{1/2} \cos\eta \qquad \text{and} \qquad v = \left(\frac{\xi}{f_{02}}\right)^{1/2} \sin\eta. \tag{4.13}$$

Clearly $\xi = f_{20}\, u^2 + f_{02}\, v^2 = F(u,v)$ and

$$\frac{1}{|\nabla F|}d\sigma = \frac{1}{2\sqrt{f_{20}f_{02}}}d\eta.$$

Thus from (4.12) it follows that

$$h(t) = \int_{\xi=t} \Phi(\xi,\eta)d\eta = \int_0^{2\pi} \Phi(t,\eta)d\eta, \tag{4.14}$$

where

$$\Phi(\xi,\eta) = \frac{1}{2\sqrt{f_{20}f_{02}}}G(u,v).$$

The second equality in (4.14) holds when t is sufficiently small. Put

$$G^*_{ij} = \frac{G_{ij}}{f_{20}^{1/2} f_{02}^{1/2}}.$$

Then $\Phi(\xi,\eta)$ has the asymptotic representation

$$\Phi(\xi,\eta) \sim \frac{1}{2\sqrt{f_{20}f_{02}}} \sum G^*_{ij}\, \xi^{(i+j)/2} \cos^i\eta \sin^j\eta, \tag{4.15}$$

as $\xi \to 0^+$. Since

$$\int_0^{2\pi} \cos^i\eta \sin^j\eta d\eta = 0 \qquad \text{for} \qquad i,j = 1,3,5,\cdots$$

and

$$\int_0^{2\pi} \cos^{2m}\eta \sin^{2n}\eta d\eta = 4\int_0^{\pi/2} \cos^{2m}\eta \sin^{2n}\eta d\eta$$
$$= 2\frac{\Gamma(m+\frac{1}{2})\Gamma(n+\frac{1}{2})}{(m+n)!},$$

we find, on inserting (4.15) in (4.14), that

$$h(t) \sim \frac{1}{\sqrt{f_{20}f_{02}}} \sum \frac{\Gamma(m+\frac{1}{2})\Gamma(n+\frac{1}{2})}{(m+n)!} G^*_{2m,2n} t^{m+n}, \tag{4.16}$$

as $t \to 0^+$. Since $h(t)$ vanishes identitically outside a small neighborhood of the origin, there is no contribution from the endpoint $t = M$, and the final result

$$I(\lambda) \sim \frac{1}{\sqrt{f_{20}f_{02}}} \sum G^*_{2m,2n} e^{i\pi(m+n+1)/2} \frac{\Gamma(m+\frac{1}{2})\Gamma(n+\frac{1}{2})}{\lambda^{m+n+1}}, \tag{4.17}$$

as $\lambda \to +\infty$, now follows from (4.11) by termwise integration in the sense of Abel summability; see Sec. 2.

Taking into account the elimination of the cross product xy and the constant term $f(0,0)$ in the Maclaurin expansion of $f(x,y)$, the leading term in (4.17) agrees with the required result (1.4) when $\sigma_f = 2$.

5. A Uniform Asymptotic Formula

As a generalization of the uniform asymptotic approximation given in Sec. 3, we consider the double integral

$$I(\lambda, \alpha) = \iint_D g(x,y) e^{i\lambda f(x,y,\alpha)} dxdy, \tag{5.1}$$

where the phase function $f(x,y,\alpha)$ has two simple stationary points $(x_+(\alpha), y_+(\alpha))$ and $(x_-(\alpha), y_-(\alpha))$ in the integration domain D, which coalesce at an interior point (x_0, y_0) in D as α approaches a critical value α_0. Problems of this nature arise in propeller acoustics; see Chapman [2] and Prentice [10]. In fact, there is an entire book written by Borovikov [1] on this topic. The material in this section is taken from Qiu and Wong [11].

We shall write $f_\alpha(x,y) \equiv f(x,y,\alpha)$. Without loss of generality, we may take $\alpha_0 = 0, (x_0, y_0) = (0,0)$ and $f(0,0,\alpha) = 0$ for all small α. Under these assumptions, it is readily seen that $(0,0)$ is a degenerate stationary point of $f_0(x,y) = f(x,y,0)$. If we put

$$a_{ij} \equiv \left.\frac{\partial^{i+j} f_0}{\partial x^i \partial y^j}\right|_{(0,0)}, \tag{5.2}$$

then the gradient of f_0 at $(0,0)$ is

$$\nabla f_0(0,0) = (a_{10}, a_{01}) = (0,0) \tag{5.3}$$

and the Hessian of f_0 at $(0,0)$ is

$$\det f_0''(0,0) = a_{20}a_{02} - a_{11}^2 = 0. \tag{5.4}$$

The Taylor expansion of $f_0(x,y)$ at the origin is given by

$$\begin{aligned} f_0(x,y) = &\frac{1}{2}(a_{20}\,x^2 + 2a_{11}xy + a_{02}\,y^2) \\ &+ \frac{1}{6}(a_{30}\,x^3 + 3a_{21}\,x^2y + 3a_{12}\,xy^2 + a_{03}\,y^3) + \cdots . \end{aligned} \tag{5.5}$$

We now recall the following result from catastrophe theory, known as the splitting lemma [9, p.95].

LEMMA. *Let $f(x,y,\alpha)$ be a smooth function of (x,y,α) defined in a neighborhood of the origin in $\mathbb{R}^2 \times \mathbb{R}$. Write $f_\alpha(x,y) = f(x,y,\alpha)$. Suppose that the Taylor expansion of $f_0(x,y)$ at $(0,0)$ is of the form* (5.5), *and that condition* (5.4) *holds with a_{20}, a_{11} and a_{02} not all zero. Then, for all small values of α, there exists a smooth and one-to-one mapping $(x,y) \to (s,t)$ given by $s = s_\alpha(x,y)$ and $t = t_\alpha(x,y)$ with $s_0(0,0) = t_0(0,0) = 0$, which depends also smoothly on α, such that*

$$f_\alpha(x,y) = \varepsilon s^2 + g_\alpha(t), \qquad \varepsilon = \pm 1,$$

where $g_\alpha(t)$ is smooth in both t and α and $g_0(0) = 0$.

Using the splitting lemma above and the cubic transformation in (3.4), it can be established that if either

$$a_{03}\,a_{20}^2 - 3a_{12}\,a_{20}\,a_{11} + 3a_{21}\,a_{20}\,a_{02} - a_{30}\,a_{02}\,a_{11} \neq 0 \tag{5.6}$$

or

$$a_{30}\,a_{02}^2 - 3a_{21}\,a_{02}\,a_{11} + 3a_{12}\,a_{20}\,a_{02} - a_{03}\,a_{20}\,a_{11} \neq 0, \tag{5.7}$$

then for all small values of α, $f_\alpha(x,y)$ can be mapped into the canonical form

$$f_\alpha(x,y) = \varepsilon u^2 + \frac{1}{3}v^3 + \xi(\alpha)v + \eta(\alpha) \tag{5.8}$$

by a smooth and one-to-one mapping $(x,y) \to (u,v)$ given by $u = u_\alpha(x,y)$ and $v = v_\alpha(x,y)$, where $\varepsilon = \operatorname{sgn} a_{20}$ if $a_{20} \neq 0$ or $\varepsilon = \operatorname{sgn} a_{02}$ if $a_{02} \neq 0$, and where $\xi(\alpha)$ and $\eta(\alpha)$ are smooth functions with $\xi(0) = \eta(0) = 0$.

Furthermore, $u_0(0,0) = v_0(0,0) = 0$. The functions $\xi(\alpha)$ and $\eta(\alpha)$ are given by

$$\xi(\alpha) = -\left[\frac{3}{4}\Big(f_\alpha(x_-(\alpha), y_-(\alpha)) - f_\alpha(x_+(\alpha), y_+(\alpha))\Big)\right]^{2/3} \tag{5.9}$$

$$\eta(\alpha) = \frac{1}{2}\left[f_\alpha(x_+(\alpha), y_+(\alpha)) + f_\alpha(x_-(\alpha), y_-(\alpha))\right]. \tag{5.10}$$

Returning to (5.1), we assume as in Sec. 4 that the support of $g(x, y)$ is contained in D. This is reasonable at least to the leading order approximation, since the contribution from the boundary is always of lower asymptotic order than that from stationary points. We also assume that α is sufficiently small so that (5.8) holds inside the support of g. Making the change of variables from (x, y) to (u, v), we have

$$I(\lambda, \alpha) = e^{i\lambda\eta(\alpha)} \iint_{\mathbb{R}^2} \varphi_0(u, v, \alpha) e^{i\lambda(\varepsilon u^2 + v^3/3 + \xi(\alpha)v)} dv du, \tag{5.11}$$

where

$$\varphi_0(u, v, \alpha) = g(x, y, \alpha)\left|\frac{\partial(x, y)}{\partial(u, v)}\right|. \tag{5.12}$$

Note that $\varphi_0(u, v, \alpha)$ vanishes identically outside a neighborhood of the origin in $\mathbb{R}^2$. To obtain the values of $\varphi_0(u, v, \alpha)$ at the stationary points $(0, \pm(-\xi(\alpha))^{1/2})$, we need to evaluate the Jacobian $|\partial(x, y)/\partial(u, v)|$ at these points. The result is

$$\left|\frac{\partial(x, y)}{\partial(u, v)}\right|^2_{(0, \pm(-\xi(\alpha))^{1/2})} = \frac{\pm 4\varepsilon(-\xi(\alpha))^{1/2}}{H_{f_\alpha}(x_\pm(\alpha), y_\pm(\alpha))}, \tag{5.13}$$

where $H_f(x, y)$ denotes the Hessian of the function f at (x, y).

Analogous to (3.9) in the one-dimensional case, we now write

$$\begin{aligned} \varphi_0(u, v, \alpha) &= a_0(\alpha) + b_0(\alpha)v \\ &\quad + (v^2 + \xi(\alpha))\psi_{02}(v, \alpha) + 2u\varepsilon\psi_{01}(u, v, \alpha). \end{aligned} \tag{5.14}$$

Substitute (5.14) in (5.11) and carry out integration term-by-term. Instead of using integration by parts as in Sec. 3, here we need Green's theorem. The two-term asymptotic expansion for the integral (5.11) is given by

$$\begin{aligned} I(\lambda, \alpha) &\sim e^{i(\lambda\eta(\alpha) + \varepsilon\pi/4)} \frac{2\pi^{3/2}}{\lambda^{5/6}} \\ &\quad \times \left[a_0(\alpha)\mathrm{Ai}(\lambda^{2/3}\xi) + \frac{b_0(\alpha)}{i\lambda^{1/3}}\mathrm{Ai}'(\lambda^{2/3}\xi)\right]. \end{aligned} \tag{5.15}$$

To obtain the coefficients $a_0(\xi)$ and $b_0(\xi)$, we put $u = 0$ and $v = \pm(-\xi(\alpha))^{1/2}$ in (5.14). This gives

$$a_0(\alpha) = \frac{1}{2}\left[\varphi_0(0,(-\xi(\alpha))^{1/2},\alpha) + \varphi_0(0,-(-\xi(\alpha))^{1/2},\alpha)\right] \tag{5.16}$$

and

$$\begin{aligned} b_0(\alpha) = \frac{1}{2(-\xi(\alpha))^{1/2}} & \left[\varphi_0(0,(-\xi(\alpha))^{1/2},\alpha)\right. \\ & \left. - \varphi_0(0,-(-\xi(\alpha))^{1/2},\alpha)\right]. \end{aligned} \tag{5.17}$$

A combination of (5.12), (5.13), (5.16) and (5.17) yields

$$a_0(\alpha) = (-\xi)^{1/4}\left[\frac{g(x_+,y_+,\alpha)}{|H_{f\alpha}(x_+,y_+)|^{1/2}} + \frac{g(x_-,y_-,\alpha)}{|H_{f\alpha}(x_-,y_-)|^{1/2}}\right]$$

and

$$b_0(\alpha) = (-\xi)^{1/4}\left[\frac{g(x_+,y_+,\alpha)}{|H_{f\alpha}(x_+,y_+)|^{1/2}} - \frac{g(x_-,y_-,\alpha)}{|H_{f\alpha}(x_-,y_-)|^{1/2}}\right].$$

References

1. V. A. Borovikov, *Uniform Stationary Phase Method*, IEE, London, 1994.
2. C. J. Chapman, *The asymptotic theory of rapidly rotating sound fields*, Proc. R. Soc. Lond. A**436** (1992), 511-526.
3. C. Chester, B. Friedman and F. Ursell, *An extension of the method of steepest descents*, Proc. Camb. Phil. Soc. **53** (1957), 599-611.
4. R. Courant and F. John, *Introduction to Calculus and Analysis*, Vol. 2, John Wiley and Sons, New York, 1974.
5. L. Hörmander, *The Analysis of Linear Partial Differential Operators*, Springer, Berlin, 1983.
6. D. S. Jones and M. Kline, *Asymptotic expansions of multiple integrals and the method of stationary phase*, J. Math. Phys. **37** (1958), 1-28.
7. J. P. McClure and R. Wong, *Justification of the stationary phase approximation in time-domain asymptotics*, Proc. Roy. Soc. Lond. A**453** (1997), 1019-1031.
8. F. W. J. Olver, *Error bounds for stationary phase approximations*, SIAM J. Math. Anal. **5** (1974), 19-29.
9. T. Poston and I. Stewart, *Catastrophe Theory and Its Applications*, Pitman, Boston, 1978.
10. P. R. Prentice, *Time-domain asymptotics* I. *General theory for double integrals*, Proc. R. Soc. Lond. A**466** (1994), 341-360.
11. W.-Y. Qiu and R. Wong, *Uniform asymptotic expansions of a double integral: coalescence of two stationary points*, Proc. R. Soc. Lond. A**456**(2000), 407-431.
12. E. C. Titchmarsh, *The Theory of Functions*, Second Edition, Oxford University Press, London, 1939.

13. G. N. Watson, *A Treatise on the Theory of Bessel Functions*, Second Edition, Cambridge University Press, London, 1944.
14. R. Wong, *Asymptotic Approximations of Integrals*, Academic Press, Boston, 1989; reprinted by SIAM, Philadelphia, 2001.

LECTURE II

METHOD OF STEEPEST DESCENT

He (Debye) found a very powerful technique for evaluating some integrals $\cdots$. (This invention) required only a clear picture of the relevant feature of the mathematical object (*not its abstraction*) and a willingness to discard that which was not essential to the task [my italics].

G. F. Carrier

1. Introduction

By name, this is probably the best known method for finding the asymptotic behavior of integrals of the form

$$I(\lambda) = \int_C g(z) e^{\lambda f(z)} dz, \tag{1.1}$$

where $f(z)$ and $g(z)$ are analytic functions, λ is a large parameter and C is a contour in the z-plane. However, it is probably also the one which is least understood by non-specialists. Excellent survey-type articles have been written on this subject; here we mention only the following two: Wyman [17] and Olver [8].

Debye devised this method in 1909 in order to derive asymptotic expansions of Bessel functions of large order. His basic idea is to deform the contour C into a new path of integration C' so that the following conditions hold:

(i) C' passes through one or more zeros of $f'(z)$;
(ii) the imaginary part of $f(z)$ is constant on C'.

To obtain a geometric interpretation of the new path of integration, we write

$$f(z) = u(x,y) + iv(x,y),$$

where $z = x + iy$ and u and v are real. If u is treated as a third axis orthogonal to both x and y, then the equation $u = u(x,y)$ defines a surface S in the (x,y,u) space. Suppose that $z_0 = x_0 + iy_0$ is a zero of $f'(z)$. Then, by the Cauchy-Riemann equation, $f'(z) = u_x - iu_y$. Thus, $f'(z_0) = 0$

implies

$$u_x(x_0, y_0) = u_y(x_0, y_0) = 0,$$

i.e., (x_0, y_0) is a critical point of $u(x, y)$. Since u is a harmonic function, $u(x, y)$ cannot have a maximum or a minimum at an interior point. Therefore, (x_0, y_0) must be a saddle point of $u(x, y)$. For this reason, we call z_0 a *saddle point* of $f(z)$.

The shape of the surface S on the (x, y) plane can be represented by drawing the level curves on which u is constant. The curves $v =$ constant are the orthogonal trajectories of the level curves, but so are the projections of the paths of steepest ascent or descent on the surface. The term *steepest descent* stems from condition (ii) above.

We suppose that it is possible to deform the original path of integration into a steepest path $v(x, y) = \text{constant} = \text{Im } f(z_0)$. On this path, we have

$$f(z) = f(z_0) - \tau, \tag{1.2}$$

where τ is real and is either monotonically increasing or monotonically decreasing. The integrand in (1.1) becomes $g(z)e^{\lambda f(z_0) - \lambda\tau}$. On a path where $\tau \to -\infty$, the integral may be divergent. For this reason we choose paths on which τ is positive and increasing. These are the paths of steepest descent from saddle points. We suppose that the original path of integration C in (1.1) can be deformed into an equivalent path consisting of paths of steepest descent through a saddle point. Then our problem is reduced to finding the asymptotic behavior of integrals of the form

$$e^{\lambda f(z_0)} \int_0^\infty g(z) \frac{dz}{dt} e^{-\lambda\tau} d\tau. \tag{1.3}$$

The above brief explanation of the method of steepest descent will be made clearer by the example of Airy's integral given in Sec. 2. The results for Airy's integral will be used to illustrate the Stokes phenomenon in Sec. 3. In Sec. 4, we present a modified version of the steepest descent method introduced by Berry and Howls [5] in 1991. The final two sections contain a brief introduction to a sub-area of asymptotics which is now known as *exponential asymptotics*.

2. The Airy Integral

This integral is defined by

$$\text{Ai}(z) = \frac{1}{2\pi i} \int_L \exp\left(\frac{1}{3}t^3 - zt\right) dt, \tag{2.1}$$

where L is any contour which begins at infinity in the sector $-\frac{1}{2}\pi < \arg t < -\frac{1}{6}\pi$ and ends at infinity in the sector $\frac{1}{6}\pi < \arg t < \frac{1}{2}\pi$. To transform it into the form given in (1.1), we first assume that z is real and positive, and then make the change of variable $t = z^{1/2}u$. This gives

$$\mathrm{Ai}(z) = \frac{z^{1/2}}{2\pi i}\int_L \exp\left\{z^{3/2}\left(\frac{1}{3}u^3 - u\right)\right\}du.$$

Once this identity has been established, the restriction on z can be removed by using analytic continuation. Put

$$\xi = \frac{2}{3}z^{3/2} \qquad \text{and} \qquad f(u) = \frac{1}{2}(u^3 - 3u) \tag{2.2}$$

so that

$$\mathrm{Ai}(z) = \frac{1}{2\pi i}\left(\frac{3}{2}\xi\right)^{1/3}\int_L e^{\xi f(u)}du. \tag{2.3}$$

The saddle points of $f(u)$ are at $u = \pm 1$. Clearly, $f(\pm 1) = \mp 1$. If we write $u = s + it$, then we have

$$f(u) = \frac{1}{2}[(s^3 - 3st^2 - 3s) + i(3s^2t - t^3 - 3t)].$$

Since $\mathrm{Im}\ f(\pm 1) = 0$, $\mathrm{Im}\ f = 0$ implies either $t = 0$ or $t^2 - 3s^2 + 3 = 0$. On $t = 0$, we have $\mathrm{Re}\ f = \frac{1}{2}(s^3 - 3s)$, which has a local minimum at $s = 1$ and a local maximum at $s = -1$. Thus, near $s = 1, t = 0$ is a steepest ascent curve. The other equation $t^2 - 3s^2 + 3 = 0$ represents two branches of a hyperbola with the asymptotes $t = \pm\sqrt{3}s$. On this hyperbola, $\mathrm{Re}\, f = -4s^3 + 3s$, which decreases for $s > 1$. From Figure 1, it is clear that the branch of the hyperbola on the right half plane is our desired path of steepest descent through the saddle point $u = 1$. (In Figure 1, arrows indicate the direction in which $\mathrm{Re}\ f$ decreases.)

Deforming the original contour L into this path of steepest descent, we can write

$$\mathrm{Ai}(z) = \frac{1}{2\pi i}\left(\frac{3}{2}\xi\right)^{1/3}\left(\int_1^{\infty e^{i\pi/3}} - \int_1^{\infty e^{-i\pi/3}}\right)e^{\xi f(u)}du. \tag{2.4}$$

In both integrals above, $f(u) - f(1)$ is real and has a maximum at $u = 1$. Also, $f(u) - f(1)$ is decreasing as u moves away from $u = 1$. Set

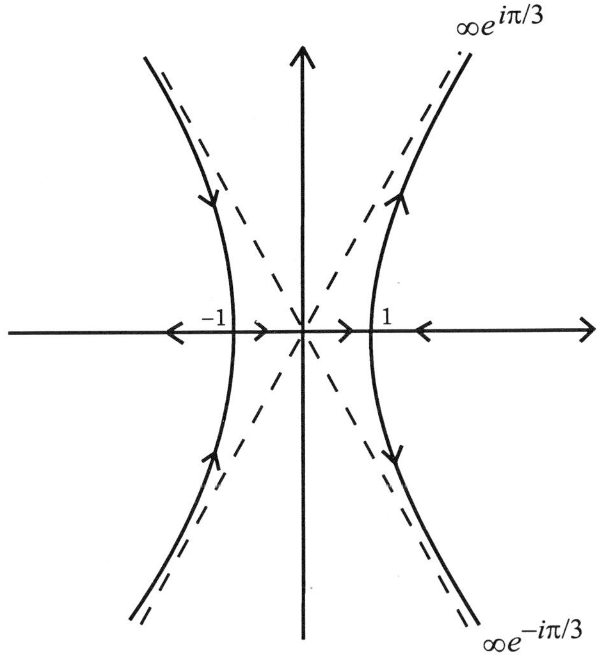

Figure 1. Steepest paths for Ai(z)

$$\tau = f(1) - f(u) = -1 + \frac{3}{2}u - \frac{1}{2}u^3 = -\frac{3}{2}(u-1)^2 - \frac{1}{2}(u-1)^3. \qquad (2.5)$$

On our steepest descent path, τ is real; cf. Eq. (1.2). From (2.5), we have

$$\pm i\left(\frac{2}{3}\tau\right)^{1/2} = (u-1)\left[1 + \frac{1}{3}(u-1)\right]^{1/2},$$

where $[\cdots]^{1/2}$ is that branch which reduces to 1 at $u = 1$. By Lagrange's formula for the reversion of series

$$u^{\pm} = 1 + \sum_{n=1}^{\infty} a_n \left[\pm i\left(\frac{2}{3}\tau\right)^{1/2}\right]^n,$$

where

$$a_n = \frac{1}{n!}\frac{d^{n-1}}{du^{n-1}}\left[1 + \frac{1}{3}(u-1)\right]^{-n/2}\Bigg|_{u=1}.$$

Since u^+ enters the first quadrant for increasing τ, we must take u^+ for the first integral in (2.4). Similarly, we take u^- for the second integral there. Equation (2.4) then becomes

$$\mathrm{Ai}(z) = \frac{1}{2\pi i}\left(\frac{3}{2}\xi\right)^{1/3} e^{-\xi}\int_0^\infty \left(\frac{du^+}{d\tau} - \frac{du^-}{d\tau}\right) e^{-\xi\tau}\, d\tau. \tag{2.6}$$

Since na_n is the coefficient of $(u-1)^{n-1}$ in the Taylor expansion of $\left[1+\dfrac{1}{3}(u-1)\right]^{-n/2}$ at $u=1$, it is easily shown that

$$a_n = \frac{(-1)^{n-1}\Gamma\left(\frac{3}{2}n - 1\right)}{n!\,\Gamma\left(\frac{1}{2}n\right)3^{n-1}}. \tag{2.7}$$

From this, it follows that

$$\frac{du^+}{d\tau} - \frac{du^-}{d\tau} = 2i\sum_{m=0}^{\infty} \frac{(-1)^m\Gamma\left(3m+\frac{1}{2}\right)\left(m+\frac{1}{2}\right)}{(2m+1)!\,\Gamma\left(m+\frac{1}{2}\right)9^m}\left(\frac{2}{3}\right)^{m+\frac{1}{2}}\tau^{m-\frac{1}{2}}.$$

Termwise integration gives

$$\mathrm{Ai}(z) \sim \frac{1}{2\pi z^{1/4}}\exp\left(-\frac{2}{3}z^{3/2}\right)\sum_{m=0}^{\infty}\frac{(-1)^m\Gamma\left(3m+\frac{1}{2}\right)}{(2m)!\,9^m} z^{-3m/2} \tag{2.8}$$

as $z \to \infty$ in $|\arg z| < \frac{1}{3}\pi$.

The sector of validity of (2.8) can be extended to $|\arg z| < \pi$ by rotating the path of integration in (2.6). This way of extending the region of validity is frequently used in asymptotics; see, e.g., [12, p.26]. However, the result in (2.8) can not be valid in a wider sector, since $\mathrm{Ai}(z)$ is a single-valued function while the factor multiplying the infinite series in (2.8) is not. In fact, from (2.1), it can be shown that $\mathrm{Ai}(z)$ has the Maclaurin expansion

$$\mathrm{Ai}(z) = 3^{-2/3}\left[\sum_{n=0}^{\infty}\frac{z^{3n}}{3^{2n}n!\,\Gamma\left(n+\frac{2}{3}\right)} - 3^{-2/3}\sum_{n=0}^{\infty}\frac{z^{3n+1}}{3^{2n}n!\,\Gamma\left(n+\frac{4}{3}\right)}\right]; \tag{2.9}$$

that is, $\mathrm{Ai}(z)$ is an entire function. The material in this section is taken from [12, Chap. II, Sec. 4], where more examples can be found on the method of steepest descent.

3. Stokes' Phenomenon

From (2.1), it can be shown that $\mathrm{Ai}(z)$ is a solution of

$$\frac{d^2w}{dz^2} - zw = 0. \tag{3.1}$$

Clearly, $\mathrm{Ai}(wz)$ and $\mathrm{Ai}(w^2z)$ are also solutions of this equation, where $w = e^{2\pi i/3}$. Using Cauchy's theorem, it can also be shown that the three solutions are connected by the relation

$$\mathrm{Ai}(z) + w\mathrm{Ai}(wz) + w^2\mathrm{Ai}(w^2z) = 0. \tag{3.2}$$

If $\pi/3 < \arg z < 5\pi/3$, then $-\pi/3 < \arg(w^{-1}z) < \pi$ and $-\pi < \arg(w^{-2}z) < \pi/3$. From (2.8) and (3.2), it follows that

$$\begin{aligned}\mathrm{Ai}(z) \sim & \frac{1}{2\pi z^{1/4}} \exp\left(-\frac{2}{3}z^{3/2}\right) \sum_{m=0}^{\infty} \frac{(-1)^m \Gamma(3m+\frac{1}{2})}{(2m)!\,9^m} z^{-3m/2} \\ & + \frac{i}{2\pi z^{1/4}} \exp\left(\frac{2}{3}z^{3/2}\right) \sum_{m=0}^{\infty} \frac{\Gamma(3m+\frac{1}{2})}{(2m)!\,9^m} z^{-3m/2}\end{aligned} \tag{3.3}$$

as $z \to \infty$ in $\pi/3 < \arg z < 5\pi/3$. If we let $u_\pm(z)$ denote the formal series

$$u_\pm(z) = \frac{1}{2\pi z^{1/4}} \exp\left(\pm\frac{2}{3}z^{3/2}\right) \sum_{m=0}^{\infty} \frac{(\pm 1)^m \Gamma(3m+\frac{1}{2})}{(2m)!\,9^m} z^{-3m/2}, \tag{3.4}$$

then (2.8) and (3.3) can be written, respectively, as

$$\mathrm{Ai}(z) \sim u_-(z) \tag{3.5}$$

as $z \to \infty$ in $-\pi < \arg z < \pi$ and.

$$\mathrm{Ai}(z) \sim u_-(z) + iu_+(z) \tag{3.6}$$

as $z \to \infty$ in $\pi/3 < \arg z < 5\pi/3$.

Note that both results (3.5) and (3.6) are valid in the common sector $\pi/3 < \arg z < \pi$, and that there is no inconsistency since the first term on the right-hand side of (3.6) is *dominant* (exponentially increasing) and the second term is *recessive* (exponentially decreasing) in the sector $\pi/3 < \arg z < \pi$. However, the roles of these two terms are interchanged in the sector $-\pi/3 < \arg z < \pi/3$; hence it is mandatory to drop $u_+(z)$ from the asymptotic expansion of $\mathrm{Ai}(z)$. By introducing a constant (coefficient) C, which is 0 for $\arg z \in (-\pi/3, \pi/3)$ and i for $\arg z \in (\pi/3, 5\pi/3)$, the two results in (3.5) and (3.6) can be combined into one, namely

$$\mathrm{Ai}(z) \sim u_-(z) + Cu_+(z) \tag{3.7}$$

as $z \to \infty$ in $\arg z \in (-\pi, 5\pi/3)$. The coefficient C is called a *Stokes multiplier*, and is domain dependent. The discontinuous change of the coefficient C, when the argument of z changes in a continuous manner, is known as *Stokes' phenomenon.*

Returning to (3.4), we let

$$S_+(z) = \frac{2}{3}z^{3/2} \qquad \text{and} \qquad S_-(z) = -\frac{2}{3}z^{3/2}.$$

It can be verified that the behavior of $u_+(z)$ and $u_-(z)$ are most unequal on the curves

$$\mathrm{Im}\{S_+(z) - S_-(z)\} = 0 \tag{3.8}$$

and they are nearly equal on the curves given by

$$\mathrm{Re}\{S_+(z) - S_-(z)\} = 0. \tag{3.9}$$

The curves given in (3.8) and (3.9) are known, respectively, as the *Stokes* and *anti-Stokes lines*. In the case of the Airy function, it is easily seen that the rays $\arg z = 0, \pm 2\pi/3$ are the Stokes lines and the rays $\arg z = \pm\pi/3, \pm\pi$ are the anti-Stokes lines.

Since $\mathrm{Ai}(z)$ on the left-hand side of (3.7) is an analytic function, it is rather undesirable to have a discontinuous coefficient C on the right-hand side of the equation. In 1989, Berry [2] gave a different interpretation of the Stokes phenomenon. In his view, if one truncates the series $u_-(z)$ at an "optimal" place, then the coefficient of the series $u_+(z)$ should be a continuous function of $\arg z$, instead of a discontinuous constant. We shall illustrate Berry's theory with the simple Airy function given in (2.1). Our approach is based on a modified version of the steepest descent method introduced by Berry and Howls [4] in 1990. The material in the next two sections is taken nearly verbatim from Wong [13]. (Permission has been obtained from Kluwer Academic Publishers.)

4. Adjacent Saddle and Adjacent Contour

In (2.3), we let

$$\theta := \arg \xi, \tag{4.1}$$

and consider the steepest descent curves

$$\Gamma_{\pm 1}(\theta): \qquad \arg\{\xi[f(\pm 1) - f(u)]\} = \arg\{e^{i\theta}[f(\pm 1) - f(u)]\} = 0, \tag{4.2}$$

i.e., curves on which $\mathrm{Im}\{\xi[f(\pm 1) - f(u)]\} = 0$ and $\mathrm{Re}\{\xi[f(\pm 1) - f(u)]\} > 0$. Deforming the contour L in (2.3) into $\Gamma_1(\theta)$, we obtain

$$\mathrm{Ai}(z) = \frac{e^{-\xi}}{2\pi i}\left(\frac{3}{2}\xi\right)^{1/3} \int_{\Gamma_1(\theta)} e^{\xi[f(u)+1]} du. \tag{4.3}$$

If we introduce the notations

$$I^{(\pm 1)}(\xi) := \xi^{1/2} \int_{\Gamma_1(\theta)} e^{\xi[f(u)\pm 1]} du, \tag{4.4}$$

then (4.3) can be written as

$$\mathrm{Ai}(z) = \frac{1}{2\pi i}\left(\frac{3}{2}\right)^{1/3} \xi^{-1/6} e^{-\xi} I^{(1)}(\xi). \tag{4.5}$$

In the integral (4.4), we make the change of variable

$$-\tau = \xi[f(u)+1]. \tag{4.6}$$

For $u \in \Gamma_1(\theta)$, τ is real and positive; cf. (4.2). As in (2.5), we now expand $f(u)$ into a Taylor series at $u = 1$. Lagrange's inversion formula again gives

$$u^{\pm} = 1 + \sum_{n=1}^{\infty} a_n \left(\pm i\sqrt{\frac{2\tau}{3\xi}}\right)^n, \tag{4.7}$$

where the coefficients a_n are given in (2.7). Note that here u is a function of τ and ξ. By breaking the integration path $\Gamma_1(\theta)$ in (4.4) at $u = 1$, we can rewrite $\mathrm{I}^{(1)}(\xi)$ as

$$\begin{aligned} \mathrm{I}^{(1)}(\xi) &= \xi^{1/2} \int_0^{\infty} \left[\frac{du^+}{d\tau} - \frac{du^-}{d\tau}\right] e^{-\tau} d\tau \\ &= \xi^{-1/2} \int_0^{\infty} \left[\frac{1}{f'(u^-(\tau))} - \frac{1}{f'(u^+(\tau))}\right] e^{-\tau} d\tau; \end{aligned} \tag{4.8}$$

cf. (2.6).

The first step in the Berry-Howls method [5] is to use Cauchy's residue theorem to represent the integrand in (4.8) as a contour integral. To see this, we let $C_1(\theta)$ be a positively oriented curve surrounding the steepest-descent path $\Gamma_1(\theta)$. Since $\Gamma_1(\theta)$ is an infinite contour, $C_1(\theta)$ actually consists of two infinite curves embracing $\Gamma_1(\theta)$; see Figure 2. We now recall the formula

$$\mathrm{Res}\left\{\frac{Q(u)}{P(u)}; u_0\right\} = \frac{Q(u_0)}{P'(u_0)},$$

where $P(u)$ and $Q(u)$ are analytic functions with $P(u_0) = 0, P'(u_0) \neq 0$ and $Q(u_0) \neq 0$. Take $P(u) \equiv \xi[-1-f(u)] - \tau, Q(u) \equiv [-1-f(u)]^{1/2}$ and $u_0 = u^{\pm}(\tau)$. Then

$$\frac{Q(u^{\pm}(\tau))}{P'(u^{\pm}(\tau))} = \mathrm{Res}\left\{\frac{Q(u)}{P(u)}; u^{\pm}(\tau)\right\} = \frac{1}{2\pi i}\int_{C^{\pm}} \frac{Q(u)}{P(u)} du, \tag{4.9}$$

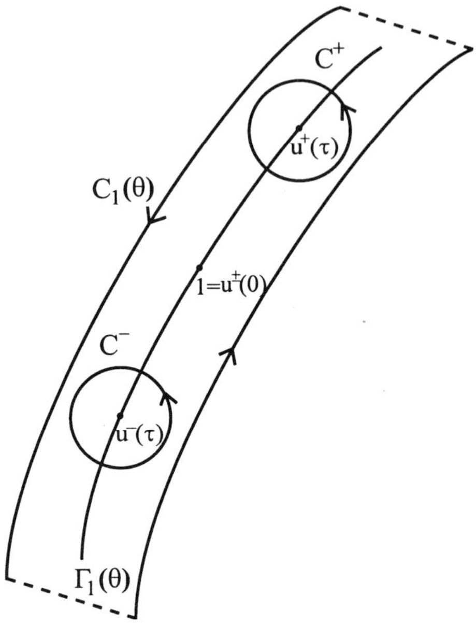

Figure 2. Contour $C_1(\theta)$.

where C^+ and C^- are the two closed contours shown in Figure 2. Since $Q(u^\pm(\tau)) = (\tau/\xi)^{1/2}$ and $P'(u^\pm(\tau)) = -\xi f'(u^\pm(\tau))$, it follows from (4.9) that

$$\frac{1}{f'(u^-(\tau))} - \frac{1}{f'(u^+(\tau))} = \frac{1}{2\pi i}\frac{\xi^{3/2}}{\tau^{1/2}}\int_{C_1(\theta)} \frac{[-1-f(u)]^{1/2}}{\xi[-1-f(u)]-\tau}du. \tag{4.10}$$

Inserting (4.10) into (4.8), we get a double integral for $\mathrm{I}^{(1)}(\xi)$. Upon interchanging the order of integration, we obtain

$$\mathrm{I}^{(1)}(\xi) = \frac{1}{2\pi i}\int_{C_1(\theta)} [-1-f(u)]^{-1/2}\int_0^\infty \frac{e^{-\tau}\tau^{-1/2}}{1-\{\tau/\xi[-1-f(u)]\}}d\tau du. \tag{4.11}$$

The geometric series

$$\frac{1}{1-x} = \sum_{s=0}^{N-1} x^s + \frac{x^N}{1-x}$$

then gives the asymptotic expansion

$$\mathrm{I}^{(1)}(\xi) = \sum_{s=0}^{N-1} c_s \xi^{-s} + R_N(\xi), \tag{4.12}$$

where

$$c_s = \frac{\Gamma\left(s+\frac{1}{2}\right)}{2\pi i} \int_{C_1(\theta)} [-1 - f(u)]^{-s-\frac{1}{2}} du \tag{4.13}$$

and

$$\begin{aligned} R_N(\xi) = \frac{\xi^{-N}}{2\pi i} \int_0^\infty e^{-\tau} \tau^{N-\frac{1}{2}} \int_{C_1(\theta)} & [-1 - f(u)]^{-N-\frac{1}{2}} \\ & \times \frac{1}{1 - \{\tau/\xi[-1 - f(u)]\}} du d\tau. \end{aligned} \tag{4.14}$$

The coefficients c_s can be evaluated exactly, and we have

$$c_s - i \frac{(-1)^s \Gamma\left(3s + \frac{1}{2}\right)}{(2s)!\, 9^s} \left(\frac{3}{2}\right)^{-s-\frac{1}{2}}. \tag{4.15}$$

The second important step in the Berry-Howls method is to consider all steepest descent paths $\Gamma_1(\theta)$ passing through $u = 1$ for different values of θ; see Figure 3. Since $f(1) - f(-1) = -2$, the path

$$\Gamma_1(\pi): \qquad \arg\{e^{i\pi}[f(1) - f(u)]\} = 0 \tag{4.16}$$

runs into the saddle point $u = -1$. Berry and Howls called $u = -1$ an *adjacent saddle* of $u = 1$, and the steepest-descent path

$$\Gamma_{-1}(\pi): \qquad \arg\{e^{i\pi}[f(-1) - f(u)]\} = 0 \tag{4.17}$$

an *adjacent contour*. Deforming the contour $C_1(\theta)$ in (4.14) into $\Gamma_{-1}(\pi)$, we obtain

$$\begin{aligned} R_N(\xi) = \frac{\xi^{-N}}{2\pi i} \int_0^\infty e^{-\tau} \tau^{N-\frac{1}{2}} \int_{\Gamma_{-1}(\pi)} & [-1 - f(u)]^{-N-\frac{1}{2}} \\ & \times \frac{1}{1 - \{\tau/\xi[-1 - f(u)]\}} du d\tau. \end{aligned} \tag{4.18}$$

In the last equation, we make the change of variable

$$\tau = t \frac{f(1) - f(u)}{f(1) - f(-1)}; \tag{4.19}$$

recall that $f(1) = -1$. Since

$$\frac{f(1) - f(u)}{f(1) - f(-1)} = 1 + \frac{f(-1) - f(u)}{f(1) - f(-1)}, \tag{4.20}$$

and since the quotient on the right-hand side is real and positive when $u \in \Gamma_{-1}(\pi)$, the quotient on the left-hand side is also real and positive for $u \in \Gamma_{-1}(\pi)$. Dingle [7] called the quantity $f(1) - f(-1)$ a *singulant*; see also [4]. Substituting (4.19) in (4.18), and making use of (4.20), we obtain

$$R_N(\xi) = \frac{\xi^{-N}}{2\pi i}(-2)^{-N-\frac{1}{2}} \int_0^\infty e^{-t} t^{N-\frac{1}{2}} \left(1 + \frac{t}{2\xi}\right)^{-1} \times \int_{\Gamma_{-1}(\pi)} \exp\left\{-t \frac{f(-1) - f(u)}{f(1) - f(-1)}\right\} du dt. \tag{4.21}$$

In the inner integral, we write $u = -w$. Since $f(u)$ is an odd function and $f(1) - f(-1) = -2$, it follows that

$$\int_{\Gamma_{-1}(\pi)} \exp\left\{-t \frac{f(-1) - f(u)}{f(1) - f(-1)}\right\} du = - \int_{\Gamma_1(0)} \exp\left\{\frac{t}{2}[f(w) - f(1)]\right\} dw.$$

The last integral can be expressed in terms of the integral $I^{(1)}(\xi)$ given in (4.4). Indeed, we have

$$\int_{\Gamma_{-1}(\pi)} \exp\left\{-t \frac{f(-1) - f(u)}{f(1) - f(-1)}\right\} du = -\left(\frac{t}{2}\right)^{-1/2} I^{(1)}\left(\frac{t}{2}\right). \tag{4.22}$$

Inserting (4.22) into (4.21) gives

$$R_N(\xi) = \frac{1}{2\pi}(-2\xi)^{-N} \int_0^\infty e^{-t} t^{N-1} \left(1 + \frac{t}{2\xi}\right)^{-1} I^{(1)}\left(\frac{t}{2}\right) dt. \tag{4.23}$$

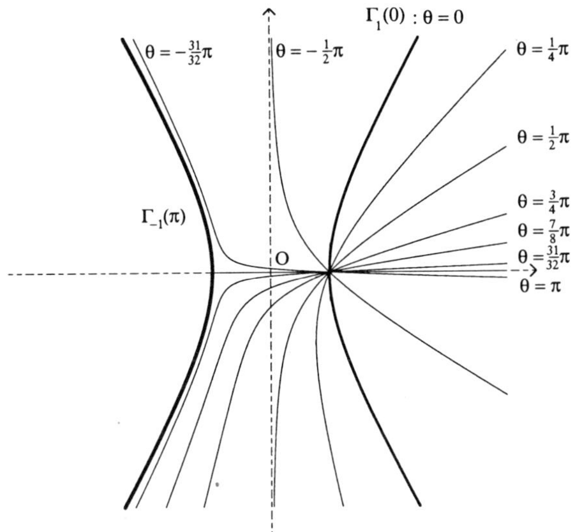

Figure 3. Contours $\Gamma_1(\theta), -\pi < \theta < \pi$

Equations (4.12) and (4.23) coupled together is known as a *resurgence formula*, since the integral $\mathrm{I}^{(1)}(\xi)$ on the left-hand side of (4.12) appears again in the remainder term $R_{\mathrm{N}}(\xi)$ given in (4.23).

5. Exponential Asymptotics

To estimate the remainder term in (4.12), we first use the double integral representation of $R_N(\xi)$ given in (4.18). For convenience, we introduce the function

$$\csc_1(\varphi) = \begin{cases} |\csc\varphi| & \text{if } 0 < |\varphi| < \dfrac{\pi}{2}, \\ 1 & \text{if } \dfrac{\pi}{2} \le |\varphi| \le \pi. \end{cases} \tag{5.1}$$

It can be readily verified that if $\zeta = |\zeta| e^{i\varphi}$ then

$$\left| \frac{1}{1-\zeta} \right| \le \csc_1(\varphi). \tag{5.2}$$

We extend $\csc_1(\varphi)$ into a 2π-periodic function by defining

$$\csc_1(\varphi + 2\pi) = \csc_1(\varphi). \tag{5.3}$$

Since this is an even function, it is unbounded at $0, \pm 2\pi, \pm 4\pi, \cdots$. Take

$\zeta = \tau/\xi[-1-f(u)]$, and note that $\arg\zeta = -\pi - \theta$. By (5.2), we have

$$\left|1 - \frac{\tau}{\xi[-1-f(u)]}\right|^{-1} \le \csc_1(\pi + \theta) \tag{5.4}$$

for $u \in \Gamma_{-1}(\pi)$. A direct application of (5.4) to (4.18) yields

$$|R_N(\xi)| \le \frac{A_N}{|\xi|^N} \csc_1(\theta + \pi), \tag{5.5}$$

where A_N is a constant given by

$$A_N = \frac{\Gamma(N+\frac{1}{2})}{2\pi} \int_{\Gamma_{-1}(\pi)} \left|[-1-f(u)]^{-N-\frac{1}{2}} du\right|. \tag{5.6}$$

There is another error estimate which is probably easier to compute. We first note that $I^{(1)}(\xi) = R_0(\xi)$; see (4.12). Hence, by (5.5),

$$\left|I^{(1)}\left(\frac{t}{2}\right)\right| \le A_0 \csc_1(\pi). \tag{5.7}$$

Next, we return to (4.23). With $\zeta = -t/2\xi$, we have from (5.2)

$$|R_N(\xi)| \le \frac{B_N}{|\xi|^N} \csc_1(\pi - \theta), \tag{5.8}$$

where

$$B_N = \frac{A_0 \Gamma(N)}{2^{N+1}\pi}. \tag{5.9}$$

Since the function $\csc_1(\varphi)$ is unbounded at $\varphi = 0, \pm 2\pi, \pm 4\pi, \cdots$, both bounds in (5.5) and (5.8) break down when $\theta = \pm\pi, \pm 3\pi, \pm 5\pi, \cdots$.

To obtain an error estimate near the Stokes lines $\theta = \pm\pi$, or equivalently, $\arg z = \pm\frac{2}{3}\pi$, we use a device suggested by Boyd [6]. Rotating the path of integration in (4.23) by an angle $\eta \in \left(-\frac{\pi}{2}, 0\right)$ gives

$$R_N(\xi) = \frac{1}{2\pi}(-2\xi)^{-N} \int_0^{\infty e^{i\eta}} e^{-t} t^{N-1} \left(1 + \frac{t}{2\xi}\right)^{-1} I^{(1)}\left(\frac{t}{2}\right) dt. \tag{5.10}$$

On the path of integration in (5.10), we have $\arg t = \eta$. Hence, by (5.5),

$$\left|I^{(1)}\left(\frac{t}{2}\right)\right| \le A_0 \csc_1(\eta + \pi) = A_0; \tag{5.11}$$

see (5.7). The equality in (5.11) follows from the fact that $\frac{\pi}{2} < \eta + \pi < \pi$. On the other hand, we also have from (5.2)

$$\left|1 + \frac{t}{2\xi}\right|^{-1} \le \csc_1(\pi + \eta - \theta). \tag{5.12}$$

A combination of (5.10), (5.11) and (5.12) yields

$$|R_N(\xi)| \le \frac{C}{(2|\xi|)^N} \csc_1(\pi + \eta - \theta) \frac{\Gamma(N)}{(\cos\eta)^N}, \tag{5.13}$$

where $C = A_0/2\pi$.

If $0 < \theta < \pi$ and θ is close to π, then for any sufficiently small $\eta \in \left(-\frac{\pi}{2}, 0\right)$ we have $-\frac{\pi}{2} < \pi + \eta - \theta < 0$. Similar, if $-\pi < \theta < 0$ and θ is close to $-\pi$, then for any sufficiently small $\eta \in \left(-\frac{\pi}{2}, 0\right)$ we have $\frac{3\pi}{2} < \pi + \eta - \theta < 2\pi$. Since the function $\csc_1(\varphi)$ defined in (5.1) and (5.3) is 2π-periodic, from (5.13) we obtain

$$|R_N(\xi)| \le \frac{C}{(2|\xi|)^N} \frac{1}{|\sin(\pm\pi + \eta - \theta)|} \frac{\Gamma(N)}{(\cos\eta)^N}, \tag{5.14}$$

where + sign is taken when $\theta \nearrow \pi$ and $-$ sign is taken when $\theta \searrow -\pi$. As $\theta \to \pm\pi$, we have $\sin(\pm\pi + \eta - \theta) \sim \sin\eta$. Since η is arbitrary, we may choose it so that

$$\cos\eta \sim \left(1 - \frac{1}{N}\right) \qquad \text{and} \qquad \sin\eta \sim \sqrt{\frac{2}{N}}.$$

Thus, for θ near $\pm\pi$, we can find a constant C_1 such that

$$|R_N(\xi)| \le \frac{C_1}{(2|\xi|)^N} \sqrt{\frac{N}{2}} \left(1 - \frac{1}{N}\right)^{-N} \Gamma(N).$$

(An explicit value for C_1 can be given if desired.) Since $\frac{1}{x}\log(1+x) \le \frac{1}{1+x}$ for negative values of x, we have $\left(1 - \frac{1}{N}\right)^{-N} \le e^{N/(N-1)}$. This together with the inequality [1]

$$\Gamma(N) < \sqrt{2\pi} N^{N-\frac{1}{2}} e^{-N+(1/12N)} \tag{5.15}$$

gives the estimate

$$|R_N(\xi)| \le C_2 e^{N(-1+\log N - \log|2\xi|)}, \tag{5.16}$$

where $C_2 = \sqrt{\pi} e^2 C_1$. The minimum value of the exponential function on the right-hand side of (5.16) is attained when

$$\frac{d}{dN}(-N + N\log N - N\log|2\xi|) = \log N - \log|2\xi| = 0.$$

Therefore, an optimal place to truncate the series in (4.12) is near

$$N = N^* := 2|\xi|.$$

With N given by this value, (5.16) yields

$$|R_N(\xi)| \leq C_2 e^{-2|\xi|}, \tag{5.17}$$

as $|\xi| \to \infty$, where $\arg\xi = \theta$ is close to the Stokes lines $\theta = \pm\pi$. Applying (5.15) to the error estimate (5.8), one readily sees that inequality (5.17) holds also for ξ away from the Stokes line. Olver [9] called the expansion (4.12) with error estimate (5.17) a *uniform, exponentially improved*, asymptotic expansion in the sector $|\xi| \leq \pi$. Optimatically truncated asymptotic expansions are also called super-asymptotic expansions by Berry and Howls [4]; see also [3].

6. Hyperasymptotics

Returning to (4.23), we now replace the function $I^{(1)}(t/2)$ by its asymptotic expansion (4.12). Termwise integration gives a series of integrals which can be expressed in terms of Dingle's *terminant function*

$$\int_0^\infty \frac{t^{k-1}e^{-t}}{1+t/\zeta}dt := 2\pi i(-\zeta)^k e^\zeta T_k(\zeta); \tag{6.1}$$

see Olver [9]. More precisely, we have

$$R_N(\xi) = ie^{2\xi}\sum_{r=0}^{N'-1}(-1)^r\frac{c_r}{\xi^r}T_{N-r}(2\xi) + R_{N,N'}(\xi), \tag{6.2}$$

where

$$R_{N,N'}(\xi) = \frac{1}{2\pi}(-2|\xi|)^{-N}\int_0^\infty e^{-t}t^{N-1}\left(1+\frac{t}{2\xi}\right)^{-1}R_{N'}\left(\frac{t}{2}\right)dt. \tag{6.3}$$

The idea of re-expanding the remainder terms in optimally truncated asymptotic series was introduced by Berry and Howls [4], and they called this theory *hyperasymptotics*.

With $\arg\zeta = \phi, u = t/|\zeta|$, and $k = |\zeta| + \alpha, \alpha$ being a positive and bounded quantity, the integral in (6.1) can be written as

$$\int_0^\infty \frac{t^{k-1}e^{-t}}{1+t/\zeta}dt = |\zeta|^k\int_0^\infty \frac{u^{\alpha-1}e^{-|\zeta|(u-\log u)}}{1+e^{-i\phi}u}du.$$

The integrand on the right-hand side has a pole at $u = -e^{i\phi}$, which coalesces with the saddle point at $u = 1$ when $\phi = \pm\pi$. An existing theory on uniform asymptotic expansions (see [12, pp. 356-358]) can now be used to show that as $|\zeta| \to \infty$,

$$T_k(\zeta) \sim \frac{1}{2}\mathrm{erfc}(Z) - \frac{i}{\sqrt{2\pi|\zeta|}}e^{-Z^2}\sum_{s=0}^\infty\left(\frac{1}{2}\right)_s g_{2s}(\phi,\alpha)\left(\frac{2}{|\zeta|}\right)^s \tag{6.4}$$

uniformly with respect to $\phi \in [-\pi+\delta, 3\pi-\delta]$, where erfc is the complementary error function, $Z := c(\phi)\sqrt{|\zeta|/2}$, and

$$\frac{1}{2}[c(\phi)]^2 := -e^{i(\phi-\pi)} + i(\phi-\pi) + 1.$$

Near $\phi = \pi$, the Taylor series of $c(\phi)$ begins

$$c(\phi) = -(\phi-\pi) - \frac{i}{6}(\phi-\pi)^2 + \frac{1}{36}(\phi-\pi)^3 + \cdots$$

The coefficients $g_{2s}(\phi, \alpha)$ can be given explicitly and the first one is

$$g_0(\phi,\alpha) = \frac{e^{i\alpha(\pi-\phi)}}{1+e^{-i\phi}} - \frac{i}{c(\phi)}.$$

This result is taken from Olver [9]. He has shown that Z lies in the sector $-\frac{1}{4}\pi < \arg Z \le 0$ when $-\pi \le \phi \le \pi$, and in the sector $0 \le \arg(-Z) \le \frac{1}{4}\pi$ when $\pi \le \phi \le 3\pi$. As ϕ increases from values below π to values above π, Z moves from the first sector to the second sector through the origin. Since $\operatorname{erfc}(Z) = O(e^{-Z^2})$ uniformly throughout the first sector and $\operatorname{erfc}(Z) = 2 + O(e^{-Z^2})$ uniformly throughout the second sector (see [12, p.42]), if $|\zeta|$ is large and fixed and ϕ increases continuously from π^- to π^+, then (6.4) shows that $T_k(\zeta)$ changes rapidly, but smoothly, from being exponentially small to being exponentially close to one.

Coupling (4.12) and (6.2) gives

$$I^{(1)}(\xi) = \sum_{s=0}^{N-1} c_s\xi^{-s} + ie^{2\xi}\sum_{r=0}^{N'-1}(-1)^r c_r \xi^{-r} T_{N-r}(2\xi) + R_{N,N'}(\xi). \quad (6.5)$$

The remainder $R_{N,N'}(\xi)$ is given in (6.3), and can be estimated as before. Of course, it is expected to be of lower order of magnitude, and hence can be neglected. Inserting (6.4) into (6.5), we obtain from (4.5)

$$\begin{aligned}\operatorname{Ai}(z) \sim \frac{(3/2)^{1/2}}{2\pi i} z^{-1/4}\Bigg[& e^{-\xi}\sum_{s=0}^{2|\xi|-1} c_s\xi^{-s} \\ & + \frac{i}{2}\operatorname{erfc}\{c(\theta)|\xi|^{1/2}\}e^{\xi}\sum_{r=0}^{N'-1}(-1)^r c_r\xi^{-r}\Bigg],\end{aligned} \quad (6.6)$$

where $\xi = \frac{2}{3}z^{3/2}$. Note that in (6.6), we have truncated the first series at an optimal place. When θ is near π, $\operatorname{erfc}\{c(\theta)|\xi|^{1/2}\}$ will have an abrupt but smooth change. In Berry's terminology, this function is called a *Stokes multiplier.* A similar result holds for θ near $-\pi$. We shall call the abrupt but smooth change *Berry's transition.*

Exponential asymptotics of other well known entire functions can be investigated in a similar manner. However, the analysis could be considerably more complicated. For instance, in [14] and [15], Wong and Zhao have studied the Berry transition of the generalized Bessel function

$$\phi(z) = \sum_{n=0}^{\infty} \frac{z^n}{\Gamma(n+1)\Gamma(\rho n + \beta)},$$

where $-1 < \rho < \infty$ and β is a real or complex number. They have also discussed the exponential asymptotics of the Mittag-Leffler function [16]

$$E_{\alpha,\beta}(z) = \sum_{n=0}^{\infty} \frac{z^n}{\Gamma(\alpha n + \beta)}, \qquad \operatorname{Re} \alpha > 0,$$

where β may again be real or complex. These two entire functions have played an important role in the study of fractional differential equations [10]; see also [11].

References

1. E. Artin, *The Gamma Function*, Holt, Reinehart and Winston, New York, 1964.
2. M. V. Berry, *Uniform asymptotic smoothing of Stokes' discontinuities*, Proc. Roy. Soc. Lond. A, **422** (1989), 7-21.
3. M. V. Berry, *Asymptotics, superasymptotics, hyperasymptotics*, in Asymptotics Beyond All Orders, H. Segur, S. Tanveer, and H. Levine (eds), Plenum, Amsterdam, 1991, 1-14.
4. M. V. Berry and C. J. Howls, *Hyperasymptotics*, Proc. Roy. Soc. Lond. A, **430** (1990), 653-668.
5. M. V. Berry and C. J. Howls, *Hyperasymptotics for integrals with saddles*, Proc. Roy. Soc. Lond. A, **434** (1991), 657-675.
6. W. G. C. Boyd, *Stieltjes transforms and the Stokes phenomenon*, Proc. Roy. Soc. Lond. A, **429** (1990), 227-246.
7. R. B. Dingle, *Asymptotic Expansions: Their Derivations and Interpretation*, Academic Press, New York, 1973.
8. F. W. J. Olver, *Why steepest descent?* SIAM Rev., **12** (1970), 228-247.
9. F. W. J. Olver, *Uniform, exponentially improved, asymptotic expansions for the generalized exponential integrals*, SIAM J. Math. Anal., **22** (1991), 1460-1474.
10. I. Podlubny, *Fractional Differential Equations*, Academic Press, New York, 1999.
11. J. Wimp, *Review of I. Podlubny's book "Fractional Differential Equations"*, SIAM Rev. **22** (2000), 766-768.
12. R. Wong, *Asymptotic Approximations of Integrals*, Academic Press, Boston, 1989.

13. R. Wong, *Exponential asymptotics*, in Special Functions 2000: Current Perspective and Future Directions, J. Bustoz et al. (eds), NATO Science Series II, Vol.30, Kluwer, 2001, 505-518.
14. R. Wong and Y.-Q. Zhao, *Smoothing of Stokes' discontinuity for the generalized Bessel function*, Proc. Roy. Soc. Lond. A, **455** (1999), 1381-1400.
15. R. Wong and Y.-Q. Zhao, *Smoothing of Stokes' discontinuity for the generalized Bessel function*, II, Proc. Roy. Soc. Lond. A, **455** (1999), 3065-3084.
16. R. Wong and Y.-Q. Zhao, *Exponential asymptotics of the Mittag-Leffler function*, Constr. Approx., **18** (2002), 355-385.
17. M. Wyman, *The method of Laplace*, Trans. Roy. Soc. Canada, **2** (1964), 227-256.

LECTURE III

WKB METHOD AND TURNING POINT

The transition from the classical physics of the late nineteenth century to the quantum mechanics of the early twentieth century is examplified by the problem of finding uniformly asymptotic solutions of the Liouville-Green (WKB) equation.

A.F. Nikiforov and V.B. Uvarov

1. Introduction

Differential equations of the type

$$y''(x) + \{\lambda^2 a(x) + b(x)\}y(x) = 0 \tag{1.1}$$

arise frequently in mathematical physics, where λ is a positive parameter. We first consider the simplest case, in which $a(x)$ is a real, positive, and twice continuously differentiable function in a given finite or infinite interval (a_1, a_2). We also assume that $b(x)$ is a continuous real- or complex-valued function. Let

$$\xi = \int a^{1/2}(x)dx, \qquad w = a^{1/4}(x)y(x). \tag{1.2}$$

It is readily verified that under this transformation, equation (1.1) becomes

$$\frac{d^2w}{d\xi^2} + \{\lambda^2 + \psi(\xi)\}w = 0, \tag{1.3}$$

where

$$\psi(\xi) = \frac{5}{16}\frac{a'^2(x)}{a^3(x)} - \frac{1}{4}\frac{a''(x)}{a^2(x)} + \frac{b(x)}{a(x)}. \tag{1.4}$$

The change of variables from (x, y) to (ξ, w) is known as the *Liouville transformation.*

If we discard ψ in (1.3), then we obtain two linearly independent solutions $e^{\pm i\lambda\xi}$. In terms of the original variables, we get

$$\begin{aligned} y(x) \sim & A a^{-1/4}(x) \exp\{i\lambda \int a^{1/2}(x)dx\} \\ & + B a^{-1/4}(x) \exp\{-i\lambda \int a^{1/2}(x)dx\}, \end{aligned} \tag{1.5}$$

where A and B are arbitrary constants. Equation (1.5) is known as the *Liouville-Green approximation*, whereas physicists refer to (1.5) as the *WKB* (or *semiclassical*) *approximation* in recognition of the work of Wentzel (1926), Kramers (1926) and Brillouin (1926). The contribution of these authors was, however, not really the construction of approximation (1.5), but the connection of exponential and oscillatory approximations across a *turning point*, i.e., a zero of $a(x)$.

In Sec. 2, we will give a rigorous proof of (1.5) from which one will learn a basic argument frequently used to establish the validity of asymptotic solutions to differential equations. From this proof, we will also see a double asymptotic feature in the Liouville-Green approximation, that is, it sometimes holds either as $\lambda \to \infty$ with x fixed, or as $x \to \infty$ with λ fixed. In Sec. 3, we introduce the Langer transformation, and present a uniform asymptotic solution in the neighborhood of a turning point. Sec. 4 deals with the case in which the coefficient functions $a(x)$ and $b(x)$ in (1.1) have, respectively, a simple and a double pole in the interval (a_1, a_2). The final section contains several examples to illustrate the usefulness of the approximations obtained in the previous sections. Most of the material for this lecture is taken from the definitive book by Olver [8].

2. Successive Approximations

The most frequently used method to prove asymptotic results for differential equations is probably the method of successive approximations. In this section, we shall illustrate this method by establishing the validity of (1.5). In (1.3) we substitute

$$w(\xi) = e^{i\lambda\xi}[1 + h(\xi)], \tag{2.1}$$

and obtain

$$h''(\xi) + i2\lambda h'(\xi) = -\psi(\xi)[1 + h(\xi)]. \tag{2.2}$$

We view (2.2) as an inhomogeneous second-order differential equation in $h(\xi)$. By the principle of variation of parameters, one can convert (2.2) into the integral equation

$$h(\xi) = -\frac{1}{i2\lambda} \int_\alpha^\xi \{1 - e^{i2\lambda(v-\xi)}\}\psi(v)[1 + h(v)]dv, \tag{2.3}$$

where α is the value of ξ at $x = a, a = a_1$ or a_2, and we assume α is finite. One can easily verify that any solution of this integral equation is also a solution of the differential equation (2.2).

Define $h_0(\xi) = 0$ and

$$h_{s+1}(\xi) = -\frac{1}{i2\lambda}\int_{\alpha}^{\xi}\{1 - e^{i2\lambda(v-\xi)}\}\psi(v)[1 + h_s(v)]dv \tag{2.4}$$

for $s = 0, 1, 2, \cdots$. Since $|1 - e^{i2\lambda(v-\xi)}| \le 2$, we have $|h_1(\xi)| \le \frac{1}{\lambda}\Psi(\xi)$, where

$$\Psi(\xi) = \int_{\alpha}^{\xi}|\psi(v)|dv. \tag{2.5}$$

Suppose that for $s = k$, we have

$$|h_k(\xi) - h_{k-1}(\xi)| \le \frac{\Psi^k(\xi)}{\lambda^k k!}. \tag{2.6}$$

Then, from (2.4) it follows that

$$h_{k+1}(\xi) - h_k(\xi) = -\frac{1}{i2\lambda}\int_{\alpha}^{\xi}\{1 - e^{i2\lambda(v-\xi)}\}\psi(v)[h_k(v) - h_{k-1}(v)]dv. \tag{2.7}$$

Hence

$$|h_{k+1}(\xi) - h_k(\xi)| \le \frac{1}{\lambda^{k+1}k!}\int_{\alpha}^{\xi}|\psi(v)|\Psi^k(v)dv = \frac{\Psi^{k+1}(\xi)}{\lambda^{k+1}(k+1)!}.$$

By induction, (2.6) holds for all $k \ge 1$. Since $\Psi(\xi)$ is bounded when ξ is finite, the series

$$h(\xi) = \lim_{n\to\infty} h_n(\xi) = \sum_{k=0}^{\infty}\{h_{k+1}(\xi) - h_k(\xi)\} \tag{2.8}$$

is uniformly convergent on any compact ξ-interval. Taking the limit as $n \to \infty$ shows that $h(\xi)$ is a solution of the integral equation (2.3), and hence a solution of the second-order differential equation (2.2). To show that $h(\xi)$ is twice continuously differentiable, we note that (2.4) gives

$$h_1'(\xi) = -\int_{\alpha}^{\xi} e^{i2\lambda(v-\xi)}\psi(v)dv, \tag{2.9}$$

and we have from (2.7)

$$h_{k+1}'(\xi) - h_k'(s) = -\int_{\alpha}^{\xi} e^{i2\lambda(v-\xi)}\psi(v)[h_k(v) - h_{k-1}(v)]dv. \tag{2.10}$$

Again since $|1 - e^{i2\lambda(v-\xi)}| \le 2$, by (2.6) we obtain

$$|h_{k+1}'(\xi) - h_k'(\xi)| \le \frac{\Psi^{k+1}(\xi)}{\lambda^k(k+1)!}, \qquad k = 0, 1, 2, \cdots,$$

which clearly shows the uniform convergence of the series $\sum\{h'_{k+1}(\xi) - h'_k(\xi)\}$ in any compact interval. Furthermore, from (2.2) we have

$$h''_{k+1}(\xi) - h''_k(\xi) = -i2\lambda[h'_{k+1}(\xi) - h'_k(\xi)] - \psi(\xi)[h_k(\xi) - h_{k-1}(\xi)].$$

Uniform convergence of $\sum\{h''_{k+1}(\xi) - h''_k(\xi)\}$ is now evident, and $h(\xi)$ is twice continuously differentiable.

In summary, we have shown that equation (1.3) is satisfied by the function $w(\xi)$ in (2.1) with $h(\xi)$ given by (2.8). Summation of (2.6) gives

$$|h(\xi)| \le \exp\left\{\frac{1}{\lambda}\Psi(\xi)\right\} - 1. \tag{2.11}$$

To express the result in terms of the original variables, we introduce the *control function*

$$F(x) = \int \left\{\frac{1}{a^{1/4}}\frac{d^2}{dx^2}\left(\frac{1}{a^{1/4}}\right) + \frac{b}{a^{1/2}}\right\}dx \tag{2.12}$$

and the notation

$$\mathcal{V}_{a,x}(F) = \int_a^x |F'(t)|dt \tag{2.13}$$

for the total variation of F over the interval (a, x). It is readily verified that $\Psi(\xi) = \mathcal{V}_{a,x}(F)$. Hence, on account of (1.2), equation (2.1) can be written as

$$y_1(x) = a^{-1/4}(x)\exp\{i\lambda \int a^{1/2}(x)dx\}[1 + \varepsilon_1(\lambda, x)] \tag{2.14}$$

with

$$|\varepsilon_1(\lambda, x)| \le \exp\left\{\frac{1}{\lambda}\mathcal{V}_{a,x}(F)\right\} - 1. \tag{2.15}$$

The same argument will yield the second linearly independent solution

$$y_2(x) = a^{-1/4}(x)\exp\{-i\lambda \int a^{1/2}(x)dx\}[1 + \varepsilon_2(\lambda, x)], \tag{2.16}$$

where $\varepsilon_2(\lambda, x)$ also satisfies (2.15), i.e.,

$$|\varepsilon_2(\lambda, x)| \le \exp\left\{\frac{1}{\lambda}\mathcal{V}_{a,x}(F)\right\} - 1. \tag{2.17}$$

For fixed x and large λ, the right-hand sides of (2.15) and (2.17) are both $O(\lambda^{-1})$. Hence, a general solution to (1.1) has the asymptotic behavior given in (1.5).

In (2.15) and (2.17), we can take $a = a_1$ and $a = a_2$, respectively. If $\mathcal{V}_{a_1,a_2}(F) < \infty$, then the O-terms obtained from (2.15) and (2.17) are uniform with respect to x.

The corresponding result for the equation

$$y''(x) - \{\lambda^2 a(x) + b(x)\}y(x) = 0 \tag{2.18}$$

is that we have two linearly independent solutions of the form

$$y_1(x) = a^{-1/4}(x)\exp\{\lambda \int a^{1/2}(x)dx\}[1 + \varepsilon_1(\lambda, x)] \tag{2.19}$$

and

$$y_2(x) = a^{-1/4}(x)\exp\{-\lambda \int a^{1/2}(x)dx\}[1 + \varepsilon_2(\lambda, x)], \tag{2.20}$$

where

$$|\varepsilon_j(\lambda, x)| \le \exp\left\{\frac{1}{2\lambda}\mathcal{V}_{a_j,x}(F)\right\} - 1 \qquad (j = 1, 2). \tag{2.21}$$

If $\mathcal{V}_{a_1,a_2}(F) < \infty$, then (2.21) gives

$$y_j(x) = a^{-1/4}(x)\exp\{(-1)^{j-1}\int a^{1/2}(x)dx\}[1 + O(\lambda^{-1})], \tag{2.22}$$

$j = 1, 2$, which holds uniformly for $x \in (a_1, a_2)$.

The proof of (2.21) is very similar to that of (2.15). The integral equation corresponding to (2.3) is

$$h(\xi) = -\frac{1}{2\lambda}\int_{\alpha_1}^{\xi}\{1 - e^{2\lambda(v-\xi)}\}\psi(v)[1 + h(v)]dv, \tag{2.23}$$

where $\xi = \alpha_1$ corresponds to $x = a_1$. Since $\xi - v \ge 0$, instead of the bound $|1 - e^{i2\lambda(v-\xi)}| \le 2$ used in proving (2.6), we can now use the estimate

$$0 \le 1 - e^{2\lambda(v-\xi)} < 1. \tag{2.24}$$

As a result, we have the extra factor $\frac{1}{2}$ in the total variation $\mathcal{V}_{a_1,x}(F)$ in (2.21). The result for $j = 2$ follows by replacing x in (2.18) by $-x$.

It is interesting to note that the error bounds in (2.21) can also be used to give asymptotic properties of the approximations (2.19) and (2.20) in the neighborhood of a singularity of the differential equation. Because of this *double* asymptotic feature, the Liouville-Green approximation is indeed a remarkably powerful tool for approximating solutions of linear second-order differential equations. To illustrate our point, we let $\lambda = 1$ in (2.18), and assume that $\mathcal{V}_{a_1,a_2}(F) < \infty$ and also

$$\int a^{1/2}(x)dx \to \infty \qquad \text{as} \quad x \to a_2^- . \tag{2.25}$$

Under these conditions, we shall show that the error term $\varepsilon_1(x) := \varepsilon_1(1, x)$ in (2.19) satisfies

$$\varepsilon_1(x) \to \quad \text{a constant} \qquad \text{as } x \to a_2^- . \tag{2.26}$$

This together with (2.19) shows that there is a solution $y_3(x)$ such that

$$y_3(x) \sim a^{-1/4}(x) \exp\left\{\int a^{1/2}(x)dx\right\}, \qquad x \to a_2^- . \tag{2.27}$$

Coupling (2.20) and (2.27), we obtain two linearly independent asymptotic solutions as $x \to a_2^-$.

We now proceed to prove (2.26). From (2.21), we know that $\varepsilon_1(x)$ is bounded in (a_1, a_2). What (2.26) says is that $\varepsilon_1(x)$ does not oscillate infinitely often as $x \to a_2$. In view of (2.25), we have $\alpha_2 = \infty$ by (1.2). Since $\mathcal{V}_{a_1,a_2}(F) < \infty$, for any given $\varepsilon > 0$, there exists a positive number $\delta \in (\alpha_1, \infty)$ such that

$$\int_\delta^\infty |\psi(v)|dv = \varepsilon.$$

Now assume that $\xi \geq \delta$, and subdivide the interval of integration in (2.9) at δ. Note that in the present case, $\lambda = 1$ and "i" is absent; cf. (2.23). Thus, we have

$$|h_1'(\xi)| \leq \int_{\alpha_1}^\delta e^{2(v-\xi)}|\psi(v)|dv + \int_\delta^\xi |\psi(v)|dv \leq e^{2(\delta-\xi)}\Psi(\delta) + \varepsilon.$$

Similarly, from (2.10) and (2.6) we obtain

$$|h_{k+1}'(\xi) - h_k'(\xi)| \leq \frac{e^{2(\delta-\xi)}\Psi^{k+1}(\delta)}{(k+1)!\,2^k} + \frac{\Psi^k(\infty)}{k!\,2^k}\,\varepsilon$$

for $k \geq 1$. The extra factor of 2^{-k} comes from the fact that we use the estimates in (2.24) for the case of eqution (2.18), instead of the bound $|1 - e^{i2\lambda(v-\xi)}| \leq 2$ for the case of equation (1.1). From (2.8), it follows that

$$|h'(\xi)| \leq 2e^{2(\delta-\xi)}\{e^{\Psi(\delta)/2} - 1\} + e^{\Psi(\infty)/2}\,\varepsilon. \tag{2.28}$$

The first term on the right-hand side tends to zero as $\xi \to \infty$, and since ε is arbitrary, (2.28) implies that $h'(\xi) \to 0$ as $\xi \to \infty$.

With $\lambda = 1$ in (2.23), it can be verified by differentiation that

$$h(\xi) = -\frac{1}{2}h'(\xi) - \frac{1}{2}\int_{\alpha_1}^{\xi} \psi(v)[1 + h(v)]dv. \tag{2.29}$$

If we let

$$l_0(\xi) = \int_{\alpha_1}^{\xi} \psi(v)dv, \qquad l_k(\xi) = \int_{\alpha_1}^{\xi} \psi(v)\{h_k(v) - h_{k-1}(v)\}dv \qquad (k \geq 1),$$

then in view of (2.8) we can rewrite (2.29) as

$$h(\xi) = -\frac{1}{2}h'(\xi) - \frac{1}{2}\sum_{k=0}^{\infty} l_k(\xi). \tag{2.30}$$

For $\xi \geq \delta$, we have from (2.6)

$$|l_k(\xi) - l_k(\delta)| \leq \frac{\Psi^k(\xi)}{k!\,2^k}\int_{\delta}^{\xi} |\psi(v)|dv, \qquad k \geq 0, \tag{2.31}$$

since $\Psi(\xi)$ is an increasing function. Coupling (2.30) and (2.31) gives

$$|h(\xi) - h(\delta)| \leq \frac{1}{2}e^{\Psi(\xi)/2}\{\Psi(\xi) - \Psi(\delta)\} + \frac{1}{2}|h'(\xi) - h'(\delta)|.$$

The right-hand side tends to zero as ξ and δ approach infinity independently. Hence, $h(\xi)$ must tend to a constant as $\xi \to \infty$; this completes the proof of (2.26).

3. Turning Point Problem

The problem of finding asymptotic solutions to the differential equation (2.18) becomes much more complicated, when the coefficient function $a(x)$ has a zero, say at $x = x_0$, in the interval (a_1, a_2). Such a point is known as a *turning point* of the differential equation. In this case, there is an ambiguity in taking the square root of the function $a(x)$, and hence the Liouville transformation (1.2) is not well-defined.

For definiteness, we assume that $a(x)$ has the same sign as $x - x_0$; i.e.,

$$a(x)(x - x_0) > 0 \qquad \text{for all} \quad x \neq x_0. \tag{3.1}$$

Instead of (1.2), we now make the change of variables

$$\begin{cases} \frac{2}{3}\zeta^{3/2} = \int_{x_0}^{x} a(t)^{1/2}dt, & x_0 \leq x, \\ \frac{2}{3}(-\zeta)^{3/2} = \int_{x}^{x_0}[-a(t)]^{1/2}dt, & x < x_0, \end{cases} \tag{3.2}$$

and

$$w(\zeta) = \left(\frac{d\zeta}{dx}\right)^{1/2} y. \tag{3.3}$$

It is easily verified that

$$\left(\frac{d\zeta}{dx}\right)^2 = \frac{a(x)}{\zeta}. \tag{3.4}$$

The transformation $(x, y) \mapsto (\zeta, w)$ was first introduced by Langer [3], under which equation (1.1) becomes

$$\frac{d^2w}{d\zeta^2} = \{\lambda^2\zeta + \psi(\zeta)\}w, \tag{3.5}$$

where

$$\psi(\zeta) = \frac{5}{16}\zeta^{-2} + \{4a(x)a''(x) - 5[a'(x)]^2\}\frac{\zeta}{16\, a^3(x)} + \frac{\zeta b(x)}{a(x)}. \tag{3.6}$$

If ψ in (3.5) is neglected, then we have the Airy equation

$$\frac{d^2w}{d\zeta^2} = \lambda^2\zeta w,$$

two linearly independent solutions of which are the Airy functions $\mathrm{Ai}(\lambda^{2/3}\zeta)$ and $\mathrm{Bi}(\lambda^{2/3}\zeta)$. Using the method of successive approximation, one can show, as in Sec. 2, that equation (3.5) has twice continuously differentiable solutions given by

$$\begin{aligned} y_1(x) &= \hat{a}^{-1/4}(x)[\mathrm{Ai}(\lambda^{2/3}\zeta) + \varepsilon_1(\lambda, x)], \\ y_2(x) &= \hat{a}^{-1/4}(x)[\mathrm{Bi}(\lambda^{2/3}\zeta) + \varepsilon_2(\lambda, x)], \end{aligned} \tag{3.7}$$

where $\hat{a}(x) = a(x)/\zeta$; see (3.3) and (3.4). To give an estimate for the error terms $\varepsilon_1(\lambda, x)$ and $\varepsilon_2(\lambda, x)$, we first introduce the error-control function

$$H(x) := -\int_0^\zeta |v|^{-1/2}\psi(v)dv. \tag{3.8}$$

In terms of the original variable, it is equivalent to

$$H(x) = \int_{x_0}^x \left\{\frac{1}{|a|^{1/4}}\frac{d^2}{dx^2}\left(\frac{1}{|a|^{1/4}}\right) - \frac{b}{|a|^{1/2}} - \frac{5|a|^{1/2}}{16|\zeta|^3}\right\}dx. \tag{3.9}$$

The *modulus function* $M(x)$ and the *weight function* $E(x)$ associated with the Airy functions $\mathrm{Ai}(x)$ and $\mathrm{Bi}(x)$ are defined by $E(x) = 1$ for $-\infty < x \le c$,

$$E(x) = \{\mathrm{Bi}(x)/\mathrm{Ai}(x)\}^{1/2}, \qquad c \le x < \infty, \tag{3.10}$$

and

$$M(x) = \{E^2(x)\mathrm{Ai}^2(x) + E^{-2}(x)\mathrm{Bi}^2(x)\}^{1/2}, \tag{3.11}$$

where $E^{-1}(x) = 1/E(x)$ and c denotes the negative root of the equation

$$\mathrm{Ai}(x) = \mathrm{Bi}(x)$$

with smallest absolute value. Numerical calculation has shown that $c = -0.36605$, correct to five decimal places. The modulus and the weight have the well known asymptotic behavior

$$M(x) \sim \pi^{-1/2}|x|^{-1/4}, \qquad x \to \pm\infty, \tag{3.12}$$

and

$$E(x) \sim 2^{1/2}\exp\left(\frac{2}{3}x^{3/2}\right), \qquad x \to +\infty. \tag{3.13}$$

The error terms $\varepsilon_1(\lambda, x)$ and $\varepsilon_2(\lambda, x)$ in (3.7) satisfy

$$\frac{|\varepsilon_1(\lambda, x)|}{M(\lambda^{2/3}\zeta)} \le \frac{E^{-1}(\lambda^{2/3}\zeta)}{\mu}\left\{\exp\left[\frac{\mu\mathcal{V}_{x,a_2}(H)}{\lambda}\right] - 1\right\} \tag{3.14}$$

and

$$\frac{|\varepsilon_2(\lambda, x)|}{M(\lambda^{2/3}\zeta)} \le \frac{E(\lambda^{2/3}\zeta)}{\mu}\left\{\exp\left[\frac{\mu\mathcal{V}_{a_1,x}(H)}{\lambda}\right] - 1\right\}, \tag{3.15}$$

where

$$\mu = \sup_{(-\infty,\infty)}\{\pi|x|^{1/2}M^2(x)\} = 1.04\cdots. \tag{3.16}$$

The above method can be extended to the more general equation

$$y''(x) = \lambda^2 a(\lambda, x)y(x), \tag{3.17}$$

where $a(\lambda, x)$ has an asymptotic expansion

$$a(\lambda, x) \sim a_0(x) + \frac{a_1(x)}{\lambda} + \frac{a_2(x)}{\lambda^2} + \cdots \tag{3.18}$$

as $|\lambda| \to \infty$, and $a_0(x)$ has a zero x_0 in the interval (a_1, a_2). Applying the Langer transformation (3.2) – (3.3) with $a(x)$ replaced by $a_0(x)$, equation (3.17) becomes

$$\frac{d^2w}{d\zeta^2} = \{\lambda^2\zeta + \lambda\phi(\zeta) + \psi(\lambda, \zeta)\}w, \tag{3.19}$$

where we put $\hat{a}(x) = a_0(x)/\zeta$,

$$\phi(\zeta) = a_1(x)/\hat{a}(x) \tag{3.20}$$

and

$$\psi(\lambda,\zeta) \sim \sum_{s=0}^{\infty} \frac{\psi_s(\zeta)}{\lambda^s}, \tag{3.21}$$

with

$$\psi_0(\zeta) = \frac{a_2(\zeta)}{\hat{a}(\zeta)} - \frac{1}{\hat{a}^{3/4}(\zeta)} \frac{d^2}{d\zeta^2}\left\{\frac{1}{\hat{a}^{1/4}(x)}\right\} \tag{3.22}$$

and $\psi_s(\zeta) = a_{s+2}(\zeta)/\hat{a}(\zeta)$ for $s \geq 1$. An asymptotic series solution to (3.19) is given by

$$\begin{aligned} w(\zeta) \sim & \mathrm{Ai}\left(\lambda^{2/3}\zeta + \frac{\Phi}{\lambda^{1/3}}\right) \sum_{s=0}^{\infty} \frac{A_s(\zeta)}{\lambda^s} \\ & + \frac{1}{\lambda^{4/3}} \mathrm{Ai}'\left(\lambda^{2/3}\zeta + \frac{\Phi}{\lambda^{1/3}}\right) \sum_{s=0}^{\infty} \frac{B_s(\zeta)}{\lambda^s}, \end{aligned} \tag{3.23}$$

where $\Phi = \Phi(\zeta)$ is defined by

$$\Phi(\zeta) = \frac{1}{2\zeta^{1/2}} \int_0^{\zeta} \frac{\phi(v)}{v^{1/2}} dv. \tag{3.24}$$

4. Simple Pole

Returning to equation (2.18), namely

$$y''(x) = \{\lambda^2 a(x) + b(x)\} y(x), \tag{4.1}$$

we now assume that $a(x)$ has a simple pole (say) at x_0 and $(x - x_0)^2 b(x)$ is analytic. For simplicity, we also assume that $a(x)$ has the same sign as $x - x_0$. In this case, Olver [7] has introduced the transformation

$$\begin{cases} \zeta^{1/2} = \displaystyle\int_{x_0}^{x} a^{1/2}(t)dt, & x \geq x_0, \\ (-\zeta)^{1/2} = \displaystyle\int_{x}^{x_0} \{-a(t)\}^{1/2} dt, & x \leq x_0, \end{cases} \tag{4.2}$$

and

$$w = \left(\frac{d\zeta}{dx}\right)^{1/2} y \tag{4.3}$$

which transforms (4.1) into the new equation

$$\frac{d^2 w}{d\zeta^2} = \left\{\frac{\lambda^2}{4\zeta} + \hat{\psi}(\zeta)\right\} w, \tag{4.4}$$

where

$$\hat{\psi}(\zeta) = \frac{b(x)}{\hat{a}(x)} + \frac{1}{\hat{a}^{1/4}(x)} \frac{d^2}{d\zeta^2}\{\hat{a}^{1/4}(x)\} \tag{4.5}$$

and $\hat{a}(x) = (d\zeta/dx)^2 = 4\zeta a(x)$. If $b(x)$ has a simple or double pole at x_0, then $\hat{\psi}(\zeta)$ has the same kind of singularity at $\zeta = 0$. Denote the value of $\zeta^2\hat{\psi}(\zeta)$ at $\zeta = 0$ by $\frac{1}{4}(\nu^2 - 1)$, and write (4.4) in the form

$$\frac{d^2w}{d\zeta^2} = \left\{\frac{\lambda^2}{4\zeta} + \frac{\nu^2 - 1}{4\zeta^2} + \frac{\psi(\zeta)}{\zeta}\right\}w \tag{4.6}$$

with $\psi(\zeta) = \zeta\hat{\psi}(\zeta) - \frac{1}{4}(\nu^2 - 1)\zeta^{-1}$. Note that $\psi(\zeta)$ is analytic at $\zeta = 0$. In terms of the original variable, $\frac{1}{4}(\nu^2 - 1)$ is the value of $(x - x_0)^2 b(x)$ at $x = x_0$, and

$$\psi(\zeta) = \frac{1 - 4\nu^2}{16\zeta} + \frac{b(x)}{4a(x)} + \frac{4a(x)a''(x) - 5a'^2(x)}{64a^3(x)}. \tag{4.7}$$

The differential equation (4.6) has a regular singularity at $\zeta = 0$, and we suppose that the range of ζ is a real interval (α_1, α_2) which contains $\zeta = 0$ and may be unbounded. We consider separately the intervals $[0, \alpha_2)$ and $(\alpha_1, 0]$.

If the term $\psi(\zeta)/\zeta$ is neglected, then (4.6) becomes

$$\frac{d^2w}{d\zeta^2} = \left\{\frac{\lambda^2}{4\zeta} + \frac{\nu^2 - 1}{4\zeta^2}\right\}w. \tag{4.8}$$

Two linearly independent solutions of (4.8) are $\zeta^{1/2}I_\nu(\lambda\zeta^{1/2})$ and $\zeta^{1/2}K_\nu(\lambda\zeta^{1/2})$. It can be demonstrated, as in the previous sections, that if $\zeta^{-1/2}\psi(\zeta)$ is absolutely integrable on $[0, \alpha_2)$, then equation (4.6) has two linearly independent solutions $w_1(\lambda, \zeta)$ and $w_2(\lambda, \zeta)$ such that

$$w_1(\lambda, \zeta) = \zeta^{1/2}I_\nu(\lambda\zeta^{1/2})[1 + O(\lambda^{-1})], \tag{4.9}$$

$$w_2(\lambda, \zeta) = \zeta^{1/2}K_\nu(\lambda\zeta^{1/2})[1 + O(\lambda^{-1})], \tag{4.10}$$

as $|\lambda| \to \infty$, where the O-terms hold uniformly with respect to $\zeta \in [0, \alpha_2)$.

When ζ is negative, two linearly independent solutions of the reduced equation (4.8) are $|\zeta|^{1/2}J_\nu(\lambda|\zeta|^{1/2})$ and $|\zeta|^{1/2}Y_\nu(\lambda|\zeta|^{1/2})$. Hence, if $\zeta^{-1/2}\psi(\zeta)$ is absolutely integrable on $(\alpha_1, 0]$, equation (4.6) has two solutions

$$w_1(\lambda, \zeta) = |\zeta|^{1/2}J_\nu(\lambda|\zeta|^{1/2})[1 + O(\lambda^{-1})], \tag{4.11}$$

$$w_2(\lambda, \zeta) = |\zeta|^{1/2}Y_\nu(\lambda|\zeta|^{1/2})[1 + O(\lambda^{-1})], \tag{4.12}$$

as $|\lambda| \to \infty$, which hold uniformly for $\zeta \in (\alpha_1, 0]$.

5. Examples

As an illustration of the results obtained in Secs. 2-4, here we give three concrete examples. The first one demonstrates that the Liouville-Green (WKB) approximation is indeed a powerful tool for approximating solutions of linear second-order differential equations. The second one provides a uniform asymptotic approximation for the polynomials orthogonal with respect to the weight $\exp(-x^4)$ on the real line. The final example deals with the Jacobi function $\varphi_\mu^{(\alpha,\beta)}(t), t > 0$.

Example 1. In an interesting paper [9] on a very abstract topic (namely, Hardy fields), M. Rosenlicht considered solutions of the equation

$$y''(x) = x^x y(x) \tag{5.1}$$

as $x \to \infty$. In view of the rapid growth of the coefficient function x^x as $x \to \infty$, it is really not easy to guess the large x-behavior of the solution $y(x)$. Let us try the results in Sec. 2, and choose $\lambda = 1, a(x) = x^x$ and $b(x) = 0$ in (2.18). The control function $F(x)$ in (2.12) is given by

$$F(x) = \frac{1}{16} \int [(1 + \ln x)^2 x^{-x/2} - 4x^{-(1+x/2)}] dx.$$

Clearly, $\mathcal{V}_{x,\infty}(F) \to 0$ as $x \to \infty$. Hence, (2.20) gives the recessive solution

$$y_2(x) \sim x^{-x/4} \exp\left(- \int x^{x/2} dx \right), \qquad x \to \infty. \tag{5.2}$$

A dominant solution is provided by (2.27), namely,

$$y_3(x) \sim x^{-x/4} \exp\left(\int x^{x/2} dx \right), \qquad x \to \infty. \tag{5.3}$$

An asymptotic expansion of the integral $\int x^{x/2} dx$ can be obtained by integration by parts.

Example 2. In [6], Nevai has studied the asymptotic behavior of the orthogonal polynomials

$$p_n(x) = \gamma_n x^n + \gamma_{n-1} x^{n-1} + \cdots, \qquad \gamma_n > 0, \tag{5.4}$$

associated with the weight function $\exp(-x^4)$ on the real line $\mathbb{R}$. These polynomials satisfy the recurrence relation

$$x p_n(x) = a_{n+1} p_{n+1}(x) + a_n p_{n-1}(x), \qquad n = 0, 1, \cdots, \tag{5.5}$$

with $p_0(x) = \gamma_0 > 0$ and $p_1(x) = \gamma_0 x/a_1$. The coefficients a_n are determined successively from the equation

$$n = 4a_n^2(a_{n+1}^2 + a_n^2 + a_{n-1}^2), \qquad n = 1, 2, \cdots, \tag{5.6}$$

where $a_0^2 = 0$ and $a_1^2 = \Gamma(\frac{3}{4})/\Gamma(\frac{1}{4})$. A two-term asymptotic expansion for a_n has been given by Lew and Quarles [4]. They showed that

$$a_n^2 = \left(\frac{n}{12}\right)^{1/2}\left[1 + \frac{1}{24n^2} + O\left(\frac{1}{n^4}\right)\right], \qquad n \to \infty. \tag{5.7}$$

If we let

$$\phi_n(x) = a_{n+1}^2 + a_n^2 + x^2, \tag{5.8}$$

then Shohat [10] and Nevai [5] independently showed that the function

$$z(x) = p_n(x)[\phi_n(x)]^{-1/2}\exp\left(-\frac{x^4}{2}\right) \tag{5.9}$$

satisfies the differential equation

$$z'' + f(n, x)z = 0, \tag{5.10}$$

where

$$\begin{aligned} f(n, x) = {} & 4a_n^2[4\phi_n(x)\phi_{n-1}(x) + 1 - 4a_n^2x^2 - 4x^4 - 2x^2\phi_n(x)^{-1}] \\ & - 4x^6 - 4x^4\phi_n(x)^{-1} - 3x^2\phi_n(x)^{-1} + 6x^2 + \phi_n(x)^{-1}. \end{aligned} \tag{5.11}$$

If we make the change of variable

$$x = \lambda^{1/4}w \qquad \text{with} \qquad \lambda = \frac{4n}{3}, \tag{5.12}$$

then equation (5.10) becomes

$$\frac{d^2z}{dw^2} = \lambda^2\left[a_0(w) + \frac{a_1(w)}{\lambda} + \frac{a_2(w)}{\lambda^2} + \cdots\right]z, \tag{5.13}$$

where $a_0(w) = (4w^6 - 3w^2 - 1) = (2w^2+1)^2(w^2-1), a_1(w) = -(1 + 2w^2)$ and

$$a_2(w) = -\frac{20w^4 - 64w^2 + 17}{9(1 + 2w^2)^2}. \tag{5.14}$$

Since $a_0(w)$ vanishes at $w = \pm 1$, we have exactly the extended form of the turning point problem mentioned in (3.17) – (3.18), and the result (3.23) can be applied. The details of this study is given in [1].

Example 3. Let α, β and μ be real numbers with $\mu > 0$ and $\alpha \neq -1, -2, \cdots$. The Jacobi function is defined by

$$\varphi_\mu^{(\alpha,\beta)}(t) = {}_2F_1\left[\frac{1}{2}(\alpha+\beta+1-i\mu), \frac{1}{2}(\alpha+\beta+1+i\mu); \alpha+1; -\sinh^2 t\right] \quad (5.15)$$

for $t > 0$, where ${}_2F_1(a, b; c; z)$ is the Gaussian hypergeometric function. This function is related to the Jacobi polynomial

$$\frac{\Gamma(\alpha+1)\Gamma(n+1)}{\Gamma(\alpha+n+1)} P_n^{(\alpha,\beta)}(x) = {}_2F_1\left[-n, n+\alpha+\beta+1; \alpha+1; \frac{1}{2}(1-x)\right]. \quad (5.16)$$

The last formula furnishes the extension of the polynomial $P_n^{(\alpha,\beta)}(x)$ to arbitrary values of the degree n. From (5.15) and (5.16), it is evident that

$$\varphi_\mu^{(\alpha,\beta)}(t) = \frac{\Gamma(\alpha+1)\Gamma(\frac{1}{2}(i\mu-\alpha-\beta+1))}{\Gamma(\frac{1}{2}(i\mu+\alpha-\beta+1))} P^{(\alpha,\beta)}_{\frac{1}{2}(i\mu-\alpha-\beta-1)}(\cosh 2t), \quad (5.17)$$

and for this reason, $\varphi_\mu^{(\alpha,\beta)}(t)$ is called the Jacobi function.

It is known that $\varphi_\mu^{(\alpha,\beta)}(t)$ is the unique, even, C^∞-function on $\mathbb{R}$ which satisfies

$$\begin{aligned} v''(t) + [(2\alpha+1)\coth t + (2\beta+1)\tanh t]v'(t) \\ + [\mu^2 + (\alpha+\beta+1)^2]v(t) = 0 \end{aligned} \quad (5.18)$$

and $v(0) = 1$; see [2, p.2]. If we set

$$u(t) = (\sinh t)^{\alpha+\frac{1}{2}}(\cosh t)^{\beta+\frac{1}{2}}\varphi_\mu^{(\alpha,\beta)}(t), \quad (5.19)$$

then it is easily verified that

$$u''(t) + \left\{\mu^2 + \frac{\frac{1}{4}-\alpha^2}{\sinh^2 t} - \frac{\frac{1}{4}-\beta^2}{\cosh^2 t}\right\}u(t) = 0. \quad (5.20)$$

When $\alpha > -\frac{1}{2}$, we also have $u(0) = 0$.

To apply the asymptotic theory of Olver discussed in Sec. 4, we restrict ourselves to the case $\alpha \geq 0$ and introduce the new variables

$$(-\zeta)^{\frac{1}{2}} = t, \qquad \zeta < 0; \qquad W(\zeta) = (-\zeta)^{\frac{1}{4}}u(t). \quad (5.21)$$

The transformed equation is given by

$$\frac{d^2W}{d\zeta^2} = \left\{\frac{\mu^2}{4\zeta} + \frac{\alpha^2-1}{4\zeta^2} + \frac{\psi(\zeta)}{\zeta}\right\}W(\zeta), \qquad \zeta < 0, \quad (5.22)$$

where

$$\psi(\zeta) = \frac{1}{4}\left\{\frac{\frac{1}{4}-\alpha^2}{\zeta} + \left[\frac{\frac{1}{4}-\alpha^2}{\sinh^2(-\zeta)^{\frac{1}{2}}} - \frac{\frac{1}{4}-\beta^2}{\cosh^2(-\zeta)^{\frac{1}{2}}}\right]\right\}. \quad (5.23)$$

Note that $\psi(\zeta)$ is analytic at $\zeta = 0$.

For negative ζ, equations (4.11) and (4.12) give two asymptotic solutions to (5.22), one involving the Bessel function $J_\alpha(\mu\sqrt{-\zeta})$ and the other involving $Y_\alpha(\mu\sqrt{-\zeta})$. To identify the function $(-\zeta)^{\frac{1}{4}}u(t)$ in (5.21) with one of these two solutions or a linear combination of them, we note that from (5.19) and (5.21) we have

$$(-\zeta)^{\frac{1}{4}}u(t) \sim (-\zeta)^{(\alpha+1)/2}, \qquad \zeta \to 0^-. \tag{5.24}$$

Since $J_\alpha(x) \sim (x/2)^\alpha/\Gamma(\alpha+1)$ for x near zero, it follows from (4.11) and (5.21) that

$$(\sinh t)^{\alpha+\frac{1}{2}}(\cosh t)^{\beta+\frac{1}{2}}\varphi_\mu^{(\alpha,\beta)}(t) \sim \frac{2^\alpha\Gamma(\alpha+1)}{\mu^\alpha}t^{\frac{1}{2}}J_\alpha(\mu t); \tag{5.25}$$

for details, see [11].

References

1. Bo Rui and R. Wong, *A uniform asymptotic formula for orthogonal polynomials associated with* $\exp(-x^4)$, J. Approx. Theory **98** (1999), 146-166.
2. T. H. Koornwinder, *Jacobi functions and analysis on noncompact semisimple Lie group*, in "Special Functions: Group Theoretical Aspects and Applications", R. A. Askey, T. H. Koornwinder, and W. Schempp, eds., D. Reidel, Dordrecht, Holland, 1984, 1-85.
3. R. E. Langer, *The asymptotic solutions of ordinary linear differential equations of the second order, with special reference to a turning point*, Trans. Amer. Math. Soc. **67** (1949), 461-490.
4. J. S. Lew and D. A. Quarles, Jr., *Nonnegative solutions of a nonlinear recurrence*, J. Approx. Theory **38** (1983), 357-379.
5. P. Nevai, *Orthogonal polynomials associated with* $\exp(-x^4)$, in "Second Edmonton Conference on Approximation Theory", Can. Math. Soc. Conf. Proc. **3** (1983), 263-285.
6. P. Nevai, *Asymptotics for orthogonal polynomials associated with* $\exp(-x^4)$, SIAM J. Math. Anal. **15** (1984), 1177-1187.
7. F. W. J. Olver, *The asymptotic solution of linear differential equations of the second order in a domain containing one transition point*, Philos. Trans. Roy. Soc. London Ser. A**249** (1957), 65-97.
8. F. W. J. Olver, *Asymptotic and Special Functions*, Academic Press, New York, 1974. Reprinted by A. K. Peters, Wellesley, 1997.
9. M. Rosenlicht, *Hardy fields*, J. Math. Anal. Appl. **93** (1983), 297-311.
10. J. Shohat, *A differential equation for orthogonal polynomials*, Duke Math. J. **5** (1939), 401-417.
11. R. Wong and Q.-Q. Wang, *On the asymptotics of the Jacobi function and its zeros*, SIAM J. Math. Anal. **23** (1992), 1637-1649.

LECTURE IV

BOUNDARY LAYER THEORY

The paper will certainly prove to be one of the most extraordinary papers of this century, and probably of many centuries

Sydney Goldstein

1. Introduction

At the third International Congress of Mathematicians (1904) in Heidelberg, Ludwig Prandtl [16] presented a short paper entitled *On fluid motion with small frictions.* In 1972, at the *Symposium on the Future of Applied Mathematics*, George Carrier [3] made the remark that "This success is probably most surprising to rigor-oriented mathematicians (or applied mathematicians) when they realize that there still exists no theorem which speaks to the validity or the accuracy of Prandtl's treatment of his boundary-layer problem; but seventy years of observational experience leave little doubt of its validity and its value". With comments like these, one can readily appreciate the important position of Prandtl's principle of boundary-layer theory in the minds of applied mathematicians.

To illustrate Prandtl's idea, K.O. Friedrichs used the simple example

$$\begin{cases} \varepsilon \dfrac{d^2u}{dx^2} + \dfrac{du}{dx} = a + 2bx \\ u(0) = 0, \qquad u(1) = 1, \end{cases} \tag{1.1}$$

in his 1942 lecture notes [5]. Friedrichs' analysis can be easily extended to the more general singularly perturbed two-point boundary-value problem

$$\varepsilon y''(x) + a(x)y'(x) + b(x)y(x) = 0, \tag{1.2a}$$

$$y(-1) = A, \qquad y(1) = B. \tag{1.2b}$$

It is now well known that if $a(x)$ is positive, then the asymptotic solution which holds uniformly in the interval $[-1, 1]$ is given by

$$\begin{aligned} y_{\text{unif}}(x) = {} & B \exp\left(\int_x^1 \frac{b(t)}{a(t)} dt\right) \\ & + \left\{A - B \exp\left(\int_{-1}^1 \frac{b(t)}{a(t)} dt\right)\right\} e^{-a(-1)(1+x)/\varepsilon}. \end{aligned} \tag{1.3}$$

This formula is provided in at least eight standard texts; see, e.g., [2, p.425], [6, pp.53 & 58], [7, p.59], [9, p.68], [10, p.421], [11, p.289], [13, p.94] and [17, p.109]. Despite what Carrier had said above, the lack of rigor in Prandtl's boundary-layer theory does raise some concern from mathematicians who believe that arguments based on purely heuristic reasoning may lead to incorrect results.

A derivation of equation (1.3) is given in Sec. 2, where we also point out that care must be taken in the use of (1.3) when exponentially small terms are involved. In Sec. 3, we consider a case in which the coefficient function $a(x)$ in (1.2a) has a zero. More precisely, we discuss the case when

$$a(x) \sim \alpha x \qquad \text{and} \qquad b(x) \sim \beta, \qquad \text{as } x \to 0, \tag{1.4}$$

where $\alpha \neq 0$ and β are constants. Our discussion will be divided into three subcases; namely, (i) $\alpha > 0$ and $\beta/\alpha \neq 1, 2, \cdots$; (ii) $\alpha < 0$ and $\beta/\alpha \neq 0, -1, -2, \cdots$; (iii) $\alpha > 0$ and $\beta/\alpha = 1, 2, \cdots$, or $\alpha < 0$ and $\beta/\alpha = 0, -1, -2, \cdots$. In case (i), as we shall see, the solution has an internal-layer behavior. The so-called Ackerberg-O'Malley reasonance refers to case (iii). Sec. 4 is devoted to the nonlinear equation

$$\varepsilon u'' + u^2 = 1, \qquad -1 < x < 1, \tag{1.5}$$

with boundary conditions

$$u(-1) = u(1) = 0. \tag{1.6}$$

2. Derivation of (1.3)

Since ε is small, it is natural to set $\varepsilon = 0$ in (1.2a) so that we obtain the *reduced* equation

$$a(x)Y'(x) + b(x)Y(x) = 0. \tag{2.1}$$

The general solution of this equation is

$$Y(x) = K \exp\left(\int_x^1 \frac{b(t)}{a(t)} dt\right),$$

K being an arbitrary constant. The boundary condition at $x = 1$ immediately suggests that $K = B$ and

$$Y(x) = B \exp\left(\int_x^1 \frac{b(t)}{a(t)} dt\right). \tag{2.2}$$

In general, $Y(x)$ can not satisfy the boundary condition at $x = -1$. Hence, the approximate solution $Y(x)$ is valid only in an interval near $x = 1$. This interval is known as the *outer region*, and $Y(x)$ is called the *outer solution.*

In the interval near $x = -1$, which is known as the *boundary-layer* (or *inner*) *region*, we make the change of variable $\xi = \frac{x+1}{\varepsilon}$ and define

$$y(x) = y(\varepsilon\xi - 1) \equiv \overline{y}(\xi).$$

Clearly, $\overline{y}(\xi)$ satisfies the new equation

$$\overline{y}''(\xi) + a(\varepsilon\xi - 1)\overline{y}'(\xi) + \varepsilon b(\varepsilon\xi - 1)\overline{y}(\xi) = 0.$$

Setting $\varepsilon = 0$ gives another simplified equation

$$y''_{BL}(\xi) + a(-1)y'_{BL}(\xi) = 0. \tag{2.3}$$

This equation can again be solved explicitly, and the general solution is given by

$$y_{BL}(\xi) = C_1 + C_2 e^{-a(-1)\xi},$$

where C_1 and C_2 are arbitrary constants. We call $y_{BL}(\xi)$ the *boundary-layer* (or *inner*) solution. The boundary condition at $x = -1$ gives

$$C_1 + C_2 = A.$$

Hence,

$$y_{BL}(\xi) = A + C_2(e^{-a(-1)\xi} - 1). \tag{2.4}$$

The first step in Prandtl's matching principle is to set the two limits

$$\lim_{x \to -1} Y(x) = B \exp\left(\int_{-1}^{1} \frac{b(t)}{a(t)} dt\right) \tag{2.5}$$

and

$$\lim_{\xi \to \infty} y_{BL}(\xi) = A - C_2 \tag{2.6}$$

equal. Thus,

$$C_2 = A - B \exp\left(\int_{-1}^{1} \frac{b(t)}{a(t)} dt\right). \tag{2.7}$$

In (2.5) and (2.6), we have made use of the assumption that $a(x)$ is positive in $[-1, 1]$. The second step in Prandtl's matching principle is to define the uniform approximate solution by

$$y_{\text{unif}}(x) := Y(x) + y_{BL}(\xi) - \text{common part};$$

that is,

$$y_{\text{unif}}(x) = B \exp\left(\int_{x}^{1} \frac{b(t)}{a(t)} dt\right) + A + C_2(e^{-a(-1)(x+1)/\varepsilon} - 1) - A + C_2,$$

thus obtaining (1.3).

If $a(x) < 0$ in $[-1,1]$, then the boundary layer region is at the right endpoint $x = 1$. A similar argument will lead to the corresponding formula

$$\begin{aligned} y_{\text{unif}}(x) &= A\exp\left(-\int_{-1}^{x} \frac{b(t)}{a(t)}dt\right) \\ &\quad + \left\{B - A\exp\left(-\int_{-1}^{1} \frac{b(t)}{a(t)}dt\right)\right\}e^{a(1)(1-x)/\varepsilon}. \end{aligned} \tag{2.8}$$

A natural question to ask now is "In what sense does $y_{\text{unif}}(x)$ approximate the true solution $y(x)$ of (1.2)?" In many books on applied mathematics (e.g., [2] and [9]), one will find the answer

$$y(x) = y_{\text{unif}}(x) + O(\varepsilon), \tag{2.9}$$

where the O-term is uniform with respect to $x \in [-1,1]$. Despite its usefulness, equation (2.8) is not entirely correct. For instance, if the boundary value B in (1.2b) is zero, then (1.3) becomes

$$y_{\text{unif}}(x) = Ae^{-a(-1)(1+x)/\varepsilon}, \tag{2.10}$$

which is exponentially small for $x > -1$, and asymptotically zero with respect to the order estimate in (2.9). The more accurate formula is

$$\begin{aligned} y(x) &= B\exp\left(\int_{x}^{1} \frac{b(t)}{a(t)}dt\right)[1+O(\varepsilon)] \\ &\quad + \frac{a(0)}{a(x)}\left\{A - B\exp\left(\int_{-1}^{1} \frac{b(t)}{a(t)}dt\right)\right\}\exp\left(\int_{-1}^{x} \frac{b(t)}{a(t)}dt\right) \\ &\quad \times \exp\left(-\frac{1}{\varepsilon}\int_{-1}^{x} a(t)dt\right)[1+O(\varepsilon)]. \end{aligned} \tag{2.11}$$

One can establish this result by using the WKB approximation given in Lecture III; it can also be found in [12]. To illustrate our point, let us consider the simple example

$$\varepsilon y'' + (3+x)y' + y = 0, \qquad y(-1) = 1, \qquad y(1) = 0. \tag{2.12}$$

Formulas (2.9) and (2.10) give

$$y_{\text{unif}}(x) = e^{-2(1+x)/\varepsilon} + O(\varepsilon). \tag{2.13}$$

In particular, we have

$$y_{\text{unif}}(0) = e^{-2/\varepsilon} + O(\varepsilon). \tag{2.14}$$

But, from (2.11) it follows that

$$y(x) = \frac{3}{2}\exp\left\{-\frac{1}{\varepsilon}\left(\frac{1}{2}x^2 + 3x + \frac{7}{2}\right)\right\}[1 + O(\varepsilon)]. \tag{2.15}$$

Both approximations (2.14) and (2.15) are exponentially small. However, (2.14) only gives

$$y_{\text{unif}}(x) = O(\varepsilon), \tag{2.16}$$

whereas from (2.15) we get

$$y(0) \sim \frac{3}{2}e^{-7/2\varepsilon}. \tag{2.17}$$

3. Internal Layers

We now turn to the case where the coefficient function $a(x)$ in (1.2a) has a zero in the interval $[-1, 1]$; that is, condition (1.4) holds. In this case, we shall see that boundary layers may occur in the interior as well as at endpoints. By using matching techniques as in Sec. 1, Bender and Orszag [2, p.458] have constructed a relatively simple asymptotic solution for the boundary-value problem (1.2a) – (1.2b). Indeed, they showed that in (1.4) if $\alpha > 0$ and $\beta/\alpha \neq 1, 2, 3, \cdots$, then

$$\begin{aligned} y_{\text{unif}}(x) = {} & \frac{\Gamma(1-\beta/\alpha)}{\sqrt{2\pi}}(\sqrt{\alpha/\varepsilon})^{\beta/\alpha}e^{-\alpha x^2/4\varepsilon} \\ & \times \left\{A\exp\left(\int_x^{-1}\left[\frac{b(t)}{a(t)} - \frac{\beta}{\alpha t}\right]dt\right)D_{\beta/\alpha-1}(x\sqrt{\alpha/\varepsilon}) \right. \\ & \left. + B\exp\left(\int_x^{1}\left[\frac{b(t)}{a(t)} - \frac{\beta}{\alpha t}\right]dt\right)D_{\beta/\alpha-1}(-x\sqrt{\alpha/\varepsilon})\right\}, \end{aligned} \tag{3.1}$$

where $D_\nu(z)$ is the parabolic cylinder function in the notation of Whittaker and Watson. In view of the asymptotic results

$$D_\nu(t) \sim t^\nu e^{-t^2/4}, \qquad D_\nu(-t) \sim \frac{\sqrt{2\pi}}{\Gamma(-\nu)}t^{-\nu-1}e^{t^2/4}, \qquad t \to \infty,$$

and

$$D_\nu(t) \sim \frac{\sqrt{\pi}2^{\nu/2}}{\Gamma(\frac{1}{2} - \frac{1}{2}\nu)}, \qquad t \to 0,$$

one readily obtains

$$y_{\text{unif}}(x) \sim A\exp\left(-\int_{-1}^{x}\frac{b(t)}{a(t)}dt\right), \qquad \frac{x}{\sqrt{\varepsilon}} \to -\infty, \tag{3.2}$$

$$y_{\text{unif}}(x) \sim B \exp\left(\int_x^1 \frac{b(t)}{a(t)} dt\right), \qquad \frac{x}{\sqrt{\varepsilon}} \to \infty \tag{3.3}$$

and

$$y_{\text{unif}}(x) = O(\varepsilon^{-\beta/2\alpha}), \qquad x = O(\sqrt{\varepsilon}). \tag{3.4}$$

Formulas (3.2) – (3.4) reveal that an *internal boundary layer* occurs when $x = O(\sqrt{\varepsilon})$, i.e., in a neighborhood of $x = 0$. A mathmatically rigorous proof of (3.1) has recently been provided by Wong and Yang [18].

For the case $\alpha < 0$ and $\beta/\alpha \neq 0, -1, -2, \cdots$, Bender and Orszag [2, p.460] have also given the leading-order uniform asymptotic solution

$$y_{\text{unif}}(x) = Ae^{-a(-1)(x+1)/\varepsilon} + Be^{a(1)(1-x)/\varepsilon} \tag{3.5}$$

to (1.2*a*) – (1.2*b*). Although this solution appears to behave like the true solution, we can conclude that this result is not correct on two accounts. First, it is evident that the approximate solution given in (3.5) depends only on the values of $a(x)$ at the endpoints of $[-1, 1]$. This means that the approximation will be the same for any coefficient function $a(x)$ if it takes the same values at the endpoints, which does not seem to be reasonable; see a remark later. Next, by applying (3.5) to the concrete problem

$$\varepsilon y'' - 2xy' + (1 + x^2)y = 0, \qquad y(-1) = 2, \qquad y(1) = 1, \tag{3.6}$$

Bender and Orszag gave the asymptotic solution

$$y_{\text{unif}}(x) = 2e^{-2(x+1)/\varepsilon} + e^{-2(1-x)/\varepsilon}. \tag{3.7}$$

The value of $y_{\text{unif}}(x)$ in (3.7) is always positive for $x \in [-1, 1]$. This is contrary to the fact that the exact solution is negative for x near the origin; see Figure 9.17 in [2, p.460].

Very recently, a more detailed study of the boundary-value problem (1.2*a*) – (1.2*b*) has been made by Wong and Yang [19], in the case when $\alpha < 0$ and $\beta/\alpha \neq 0, -1, -2, \cdots$. They have, in fact, given a rigorous derivation of a uniform asymptotic solution in the whole interval $[-1, 1]$. Unlike equation (3.5), their asymptotic formula involves parabolic cylinder functions and the values of the coefficient functions in the entire interval $[-1, 1]$. A crucial step in their derivation is to seek an approximate equation for (1.2a). Their uniform asymptotic formula shows that the true solution has boundary layers near the two endpoints $x = -1$ and $x = 1$, and decays exponentially when the independent variable x is away from the boundary layer regions. Comparing their result with Bender and Orszag's formula (3.5), it is found that (3.7) actually gives only the behavior of the exact solution inside the boundary layers. Away from the boundary layers, the approximate solution in (3.7) is incorrect even though it is exponentially decaying.

When the uniform approximation given in [19] is applied to the concrete example (3.6), we get

$$y(0) \sim -\frac{6\sqrt{2\pi}}{\Gamma(\frac{1}{4})} e^{1/4} \varepsilon^{-3/4} e^{-1/\varepsilon}, \tag{3.8}$$

whereas Bender and Orszag's result (3.7) gives

$$y_{\text{unif}}(0) = 3e^{-2/\varepsilon}. \tag{3.9}$$

This example again illustrates the fact that the heuristic method of matched asymptotics may lead to incorrect results, when exponentially small terms are involved.

As a follow-up to their first two papers [18] and [19], Wong and Yang [20] also studied the exceptional cases (i) $\beta/\alpha = 1, 2, 3, \cdots$ when $\alpha > 0$ and (ii) $\beta/\alpha = 0, -1, -2, \cdots$ when $\alpha < 0$. Uniform asymptotic solutions have been constructed even in these cases. Analyzing the asymptotic behavior of the solutions shows that the solutions do not exhibit the bahavior described in their earlier papers. When $\beta/\alpha = 1, 2, 3, \cdots$ and $\alpha > 0$, the solution grows exponentially in a subinterval of $[-1, 1]$; cf. (3.4). When $\beta/\alpha = 0, -1, -2, \cdots$ and $\alpha < 0$, the solution does not vanish exponentially in a subinterval of $[-1, 1]$; cf. (3.5). This phenomenon was first observed by Ackerberg and O'Malley [1], and they called it a "resonance" phenomenon. Summarizing the results of Wong and Yang in their three papers shows that the conditions in the two exceptional cases are necessary and sufficient for resonance.

4. Carrier-Pearson Equation

The power of boundary-layer theory (the method of matched asymptotics) is that it works equally well for nonlinear differential equations. A simple, yet nontrivial, example is the Carrier-Pearson equation [4]

$$\varepsilon u'' + u^2 = 1, \qquad -1 < x < 1, \tag{4.1}$$

with boundary conditions

$$u(-1) = u(1) = 0. \tag{4.2}$$

The outer solutions obtained by setting $\varepsilon = 0$ in (4.1) are $u_{\text{out},-} = -1$ and $u_{\text{out},+} = 1$, neither of which satisfies the boundary conditions at $x =$

± 1. Therefore, there must exist boundary layers at $x = \pm 1$. Let us examine first the boundary layer at $x = 1$. When we substitute the inner variables

$$\mathrm{X} = \frac{(1-x)}{\delta(\varepsilon)}, \qquad U_{\text{in, right}}(\mathrm{X}) = u(x),$$

into (4.1) and (4.2), we obtain

$$\frac{d^2 U_{\text{in, right}}(\mathrm{X})}{d\mathrm{X}^2} + \frac{\delta^2}{\varepsilon}(U_{\text{in, right}}^2 - 1) = 0 \tag{4.3}$$

and

$$U_{\text{in, right}}(0) = 0. \tag{4.4}$$

To balance the order terms in (4.3), we take $\delta(\varepsilon) = \sqrt{\varepsilon}$. From here on we proceed to solve (4.3) and (4.4), and have the solution asymptotically matched with one (or both) of the outer solutions.

We claim that $U_{\text{in, right}}$ cannot approach $u_{\text{out},+}$ as $\mathrm{X} \to \infty$. Suppose so, and let $U_{\text{in, right}}(\mathrm{X}) = 1 + w(\mathrm{X})$. From (4.3) and (4.4), we have

$$w'' + 2w + w^2 = 0 \tag{4.5}$$

and

$$w(0) = -1, \qquad w(\infty) = 0. \tag{4.6}$$

When w tends to zero, equation (4.5) can be approximated by $w'' + 2w = 0$, whose solutions oscillate as $\mathrm{X} \to \infty$ and cannot tend to 0. This confirms our claim. Thus, we put $U_{\text{in, right}}(\mathrm{X}) = u_{\text{out},-} + w(\mathrm{X}) = -1 + w(\mathrm{X})$. In a similar manner, we have

$$w'' - 2w + w^2 = 0, \tag{4.7}$$

$w(0) = 1$ and $w(\infty) = 0$. Approximating (4.7) by $w'' - 2w = 0$, we anticipate w to decay exponentially as $\mathrm{X} \to \infty$. Hence, we may assume $w'(\infty) = 0$. Multiplying both sides of equation (4.7) by w' and integrating, we obtain

$$(w')^2 - 2w^2 + \frac{2}{3}w^3 = \text{constant}.$$

Since both w' and w tend to zero as $\mathrm{X} \to \infty$, we conclude that the constant is equal to zero and

$$(w')^2 - 2w^2 + \frac{2}{3}w^3 = 0,$$

which is equivalent to the separable first-order equation

$$\frac{dw}{\sqrt{2w^2 - \frac{2}{3}w^3}} = \pm d\mathrm{X}.$$

Integrating this equation gives

$$-\sqrt{2}\tanh^{-1}\sqrt{1 - \frac{1}{3}w} = \pm\mathrm{X} + C.$$

The constant C is determined by the requirement that $w(0) = 1$. Returning to the original variables, we obtain

$$U_{\text{in, right}} = -1 + 3\,\mathrm{sech}^2\left(\pm\frac{1-x}{\sqrt{2\varepsilon}} + \ln(\sqrt{3} + \sqrt{2})\right),$$

which is valid near $x = 1$.

The same argument applies at $x = -1$, and we have

$$U_{\text{in, left}} = -1 + 3\,\mathrm{sech}^2\left(\pm\frac{1+x}{\sqrt{2\varepsilon}} + \ln(\sqrt{3} + \sqrt{2})\right).$$

Matching the outer solution $u_{\text{out},-}$ with $U_{\text{in, right}}$ and $U_{\text{in, left}}$ gives

$$\begin{aligned} u_{\text{unif}}(x) = -1 &+ 3\,\mathrm{sech}^2\left(\pm\frac{1-x}{\sqrt{2\varepsilon}} + \ln(\sqrt{3} + \sqrt{2})\right) \\ &+ 3\,\mathrm{sech}^2\left(\pm\frac{1+x}{\sqrt{2\varepsilon}} + \ln(\sqrt{3} + \sqrt{2})\right) \end{aligned} \tag{4.8}$$

valid over the entire interval $[-1, 1]$.

An alternative, but equivalent, form of (4.8) can also be found in [4]; that is,

$$u_{\text{unif}}(x) = -1 + \frac{12e^{p_1}}{(1 + e^{p_1})^2} + \frac{12e^{p_2}}{(1 + e^{p_2})^2}, \tag{4.9}$$

where

$$p_1 = \pm\sqrt{\frac{2}{\varepsilon}}(1 - x) + 2\ln(\sqrt{3} + \sqrt{2}) \tag{4.10}$$

and

$$p_2 = \pm\sqrt{\frac{2}{\varepsilon}}(1 + x) + 2\ln(\sqrt{3} + \sqrt{2}). \tag{4.11}$$

Note that the solution in (4.8) is not unique. There are in fact four different solutions, depending on the two choices of plus and minus signs in the boundary layer solutions; see Figure 1.

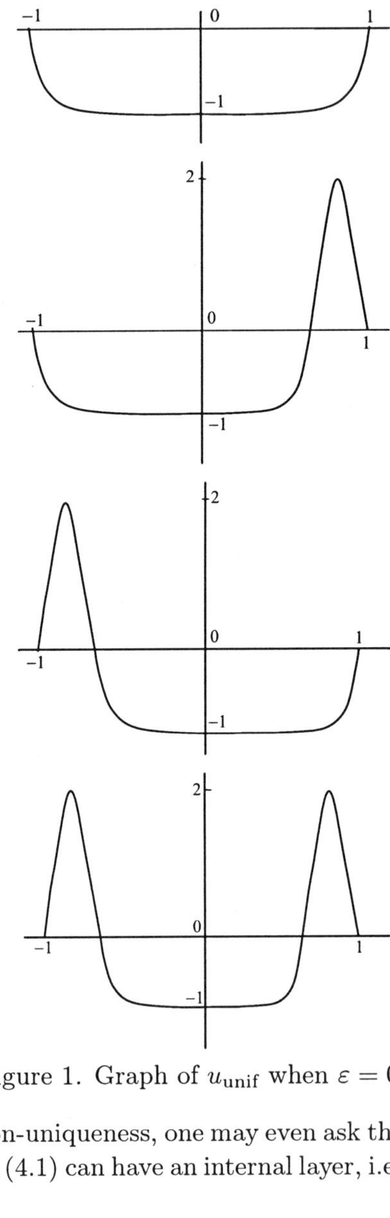

Figure 1. Graph of u_{unif} when $\varepsilon = 0.01$

In view of the non-uniqueness, one may even ask the question: whether a solution of equation (4.1) can have an internal layer, i.e., a narrow region not adjacent to a boundary in which the solution ascends or descends steeply.

Assume that there is an internal layer at $x_0 \in (-1,1)$, where $1-|x_0| \gg \sqrt{\varepsilon}$. It can be shown, as in (4.3), the thickness of such a layer is $\sqrt{\varepsilon}$. Stretching the variable near $x = x_0$ by the transformation

$$\zeta = \frac{x - x_0}{\sqrt{\varepsilon}}, \qquad q(\zeta) = 1 + u(x),$$

we have

$$q'' - 2q + q^2 = 0, \qquad |\zeta| < \infty, \tag{4.12}$$

with

$$q(-\infty) = q(\infty) = 0. \tag{4.13}$$

Solving (4.12) and (4.13) gives

$$q(\zeta) = \frac{12e^{\sqrt{2}\zeta}}{(1+e^{\sqrt{2}\zeta})^2} = 3\,\mathrm{sech}^2\left(\frac{\zeta}{\sqrt{2}}\right)$$

or, equivalently,

$$u = -1 + 3\,\mathrm{sech}^2\left(\frac{x - x_0}{\sqrt{2\varepsilon}}\right) \tag{4.14}$$

for x near x_0. The function $q(\zeta)$ takes the value 3 at $\zeta = 0$, decays to zero with exponential rapidity as $|\zeta| \to \infty$; thus, it behaves like a *spike* near x_0 for sufficiently small ε. Matching (4.14) with the outer solution and the two inner solutions near $x = \pm 1$, we get a composite formula

$$\begin{aligned} u_{\mathrm{unif}}(x) = {}& -1 + 3\,\mathrm{sech}^2\left(\pm\frac{1-x}{\sqrt{2\varepsilon}} + \ln(\sqrt{3}+\sqrt{2})\right) \\ & + 3\,\mathrm{sech}^2\left(\pm\frac{1+x}{\sqrt{2\varepsilon}} + \ln(\sqrt{3}+\sqrt{2})\right) \\ & + 3\,\mathrm{sech}^2\left(\frac{x-x_0}{\sqrt{2\varepsilon}}\right). \end{aligned} \tag{4.15}$$

Formula (4.15) *appears* to be a valid approximation for x in the entire interval. But, by using phase plane analysis, Carrier and Pearson [4, p.204] showed that (4.15) can approximate an exact solution only if

$$x_0 = 0.$$

Thus, for most values of x_0, the solutions given in (4.15) cannot be valid, and they are called *spurious solutions*.

In a recent paper, Ou and Wong [14] have investigated (4.1) – (4.2) from a rigorous point of view. By using a "shooting method", they proved that the formal solutions in (4.15) obtained from the method of matched asymptotics approximate true solutions with an exponentially small error. The so-called spurious solutions turn out to be approximations of true solutions, when the locations of their "spikes" are properly assigned. They also gave an estimate for the maximum number of spikes that these solutions can have.

As a continuation of [14], Ou and Wong extended their results in [15] to include the singularly perturbed two-point problem

$$\varepsilon u'' + Q(u) = 0, \qquad -1 < x < 1, \tag{4.16}$$

with boundary conditions

$$u(-1) = u(1) = 0 \tag{4.17}$$

or

$$u'(-1) = u'(1) = 0, \tag{4.18}$$

where ε is a small positive parameter. The nonlinear term $Q(u)$ vanishes at $s_-, 0, s_+$ and nowhere else in $[s_-, s_+]$, with $s_- < 0 < s_+$. Furthermore, they assumed that $Q'(s_\pm) < 0$, $Q'(0) > 0$ and

$$\int_{s_-}^{s_+} Q(s)ds = 0. \tag{4.19}$$

Simple examples of functions satisfying these conditions are $Q(u) = u(1 - u^2)$ and $Q(u) = \sin \pi u$ for $u \in [-1, 1]$.

Equation (4.16) can be considered as the equation of motion of a nonlinear spring with spring constant large compared to the mass. It is also the steady state version of many partial differential equations arising from physics and biochemistry. Unlike the case $Q(u) = u^2 - 1$, now the solutions exhibit a new phenomenon, known as the *shock* layer, i.e., solutions vary rapidly from one value to another in a very short interval.

References

1. R. C. Ackerberg and R. E. O'Malley, Jr., *Boundary layer problems exhibiting resonance*, Stud. Appl. Math. **49** (1970), 277-295.
2. C. M. Bender and S. A. Orszag, *Advanced Mathematical Methods for Scientists and Engineers*, McGraw-Hill, New York, 1978.
3. G. F. Carrier, *Heuristic Reasoning in Applied Mathematics*, (Special issue : Symposium on the "Future of Applied Mathematics"), Quart. Appl. Math., 1972, 11-15.

4. G. F. Carrier and C. E. Pearson, *Ordinary Differential Equations*, Blaisdell Pub. Co., Waltham, MA, 1968. (Reprinted in SIAM *Classics in Applied Mathematics* series, Vol.6, SIAM, Philadelphia, 1991.)
5. K. O. Friedrichs, *Fluid Dynamics*, Brown University, 1942. (Reprinted in Springer, 1971.)
6. J. Kevorkian and J. D. Cole, *Perturbation Methods in Applied Mathematics*, Springer-Verlag, New York, 1981.
7. P. A. Lagerstrom, *Matched Asymptotic Expansions : Ideas and Techniques*, Springer-Verlag, New York, 1988.
8. C. G. Lange, *On spurious solutions of singular perturbation problems*, Stud. Appl. Math. **68** (1983), 227-257.
9. J. David Logan, *Applied Mathematics*, John Wiley and Sons, New York, 1987.
10. J. A. Murdock, *Perturbation : Theory and Methods*, John Wiley & Sons, New York, 1991.
11. A. H. Nayfeh, *Introduction to Perturbation Techniques*, John Wiley & Sons, New York, 1981.
12. R. E. O'Malley, Jr., *Topics in singular perturbations*, Adv. Math. **2** (1968), 365-470.
13. R. E. O'Malley, Jr., *Singular Perturbation Methods for Ordinary Differential Equations*, Springer-Verlag, New York, 1991.
14. C. H. Ou and R. Wong, *On a two-point boundary value problem with spurious solutions*, Stud. Appl. Math. **111** (2003), 377-408.
15. C. H. Ou and R. Wong, *Shooting method for nonlinear singularly perturbed boundary-value problems*, Stud. Appl. Math., to appear.
16. L. Prandtl, *Über Flüssigkeits - bewegung bei kleiner Reibung*, Verhandlungen, III. Int. Math. Kongresses, Tuebner, Leipzig, 1905, 484-491.
17. J. G. Simmonds and J. E. Mann, *A First Look at Perturbation Theory*, Robert E. Krieger Publishing Co., Malabar, Florida, 1986.
18. R. Wong and H. Yang, *On an internal layer problem*, J. Comp. Appl. Math. **144** (2002), 301-323.
19. R. Wong and H. Yang, *On a boundary-layer problem*, Stud. Appl. Math. **108** (2002), 369-398.
20. R. Wong and H. Yang, *On the Ackerberg-O'Malley Resonance*, Stud. Appl. Math. **110** (2003), 157-179.

LECTURE V

DIFFERENCE EQUATIONS AND ORTHOGONAL POLYNOMIALS

Difference equations might be a handy and practical means to compute differential equations, but they are considerably more complicated to analyze.

Arieh Iserles

1. Introduction

Orthogonal polynomials play an important role in many branches of mathematical physics; for instance, quantum mechanics, scattering theory, and statistical mechanics. A major topic in orthogonal polynomials is the study of their asymptotic behavior as the degree grows to infinity. Since the classical orthogonal polynomials (Hermite, Laguerre, and Jacobi) all satisfy a second-order linear differential equation, their asymptotic behavior can be obtained from the WKB approximation or the turning point theory (Lecture III). For discrete orthogonal polynomials (e.g., Charlier, Meixner, and Krawtchouk), one can use their generating function to obtain a Cauchy integral representation and then apply the steepest descent method or its extensions (Lecture II). However, there are orthogonal polynomials that neither satisfy any differential equation nor have integral representations. A powerful method, known as the steepest descent method for the Riemann-Hilbert problem, has recently been developed that can be applied to such polynomials. Two papers that deserve special mention are Deift & Zhou [10], and Bleher & Its [5]. In fact, there is a whole group of people now working in this area, which includes, in addition to the authors of the two above mentioned papers, Kuijlaars, Kriecherbauer, McLaughlin, Vanlessen, Venakides and Van Assche; see, e.g., [8, 9, 12, 13, 14].

However, in our view, a more desirable approach to derive asymptotic expansions for orthogonal polynomials is to develop an asymptotic theory for linear second order difference equations, in the same way as Langer, Cherry, Olver and others have done for linear second-order differential equations (Lecture III). Our view is based on the fact that any sequence of orthogonal polynomials satisfies the three-term recurrence relation

$$p_{n+1}(x) = (a_n x + b_n)p_n(x) - c_n p_{n-1}(x), \qquad n = 1, 2, \cdots, \tag{1.1}$$

where a_n, b_n and c_n are constants; see [16, p.42]. If x is a fixed number, then this recurrence relation is equivalent to the second-order linear difference equation

$$y(n+2) + n^p a(n) y(n+1) + n^q b(n) y(n) = 0, \tag{1.2}$$

where p and q are integers. When the coefficients $a(n)$ and $b(n)$ have asymptotic expansions of the form

$$a(n) = \sum_{s=0}^{\infty} \frac{\alpha_s}{n^s}, \qquad b(n) = \sum_{s=0}^{\infty} \frac{\beta_s}{n^s}, \tag{1.3}$$

with $\alpha_0 \neq 0$ and $\beta_0 \neq 0$, asymptotic solutions to equation (1.2) have been given by Birkhoff [3], Birkhoff and Trjitzinsky [4]. But, their papers have been considered far too complicated and even impenetrable. A more accessible approach to these results has been given in more recent years by Wong and Li [23, 24]. However, the results in all these papers can be applied to (1.1) only when x is a fixed number. When x is a parameter and allowed to vary, not much work has been done in this area until just recently. In a series of papers [19, 20, 21], Wang and Wong have derived asymptotic expansions for the solutions to (1.1), which hold uniformly for x in infinite intervals. They first define a sequence $\{K_n\}$ recursively by $K_{n+1}/K_{n-1} = c_n$, with K_0 and K_1 depending on the particular sequence of polynomials. Then they put $A_n \equiv a_n K_n / K_{n+1}, B_n \equiv b_n K_n / K_{n+1}$ and $P_n(x) \equiv p_n(x)/K_n$, so that (1.1) becomes

$$P_{n+1}(x) - (A_n x + B_n) P_n(x) + P_{n-1}(x) = 0. \tag{1.4}$$

The coefficients A_n and B_n are assumed to have asymptotic expansions of the form

$$A_n \sim n^{-\theta} \sum_{s=0}^{\infty} \frac{\alpha_s}{n^s} \qquad \text{and} \qquad B_n \sim \sum_{s=0}^{\infty} \frac{\beta_s}{n^s}, \tag{1.5}$$

where θ is a real number and $\alpha_0 \neq 0$.

In this lecture, we summarize the results in [23, 24] and [20, 21]. More precisely, in Sec. 2 we present the results for equation (1.2) when $p = q = 0$, and discuss the general case in Sec. 3. Equation (1.4) is studied in Sec. 4 when $\theta \neq 0$, and in Sec. 5 when $\theta = 0$. Interested readers are referred to the original papers for proofs of the results.

2. Normal and Subnormal Series

When $p = q = 0$, equation (1.2) becomes

$$y(n+2) + a(n)y(n+1) + b(n)y(n) = 0. \tag{2.1}$$

Asymptotic solutions to this equation are classified by the roots of the *characteristic equation*

$$\rho^2 + a_0\rho + b_0 = 0. \tag{2.2}$$

Two possible values of ρ are

$$\rho_1, \rho_2 = -\frac{1}{2}a_0 \pm \left(\frac{1}{4}a_0^2 - b_0\right)^{1/2}. \tag{2.3}$$

If $\rho_1 \neq \rho_2$, i.e., $a_0^2 \neq 4b_0$, then Birkhoff [2] showed that (2.1) has two linearly independent solutions $y_j(n), j = 1, 2$, such that

$$y_j(n) \sim \rho_j^n n^{\alpha_j} \sum_{s=0}^{\infty} \frac{c_{s,j}}{n^s}, \qquad n \to \infty, \tag{2.4}$$

where

$$\alpha_j = -\frac{a_1\rho_j + b_1}{2\rho_j^2 + \rho_j a_0} = \frac{a_1\rho_j + b_1}{a_0\rho_j + 2b_0}, \tag{2.5}$$

$c_{0,j} = 1$ and

$$\begin{aligned} \rho_j(a_0 + 2\rho_j)sc_{s,j} = \sum_{r=1}^{s} \Bigg[& \rho_j^2 2^{r+1} \binom{\alpha_j + r - s}{r+1} \\ & + \rho_j \sum_{q=0}^{r+1} \binom{\alpha_j + r - s}{r+1-q} a_q + b_{r+1} \Bigg] c_{s-r,j}, \end{aligned} \tag{2.6}$$

$s = 1, 2, \cdots$. This construction fails when and only when $\rho_1 = \rho_2$, i.e., when $a_0^2 = 4b_0$. The series in (2.4) are known as *normal series* or *normal solutions.*

If $\rho_1 = \rho_2$ but their common value $\rho = -\frac{1}{2}a_0$ is not a root of the auxiliary equation

$$a_1\rho + b_1 = 0, \tag{2.7}$$

i.e., $2b_1 \neq a_0a_1$, then Adams [1] showed that (2.1) has two linearly independent solutions $y_j(n), j = 1, 2$, such that

$$y_j(n) \sim \rho^n \exp((-1)^j \gamma\sqrt{n}) n^{\alpha} \sum_{s=0}^{\infty} (-1)^{js} \frac{c_s}{n^{s/2}}, \tag{2.8}$$

where

$$\gamma = 2\sqrt{\frac{a_0 a_1 - 2b_1}{2b_0}}, \tag{2.9}$$

$$\alpha = \frac{1}{4} + \frac{b_1}{2b_0}, \tag{2.10}$$

and $c_0 = 1$. Series of the form (2.8) are called *subnormal series* or *subnormal solutions.* Higher coefficients can be determined by formal substitution.

When the double root of the characteristic equation (2.2) satisfies the auxiliary equation (2.7), i.e., when $2b_1 = a_0 a_1$, we have three (exceptional) cases to consider, depending on the values of the zeros $\alpha_1, \alpha_2(\text{Re } \alpha_2 \geq \text{Re } \alpha_1)$ of the *indicial polynomial*

$$q(\alpha) = \alpha(\alpha - 1)\rho^2 + (a_1\alpha + a_2)\rho + b_2. \tag{2.11}$$

Case (i) : $\alpha_2 - \alpha_1 \neq 0, 1, 2, \cdots$. In this case, (2.1) has independent solutions $y_j(n), j = 1, 2$, of the form

$$y_j(n) \sim \rho^n n^{\alpha_j} \sum_{s=0}^{\infty} \frac{c_{s,j}}{n^s}, \qquad n \to \infty, \tag{2.12}$$

with $c_{0,j} = 1$.

Case (ii) : $\alpha_2 - \alpha_1 = 1, 2, \cdots$. Here, (2.12) applies only in the case of $j = 1$. A second independent solution is given by

$$y_2(n) \sim \rho^n n^{\alpha_2} {\sum_{s=0}^{\infty}}' \frac{d_s}{n^s} + c(\log n) y_1(n), \tag{2.13}$$

where the prime on $\sum$ denotes that the term for $s = \alpha_2 - \alpha_1$ is absent. The coefficients c and d_s can be determined by formal substitution, beginning with $d_0 = 1$.

Case (iii) : $\alpha_2 = \alpha_1$. As in case (ii), (2.12) again gives only one solution $y_1(n)$. The second solution is given by

$$y_2(n) \sim \rho^n n^{\alpha_2} \sum_{s=1}^{\infty} \frac{d_s}{n^s} + (\log n) y_1(n). \tag{2.14}$$

3. Equation (1.2) with p and $q \neq 0$.

Many orthogonal polynomials satisfy difference equations of the form (1.2), but not of the form (2.1). For example, the recurrence relation for the Charlier polynomials is

$$C_{n+1}^{(a)}(x) + (n + a - x)C_n^{(a)}(x) + anC_{n-1}^{(a)}(x) = 0, \qquad a \neq 0, \tag{3.1}$$

and the recurrence relation for the Bessel polynomials $y_n(x)$ is

$$y_{n+1}(x) = (2n+1)xy_n(x) + y_{n-1}(x); \tag{3.2}$$

see [6, Chap.VI].

If the exponents p and q in (1.2) are related in the manner $q = 2p$, then (1.2) can be reduced to (2.1) by using the transformation

$$x(n) = [(n-2)!]^{\mu} y(n). \tag{3.3}$$

Indeed, substitution of (3.3) in (1.2) gives the equivalent equation

$$x(n+2) + n^{p+\mu} a(n) x(n+1) + n^{q+2\mu} b^*(n) x(n) = 0, \tag{3.4}$$

where

$$b^*(n) = b(n)\left(1 - \frac{1}{n}\right)^{\mu} = \sum_{s=0}^{\infty} \frac{b_s^*}{n^s}. \tag{3.5}$$

(Note that the constant term in the expansion (3.5) is again not zero.) If $q = 2p$ then, by taking $\mu = -p$, (3.4) becomes an equation of the same form as (2.1). Thus, our discussion of equation (1.2) consists of only two cases, namely, (i) $q < 2p$ and (ii) $q > 2p$. We set $k \equiv 2p - q$. Each of these two cases has two or three subcases, depending on the values of k.

In case (i), i.e., when $k > 0$, it has been shown in [23] that equation (1.2) has an asymptotic solution of the form

$$y_1(n) \sim [(n-2)!]^p \rho^n n^{\alpha} \sum_{s=0}^{\infty} \frac{c_s}{n^s}, \tag{3.6}$$

where $\rho = -a_0, \alpha = a_1/a_0$ if $k > 1$, and

$$\alpha = \frac{1}{a_0}\left(a_1 - \frac{b_0}{a_0}\right) \qquad \text{if} \quad k = 1. \tag{3.7}$$

A second independent solution is given by

$$y_2(n) = [(n-2)!]^{q-p} \rho^n n^{\alpha} \sum_{s=0}^{\infty} \frac{d_s}{n^s}, \tag{3.8}$$

where $\rho = -b_0/a_0$,

$$\alpha = \frac{b_0}{a_0^2} - \frac{a_1}{a_0} + \frac{b_1}{b_0} - p + q \qquad \text{if} \quad k = 1, \tag{3.9}$$

and

$$\alpha = \frac{b_1}{b_0} - \frac{a_1}{a_0} - p + q \qquad \text{if} \quad k > 1. \tag{3.10}$$

Recursive formulas can be obtained for the coefficients c_s and d_s by formal substitution.

In case (ii), i.e., when $k < 0$, we have three subcases to consider depending on whether $k = -1$, or $k \leq -3$ and is odd, or $k \leq -2$ and is even. If $k = -1$, then two asymptotic solutions are of the form

$$y_j(n) \sim [(n-2)!]^{q/2} \rho_j^n \exp(\gamma_j \sqrt{n}) n^{\alpha} \sum_{s=0}^{\infty} \frac{c_{s,j}}{n^{s/2}}, \tag{3.11}$$

$j = 1, 2$, where $\rho_j^2 = -b_0, \gamma_j = -a_0/\rho_j$ and

$$\alpha = \frac{b_1}{2b_0} + \frac{q}{4}. \tag{3.12}$$

If $k \leq -3$ and k is odd, then the exponential factor in (3.11) is absent and equation (3.11) becomes

$$y_j(n) \sim [(n-2)!]^{q/2} \rho_j^n n^{\alpha} \sum_{s=0}^{\infty} \frac{c_{s,j}}{n^{s/2}}, \tag{3.13}$$

where ρ and α are as given in (3.11) and (3.12). If $k \leq -2$ and k is even, then coefficients of odd powers of $n^{-\frac{1}{2}}$ all vanish and (3.11) simplifies to

$$y_j(n) \sim [(n-2)!]^{q/2} \rho_j^n n^{\alpha_j} \sum_{s=0}^{\infty} \frac{c_{s,j}}{n^s} \tag{3.14}$$

with $\rho_j^2 = -b_0 (j = 1, 2), \alpha_j = \alpha$ given in (3.12) if $k = -4, -6, \cdots$, and

$$\alpha_j = -\frac{1}{2\rho_j^2}\left(b_1 + \frac{1}{2}qb_0 + \rho_j a_0\right) \qquad \text{if} \quad k = -2. \tag{3.15}$$

4. Airy-type Expansion

Let τ_0 be a constant, and put $\nu := n + \tau_0$. Clearly, the expansions in (1.5) can be recast in the form

$$A_n \sim \nu^{-\theta} \sum_{s=0}^{\infty} \frac{\alpha'_s}{\nu^s} \qquad \text{and} \qquad B_n \sim \sum_{s=0}^{\infty} \frac{\beta'_s}{\nu^s}. \tag{4.1}$$

In (1.4), we now let $x = \nu^{\theta} t$ and $P_n = \lambda^n$. Substituting (4.1) into (1.4) and letting $n \to \infty$ (and hence $\nu \to \infty$), we obtain the characteristic equation

$$\lambda^2 - (\alpha'_0 t + \beta'_0)\lambda + 1 = 0. \tag{4.2}$$

The roots of this equation are given by

$$\lambda = \frac{1}{2}\left[(\alpha'_0 t + \beta'_0) \pm \sqrt{(\alpha'_0 t + \beta'_0)^2 - 4}\right], \tag{4.3}$$

and they coincide when $t = t_\pm$, where

$$\alpha_0' t_\pm + \beta_0' = \pm 2. \tag{4.4}$$

The values $t_\pm$ play an important role in the asymptotic theory of the three-term recurrence relation (1.4), and they correspond to the transition points (i.e., turning points and poles) occurring in differential equations; cf. [15, p.362]. For this reason, we shall also call them *transition points*. Since t_+ and t_- have different values, we may restrict ourselves to just the case $t = t_+$. For t near t_+, we try a formal series solution to (1.4) in the form

$$P_n(x) = \sum_{s=0}^{\infty} \chi_s(\xi)\delta^s, \tag{4.5}$$

where δ is a small quantity depending on ν (e.g., a power of ν^{-1}) and ξ depends on x and ν. This particular form of expansion was suggested by Costin and Costin [7]. In terms of the exponent θ in (4.1) and the transition point t_+, we have three cases to consider; namely, (i) $\theta \neq 0$ and $t_+ \neq 0$; (ii) $\theta \neq 0$ and $t_+ = 0$; and (iii) $\theta = 0$.

In this section, we shall consider only the first case, namely, case (i). For simplicity, we assume $\theta > 0$. The analysis for the case $\theta < 0$ is very similar; for an important example with $\theta = -1$, the interested reader is referred to [19]. In case (i), we choose

$$\tau_0 = -\frac{(\alpha_1 t_+ + \beta_1)}{(2-\beta_0)\theta} \tag{4.6}$$

so that

$$\alpha_1' t_+ + \beta_1' = 0. \tag{4.7}$$

Also, in (4.5), we choose $\delta = \nu^{-\frac{1}{3}}$ and $\xi = \delta^{-2}\zeta(t)$, where $\zeta(t)$ is an increasing function with $\zeta(t_+) = 0$. Substituting (4.5) into (1.4), we find that χ_0 satisfies the Airy equation

$$\chi''(\xi) = \Theta^3 \xi \chi(\xi), \tag{4.8}$$

where $\Theta^3 = \alpha_0'/\theta^2 t_+^2 \zeta'^3(t_+)$, and that each $\chi_s, s = 1, 2, \cdots$, is a solution of an inhomogeneous Airy equation. (For details, see [18, Chap.3].) This suggests that instead of (4.5), we might as well try the more accurate formal series solution

$$\begin{aligned} P_n(\nu^\theta t) = \nu^{\frac{1}{6}} \left(\frac{4\zeta}{(\alpha_0' t + \beta_0')^2 - 4} \right)^{\frac{1}{4}} \Bigg[\mathrm{Ai}(\nu^{\frac{2}{3}}\zeta) \sum_{s=0}^{\infty} \frac{A_s(\zeta)}{\nu^s} \\ + \frac{\mathrm{Ai}'(\nu^{\frac{2}{3}}\zeta)}{\nu^{\frac{1}{3}}} \sum_{s=0}^{\infty} \frac{B_s(\zeta)}{\nu^s} \Bigg], \end{aligned} \tag{4.9}$$

which we have encountered in the differential equation theory [15, p.409] and the integral approach [22, p.370]. It turns out that this form of solution is not sufficiently general, unless

$$\alpha_1' = \beta_1' = 0. \tag{4.10}$$

It is interesting to note that this condition holds in most of the classical cases. In fact, in [11] Dingle and Morgan have assumed that $\alpha_{2s+1} = \beta_{2s+1} = 0$ for $s = 0, 1, 2, \cdots$. For a more general form of the solution, we refer the reader to [20]. To show that (4.9) is indeed an asymptotic expansion, we need first to determine the function $\zeta(t)$ in the argument of the Airy function. To do this, we first substitute (4.9) in (1.4), then match the coefficients of Ai and Ai′, and finally let $\nu \to \infty$. This leads us to

$$\begin{aligned} \frac{2}{3}[\zeta(t)]^{\frac{3}{2}} = \alpha_0' \, t^{1/\theta} \int_{t_+}^{t} \frac{s^{-1/\theta}}{\sqrt{(\alpha_0' s + \beta_0')^2 - 4}} \, ds \\ - \log \frac{\alpha_0' t + \beta_0' + \sqrt{(\alpha_0' t + \beta_0')^2 - 4}}{2} \end{aligned} \tag{4.11}$$

if $t \geq t_+$, and

$$\begin{aligned} \frac{2}{3}[-\zeta(t)]^{\frac{3}{2}} = \cos^{-1} \frac{\alpha_0' t + \beta_0'}{2} \\ - \alpha_0' \, t^{1/\theta} \int_{t}^{t_+} \frac{s^{-1/\theta}}{\sqrt{4 - (\alpha_0' s + \beta_0')^2}} \, ds \end{aligned} \tag{4.12}$$

if $t < t_+$. To make the presentation simpler, we have assumed in [20] that $t_- < 0 < t_+$. This assumption is equivalent to the condition $|\beta_0| < 2$.

The second solution, independent of (4.9), is given by

$$\begin{aligned} Q_n(\nu^\theta t) \sim \nu^{\frac{1}{6}} \left(\frac{4\zeta}{(\alpha_0' t + \beta_0')^2 - 4} \right)^{\frac{1}{4}} \Bigg[\mathrm{Bi}(\nu^{\frac{2}{3}} \zeta) \sum_{s=0}^{\infty} \frac{A_s(\zeta)}{\nu^s} \\ + \frac{\mathrm{Bi}'(\nu^{\frac{2}{3}} \zeta)}{\nu^{\frac{1}{3}}} \sum_{s=0}^{\infty} \frac{B_s(\zeta)}{\nu^s} \Bigg]. \end{aligned} \tag{4.13}$$

The coefficients $A_s(\zeta)$ and $B_s(\zeta)$ are determined successively from some recursive formulas, beginning with $A_0(\zeta) = 1$ and $\zeta^{\frac{1}{2}} B_0(\zeta) = 0$.

5. Bessel-type Expansion

We now consider case (iii), i.e., $\theta = 0$ in (4.1). As in Sec. 4, we let τ_0 be a constant and define $N := n + \tau_0$. The characteristic equation (4.2) is obtained in the same manner, except that t is replaced by x. The characteristic roots again coincide when $x = x_\pm$, where $\alpha_0 x_\pm + \beta_0 = \pm 2$. For x near x_+, we try a formal series solution of the form

$$P_n(x) = \sum_{s=0}^{\infty} \chi_s(\xi) N^{-s}, \tag{5.1}$$

where ξ depends on x and N. In the present case, we choose $\xi = N\zeta^{1/2}(x)$, where $\zeta(x)$ is an increasing function with $\zeta(x_+) = 0$. Substituting (5.1) into (1.4), one finds that $\chi_0(\xi)$ satisfies the Bessel equation

$$\frac{d^2\chi_0}{d\xi^2} = \left(\frac{\alpha_0'}{\zeta'(0)} + \frac{\alpha_2' x_+ + \beta_2'}{\xi^2}\right)\chi_0.$$

Thus, it follows that $\chi_0(\xi)$ can be expressed in terms of either the Bessel functions $J_\nu(\xi)$ and $Y_\nu(\xi)$, or the modified Bessel functions $I_\nu(\xi)$ and $K_\nu(\xi)$. That is, there are constants C_1 and C_2 such that

$$\chi_0(\xi) = C_1 \xi^{1/2} J_\nu(\xi) + C_2\, \xi^{1/2} Y_\nu(\xi) \qquad \text{if} \quad \alpha_0' < 0$$

and

$$\chi_0(\xi) = C_1 \xi^{1/2} I_\nu(\xi) + C_2\, \xi^{1/2} K_\nu(\xi) \qquad \text{if} \quad \alpha_0' > 0,$$

where

$$\nu = \left(\alpha_2' x_+ + \beta_2' + \frac{1}{4}\right)^{1/2} \tag{5.2}$$

and we take the square root with nonnegative real part. Moreover, each of the subsequent coefficient functions $\chi_s(\xi), s = 1, 2, \cdots$, in (5.1) satisfies an inhomogeneous Bessel equation. This suggests that instead of (5.1), we might as well try the formal series solution of the form

$$P_n(x) = N^{\frac{1}{2}} Z_\nu(N\zeta^{\frac{1}{2}}) \sum_{s=0}^{\infty} \frac{\overline{A}_s(\zeta)}{N^s} + N^{\frac{1}{2}} \zeta^{\frac{1}{2}} Z_{\nu+1}(N\zeta^{\frac{1}{2}}) \sum_{s=0}^{\infty} \frac{\overline{B}_s(\zeta)}{N^s} \tag{5.3}$$

motivated from the difference equation theory [15, p.441]. In (5.3), $Z_\nu(\xi)$ can be any solution of the modified Bessel equation

$$y'' + \frac{1}{x} y' - \left(1 + \frac{\nu^2}{x^2}\right) y = 0. \tag{5.4}$$

The function $\zeta(x)$ in (5.3) can be determined by subtituting (5.3) in (1.4), matching the coefficients of Z_ν and $Z_{\nu+1}$, and letting $n \to \infty$. The result is

$$\zeta^{\frac{1}{2}}(x) = \cosh^{-1}\left(\frac{\alpha_0' x + \beta_0'}{2}\right). \tag{5.5}$$

In [21], it was shown that when $\theta = 0$ in (4.1), equation (1.4) has a pair of linearly independent solutions

$$P_n(x) \sim \left(\frac{4\zeta}{(\alpha_0' x + \beta_0')^2 - 4}\right)^{\frac{1}{4}} \left[N^{\frac{1}{2}} I_\nu(N\zeta^{\frac{1}{2}}) \sum_{s=0}^{\infty} \frac{A_s(\zeta)}{N^s} + N^{\frac{1}{2}} \zeta^{\frac{1}{2}} I_{\nu-1}(N\zeta^{\frac{1}{2}}) \sum_{s=0}^{\infty} \frac{B_s(\zeta)}{N^s}\right] \tag{5.6}$$

and

$$Q_n(x) \sim \left(\frac{4\zeta}{(\alpha_0' x + \beta_0')^2 - 4}\right)^{\frac{1}{4}} \left[N^{\frac{1}{2}} K_\nu(N\zeta^{\frac{1}{2}}) \sum_{s=0}^{\infty} \frac{A_s(\zeta)}{N^s} + N^{\frac{1}{2}} \zeta^{\frac{1}{2}} K_{\nu-1}(N\zeta^{\frac{1}{2}}) \sum_{s=0}^{\infty} \frac{B_s(\zeta)}{N^s}\right], \tag{5.7}$$

where ν is given in (5.2) and

$$N = n + \tau_0 = n - \frac{\alpha_3 x_+ + \beta_3}{2(\nu^2 - \frac{1}{4})}. \tag{5.8}$$

References

1. C. R. Adams, *On the irregular cases of linear ordinary difference equations*, Trans. Amer. Math. Soc. **30** (1928), 507-541.
2. G. D. Birkhoff, *General theory of linear difference equations*, Trans. Amer. Math. Soc. **12** (1911), 243-284.
3. G. D. Birkhoff, *Formal theory of irregular linear difference equations*, Acta Math. **54** (1930), 205-246.
4. G. D. Birkhoff and W. J. Trjitzinsky, *Analytic theory of singular difference equations*, Acta Math. **60** (1932), 1-89.
5. P. Bleher and A. Its, *Semiclassical asymptotics of orthogonal polynomials, Riemann-Hilbert problem, and universality in the matrix model*, Ann. Math. **150** (1999), 185-266.
6. T. S. Chihara, *An Introduction to Orthogonal Polynomials*, Gordon and Breach, New York, 1978.
7. O. Costin and R. Costin, *Rigorous WKB for finite-order linear recurrence relations with smooth coefficients*, SIAM J. Math. Anal. (1996), 110-134.
8. P. Deift, T. Kriecherbauer, K. T-R McLaughlin, S. Venakides, and X. Zhou, *Uniform asymptotics for polynomials orthogonal with respect to varying exponential weights and applications to universality questions in random matrix theory*, Comm. Pure Appl. Math. **52** (1999), 1335-1425.
9. P. Deift, T. Kriecherbauer, K. T-R McLaughlin, S. Venakides, and X. Zhou, *Strong asymptotics of orthogonal polynomials with respect to exponential weights*, Comm. Pure Appl. Math. **52** (1999), 1491-1552.

10. P. Deift and X. Zhou, *A steepest descent method for oscillatory Riemann-Hilbert problems, Applications for the MKdV equation*, Ann. Math. **137** (1993), 295-368.
11. R. B. Dingle and G. J. Morgan, *WKB methods for difference equtions* I, Appl. Sci. Res. **18** (1967), 221-237.
12. T. Kriecherbauer and K. T-R McLaughlin, *Strong asymptotics of polynomials orthogonal with respect to Freud weights*, Internat. Math. Res. Notes (1999), 299-333.
13. A. B. J. Kuijlaars and K. T-R McLaughlin, *Riemann-Hilbert analysis for Laguerre polynomials with large negative parameter*, Comput. Math. Funct. Theory **1** (2001), 205-233.
14. A. B. J. Kuijlaars and M. Vanlessen, *Universality for eigenvalue correlations from the modified Jacobi unitary ensemble*, Internat. Math. Res. Notices (2002), 1575-1600.
15. F. W. J. Olver, *Asymptotics and Special Functions*, Academic Press, New York, 1974. (Reprinted by A. K. Peters Ltd., Wellesley, 1997).
16. G. Szegö, *Orthogonal Polynomials*, Fourth edition, Colloquium Publications, Vol 23, Amer. Math. Soc., Providence R. I., 1975.
17. W. Van Assche, J. S. Geromino, and A. B. J. Kuijlaars, *Riemann-Hilbert problems for multiple orthogonal polynomials*, pp.23-50 in "NATO ASI Special Functions 2000" (J. Bustoz et. al., eds.), Kluwer Academic Publishers, Dordrecht 2001.
18. Z. Wang, *Asymptotic expansions for second order linear difference equations with a turning point*, Ph.D. Thesis, City University of Hong Kong, 2001.
19. Z. Wang and R. Wong, *Uniform asymptotic expansion of $J_\nu(\nu a)$ via a difference equation*, Numer. Math. **91** (2002), 147-193.
20. Z. Wang and R. Wong, *Asymptotic expansions for second-order linear difference equations with a turning point*, Numer. Math. **94** (2003), 147-194.
21. Z. Wang and R. Wong, *Linear difference equations with transition points*, Math. Comp., to appear.
22. R. Wong, *Asymptotic Approximations of Integrals*, Academic Press, Boston, 1989. (Reprinted by SIAM, Philadelphia, PA, 2001.)
23. R. Wong and H. Li, *Asymptotic expansions for second-order linear difference equations*, J. Comput. Appl. Math. **41** (1992), 65-94.
24. R. Wong and H. Li, *Asymptotic expansions for second-order linear difference equations* II, Stud. Appl. Math. **87** (1992), 289-324.

A PERTURBATION MODEL FOR THE GROWTH OF TYPE III-V COMPOUND CRYSTALS

C. SEAN BOHUN,* IAN FRIGAARD,† HUAXIONG HUANG,‡ & SHUQING LIANG§

In this paper we describe the basic idea of a semi-analytical approach for computing the temperature and thermal stress inside a type III-V compound grown with the Czochralski technique. An analysis of the growing conditions indicates that the crystal growth occurs on the conductive time scale. A perturbation method for the temperature field is developed using the Biot number as a (small) expansion parameter whose zeroth order solution is one-dimensional (in the axial direction) and is obtained for a cylindrical and a conical crystal. Under typical growth conditions, a parabolic temperature profile in the radial direction is shown to arise naturally as the first order correction. As a result, the thermal stress is obtained explicitly and its magnitude is shown to depend on the zeroth order temperature and Biot number.

1. Introduction

The Czochralski method is the most popular technique for growing the single large crystals used by the semi-conductor and related industries. By dipping a small seed crystal into a pool of molten material in the crucible and carefully controlling the heat balance inside the grower, a large crystal can be grown by pulling the crystal away from the melt in a slow and steady fashion. The pulling rod and the crucible are normally rotated in the opposite directions during the growth period. Delicate control is often needed to maintain the crystal quality and a slight change of growth condition may result in defect formation inside the crystal. With care, a

*department of mathematics, pennsylvania state university, mont alto, pa 17237. csb15@psu.edu
†department of mathematics, university of british columbia, vancouver, bc v6t 1z2. frigaard@math.ubc.ca
‡department of mathematics and statistics, york university, toronto, ontario m3j 1p3. hhuang@yorku.ca
§department of mathematics and statistics, york university, toronto, ontario m3j 1p3. sqliang@mathstat.yorku.ca

pure ('defect free') crystal can be obtained routinely when the size of the crystal does not exceed a critical size.

Due to the seemingly complex nature of the thermal, structural and dynamical coupling of the molten material, the crystal, the crucible, the gas chamber and other parts of the growth, considerable efforts have been devoted to the laboratory experiments and to modelling and simulations of the growth environment over the past several decades. As a result, there exists an extensive literature, mostly in engineering fields. These studies cover a wide spectrum of areas, from decoupled one or two dimensional simulations to fully coupled three-dimensional computations [3,4,9,10,13,14,15]. Most of the studies rely heavily on computer simulation since the fully coupled system can not be solved otherwise. These investigations have generated useful information including temperature distribution, crystal-melt interface shape, and melt flow patterns inside the crucible. By comparison, until recently much less attention has been paid to the modelling of defects inside the crystal and main factors which determine the formation of defects [16].

In this paper, we present a perturbation approach to study the temperature field inside the crystal and related thermal stress. It is believed that the defects formation can be related to the excessive thermal stress above some critical value (see [11,18,8,1,7,15,19] and references therein). Therefore, some analysis on the growth factors which determine the stress level will be extremely useful for crystal growers using different operating conditions for less well-known type III-V crystals such as the Indium Antimonide (InSb) compound. While the basic solution remains the same for these compounds, we will focus on InSb crystals in the remainer of the paper.

By examining the physical process and parameter values of the growth environment closely, we are able to identify the main features associated with InSb crystals. In particular, we found that the temperature field is dominated by the lateral flux through the crystal-gas surface, characterized by the non-dimensional Biot number. The value of the Biot number is small under the growth condition for InSb crystals, which suggests an asymptotic expansion of the solution with respect to the Biot number. As a result, analytical solutions could be obtained for the pseudo-steady case, which is another main feature of the temperature field inside the crystal. This is similar to the growth of other crystals where the pseudo-steady assumption has been discussed in detail [6]. Even for the fully unsteady case, the asymptotic expansion results in a system of one-dimensional equations and the thermal stress can be obtained explicitly in an analytical form, under the plane strain assumption. To simplify the presentation, we have used a simplified model for the melt and gas flows. However, the asymptotic solution developed here is still valid and can be incorporated with more realistic models for the melt and gas flows. Compared to most of the previous work, the results of this study are different since an explicit form for the stress is obtained. Formulated in a non-dimensional form, the dependence of the stress level on the Biot number is useful for crystal growers when larger crystals are grown. Since the Biot number is proportional to the product of the heat transfer coefficient and the mean crystal radius, it is obvious that one should try to reduce the heat flux via the lateral surface when a crystal of larger radius is grown.

The rest of the paper is organized as follows. In Section 2, we will present the mathematical model and dimensional analysis. Asymptotic solutions are given in Section 3. Thermal stress is discussed in Section 4. We will conclude the paper with a brief summary and discussion on future directions in Section 5.

2. Mathematical Model and Dimensional Analysis

The overall aim of the paper is to derive a realistic but simplified model of InSb crystal growth. Figure 1 illustrates the profile of a typical crystal and the coordinate system, which is fixed to the top of the growing crystal.

Within the crystal Ω, the temperature $T(\mathbf{x}, t)$ satisfies the heat equation

$$\rho_s c_s \frac{\partial T}{\partial t} = k_s \sum_j \frac{\partial^2 T}{\partial x_j^2}, \qquad \mathbf{x} \in \Omega,\ t > 0 \tag{1}$$

where ρ_s, c_s and k_s are the density, specific heat and thermal conductivity of the crystal, (i.e. the solid phase), respectively. The lateral surface of the crystal is denoted Γ_g and is subjected to cooling from the circulating chamber gases and radiative heat loss. Although radiation is not insignificant, for simplicity we model both effects through a simple Newtonian cooling law:

$$-k_s \frac{\partial T}{\partial \mathbf{n}} = h_{gs}(T - T_g), \qquad \mathbf{x} \in \Gamma_g. \tag{2}$$

Here we assume that the heat transfer coefficient, h_{gs}, incorporates both convective and radiative heat transfer, (the latter via linearisation). The top of the crystal is fixed at $z = 0$ where we also invoke a Newtonian cooling law

$$k_s \frac{\partial T}{\partial z} = h_{ch}(T - T_{ch}), \tag{3}$$

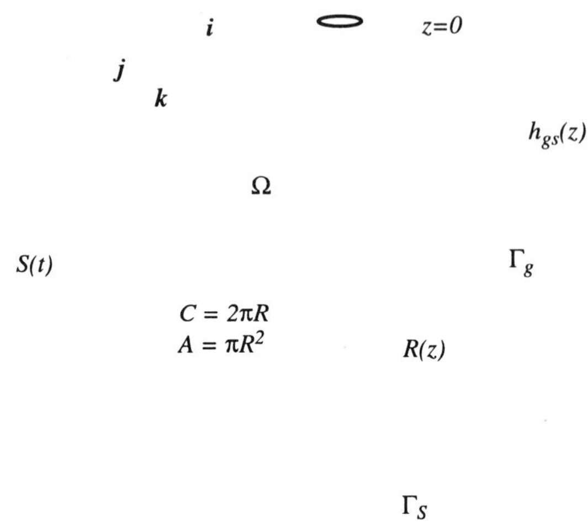

Figure 1. *Shown is a typical crystal at some time t during a growth run with a newly solidified portion at $z = S(t)$. The coordinate system is chosen so that the top of the crystal remains at $z = 0$ and the solidification front grows downwards in the positive z direction. The radial profile is given by $R(z)$ and the crystal length is $S(t)$. Finally, the heat transfer coefficient $h_{gs}(z)$ may be a function of the axial position z.*

in the case that the radius at $z = 0$ is assumed to be non-zero. Here h_{ch} represents the heat transfer coefficient for the seed-chuck connection and T_{ch} is the chuck temperature.

The crystal-melt interface is denoted Γ_S and is where $T = T_m$, the melting temperature. The interface of the phase transition is thus implicitly defined from the temperature field. Explicity we denote the melting isotherm by

$$z - S(\mathbf{x}, t) = 0, \qquad \mathbf{x} \in \Gamma_S. \tag{4}$$

The motion of the interface of the phase transition is governed by the Stefan condition

$$\rho_s L \frac{\partial S}{\partial t} = k_s \frac{\partial T}{\partial \mathbf{n}}\bigg|_{z \to S^-} - q_l \tag{5}$$

where L is the latent heat and q_l is the heat flux from the melt. The speed $\partial S / \partial t$ above is the speed at which S moves in the direction of the outward normal $\mathbf{n}$.

Data	Symbol	Value
Growing Properties		
Mean crystal radius	$\overline{R}$	0.03 m
Final crystal length	Z	0.30 m
Characteristic growth rate	$\overline{v}$	3×10^{-6} m/s
Ambient gas temperature	T_g	600 K
Solid Properties at $T = T_m$		
Melting temperature	T_m	798.4 K
Density	ρ_s	5.64×10^3 kg/m^3
Thermal conductivity	k_s	4.57 W/m K
Heat capacity	$\rho_s c_s$	1.5×10^6 J/m^3 K
Latent heat of fusion	L	2.3×10^5 J/kg
Heat Transfer Coefficients		
Crystal-Gas	h_{gs}	$1 - 4$ W/m^2 K

2.1. *Non-dimensionalization*

Typical values are shown in Table 1. Using these values we can deduce that the main thermal gradients are in the radial direction, due to gas cooling at the surface, and that these drive the vertical thermal gradients. The conductive timescale is generally much shorter than the growth timescale, and this indicates that the process is pseudo-steady.

These observations motivate our scaling below. For simplicity, we start by assuming an axisymmetric model, although the crystal cross-section is not in fact circular. The other assumptions that we make here, for simplicity only, are that the heat transfer coefficient h_{gs} and the gas temperature T_g are constant. In reality there will be local variations along the crystal surface, but in any case these require a more detailed analysis of the gas flows in order to be properly evaluated.

We define the Biot number by

$$\epsilon = \frac{h_{gs}\overline{R}}{k_s} \tag{6}$$

and using the parameter values in Table 2, we find $\epsilon \simeq 0.026 \ll 1$. We seek an asymptotic expansion in terms of ϵ. With this in mind we adopt the following scalings:

$$r = \overline{R}\hat{r}, \quad \epsilon^{1/2}z = \overline{R}\hat{z}, \quad R(z) = \overline{R}\hat{R}(\hat{z}), \quad \epsilon^{1/2}S(r,t) = \overline{R}\hat{S}(\hat{r},\hat{t}),$$

$$T = T_g + \Delta T\Theta, \quad \Delta T = T_m - T_g,$$

$$t = \frac{\mathrm{St}\overline{R}^2\rho_s c_s}{k_s\epsilon}\hat{t}, \quad \mathrm{St} = \frac{L}{c_s\Delta T}.$$

Here variables with hats ($\hat{}$) are the non-dimensional ones. In terms of these variables the heat equation in the crystal (1) becomes

$$\frac{\epsilon}{\mathrm{St}}\Theta_t = \frac{1}{r}(r\Theta_r)_r + \epsilon\Theta_{zz}, \qquad \mathbf{x} \in \Omega,\ t > 0 \tag{7}$$

with boundary conditions (2)-(4)

$$\begin{aligned} -\Theta_r + \epsilon\Theta_z R'(z) &= \epsilon\left[1 + \epsilon(R'(z))^2\right]^{1/2}\Theta, & \mathbf{x} &\in \Gamma_g, \\ \Theta &= 1, & \mathbf{x} &\in \Gamma_S, \\ \Theta_z &= \delta(\Theta - \Theta_{ch}), & z &= 0, \end{aligned} \tag{8}$$

where $\delta = \epsilon^{1/2}h_{ch}/h_{gs}$. The hats have been dropped for brevity. The crystal/melt interface advances according to the Stefan condition (5) which in non-dimensional coordinates becomes

$$\Theta_z - \frac{1}{\epsilon}S_r\Theta_r = \left(1 + \frac{S_r^2}{\epsilon}\right)^{1/2}(\gamma + S_t), \tag{9}$$

where

$$\gamma = \frac{q_l}{\dfrac{\epsilon^{1/2}k_s\Delta T}{\overline{R}}}, \tag{10}$$

which is the non-dimensional heat flux in the liquid across the crystal/melt interface. Typically the value of γ is determined by the characteristics of the melt, which is controlled by the growth conditions such as the heater temperature, crucible shape, etc. To simplify the discussion here, we will assume that the value of γ can be chosen. In [2], a simple model is used to determine γ and work is currently underway to develop a fully-coupled model.

Note that we have chosen the rate of solidification to define the characteristic time scale. The Stefan number, St, gives the ratio of this characteristic solidification time scale to the time scale associated with conductive heat loss through the crystal side surface. Based on the parameter values in Table 2, we have St $\approx$ 4.3, which suggests that the conductive scale is small and the temperature inside the crystal is steady on the growth time scale.

3. Perturbation Solution

We now seek to approximate the scaled model in Section 2.1 via a straightforward perturbation expansion. In turn, this perturbation model will form the basis for a numerical solution. Since St is neither small nor too large under the current growth conditions it is retained as a parameter. Equations (7) & (8) strongly suggest that the temperature Θ is independent of r to leading order. If true then the crystal/melt interface S is also independent of r to leading order, and we see that this is consistent in (9) with the growth being driven primarily by the vertical gradients. These observations motivate the following approximations:

$$\begin{aligned} \Theta &\sim \Theta_0(z,t) + \epsilon\Theta_1(r,z,t) + \epsilon^2\Theta_2(r,z,t) + \dots \\ S &\sim S_0(t) + \epsilon S_1(r,t) + \epsilon^2 S_2(r,t) + \dots \ . \end{aligned} \tag{11}$$

We substitute them into the scaled model, expand in powers of ϵ, simplify and collect terms. The resulting field equations to first order are:

$$\frac{1}{\mathrm{St}}\Theta_{0,t} = \frac{1}{r}\frac{\partial}{\partial r}(r\Theta_{1,r}) + \Theta_{0,zz}, \qquad \mathbf{x} \in \Omega,\ t > 0, \tag{12}$$

$$\frac{1}{\mathrm{St}}\Theta_{1,t} = \frac{1}{r}\frac{\partial}{\partial r}(r\Theta_{2,r}) + \Theta_{1,zz}, \qquad \mathbf{x} \in \Omega,\ t > 0, \tag{13}$$

where the boundary condition on the lateral surface becomes

$$\Theta_{1,r} - R'\Theta_{0,z} = -\Theta_0, \qquad r = R(z), \tag{14}$$

$$\Theta_{2,r} - R'\Theta_{1,z} + \frac{1}{2}{R'}^2\Theta_0 = -\Theta_1, \qquad r = R(z). \tag{15}$$

Continuing this procedure for the remaining conditions, at the top of the crystal one has

$$\begin{aligned} \Theta_{0,z} &= \delta(\Theta_0 - \Theta_{ch}), \qquad z = 0, \\ \Theta_{1,z} &= \delta\Theta_1, \qquad z = 0, \end{aligned}$$

and at the solid-liquid interface

$$\begin{aligned} \Theta_0 &= 1, \qquad z = S_0(t), \\ S_1\Theta_{0,z} + \Theta_1 &= 0, \qquad z = S_0(t). \end{aligned}$$

Finally, the evolution of the interface is governed by

$$\gamma + S_{0,t} = \Theta_{0,z}\big|_{z=S_0(t)}, \qquad S_0(0) = Z_{\text{seed}}, \tag{16}$$

$$S_{1,t} = \left(\Theta_{1,z} + S_1\Theta_{0,zz} + \frac{1}{2}\frac{\Theta_{1,r}^2}{\Theta_{0,z}}\right)_{z=S_0(t)}, \qquad S_1(r,0) = 0. \tag{17}$$

We note that Z_{seed} is the non-dimensional length of the seed. In addition there will be symmetry conditions at $r = 0$ for Θ_k, S_k, $k = 0, 1$.

3.1. *Resolution of the zeroth order model*

Integrating (12) once and imposing the symmetry condition $\Theta_{1,r} = 0$ at $r = 0$, we have:

$$\frac{r}{2}\left(\frac{1}{\mathrm{St}}\Theta_{0,t} - \Theta_{0,zz}\right) = \Theta_{1,r},$$

and applying (14) at $r = R$ gives the zeroth order problem:

$$\frac{1}{\mathrm{St}}\Theta_{0,t} = \Theta_{0,zz} + \frac{2}{R}\left(R'\Theta_{0,z} - \Theta_0\right), \qquad 0 < z < S_0(t),\ t > 0, \quad (18)$$

$$\Theta_{0,z} = \delta(\Theta_0 - \Theta_{ch}), \qquad z = 0, \quad (19)$$

$$\Theta_0 = 1, \qquad z = S_0(t) \quad (20)$$

where the advance of $S_0(t)$ is coupled to the thermal gradients via (16).

Equation (18) is parabolic and involves only the heat fluxes along the length of the crystal. With the chosen expansion we see that at zeroth order the temperature field has no radial dependence. In addition, we can see that the thermal gradients, as discussed previously, are caused by cooling effects at the surface. Also notice that expression (16) illustrates that the chosen time scale balances the growth. The resultant appearance of $1/\mathrm{St} < 1$ in (18) suggests that thermal transients in the bulk of the crystal are not as important as the growth transient. The limit as $\mathrm{St} \to \infty$ leads naturally to a pseudo-steady leading order model, in which time dependency only enters the thermal model through the growth, i.e. we also solve:

$$\Theta_{0,zz} + \frac{2}{R}\left(R'\Theta_{0,z} - \Theta_0\right) = 0, \qquad 0 < z < S_0(t) \quad (21)$$

with the boundary conditions (19) & (20) with the growth of $S_0(t)$ given by (16) as the pseudo-steady limit.

We note that properly it is necessary to close the model by relating growth in S to that in R. To do this we must model the crystal withdrawal from the crucible, formation of the meniscus in the holm region and coupling of S and R. It has been shown in [17] that the growth angle is related to the capillary height for large Bond number growth. In principle, crystals with desirable shapes can be grown by adjusting the pulling rate. Therefore, to simplify the computation, we impose a geometry $R(z)$ on the model. This approach has the advantage of allowing us to investigate the thermal fields and associated stresses that develop for a particular observed shape. We start by exploring two special cases for which an analytic solution may be computed to the pseudo-steady model.

3.1.1. *Constant radius crystals*

In this case we take $R(z) = 1$ and (21) becomes simply

$$0 = \Theta_{0,zz} - 2\Theta_0, \qquad 0 < z < S_0(t)$$

with boundary conditions (19) & (20). Solving for Θ_0 gives:

$$\Theta_0(z) = \frac{\sqrt{2}\cosh\sqrt{2}z + \delta\sinh\sqrt{2}z + \delta\Theta_{ch}\sinh\sqrt{2}(S_0 - z)}{\sqrt{2}\cosh\sqrt{2}S_0 + \delta\sinh\sqrt{2}S_0}. \tag{22}$$

The crystal grows at a rate governed by the Stefan condition (16):

$$\gamma + S_{0,t} = \sqrt{2}\frac{\sqrt{2}\sinh\sqrt{2}S_0 + \delta\cosh\sqrt{2}S_0 - \delta\Theta_{ch}}{\sqrt{2}\cosh\sqrt{2}S_0 + \delta\sinh\sqrt{2}S_0}.$$

The insulated chuck ($\delta = 0$) and the cold chuck ($\Theta_{ch} = 0$) cases are easily extracted.

3.1.2. *Conical crystals*

One source of ambiguity in the constant radius model above is the need to specify the chuck temperature and heat transfer coefficient. In the case of a conical crystal, (which is closer to reality), this ambiguity is less prominent. We assume $R(z) = R_{\text{seed}} + \alpha z$ where $\arctan\alpha \simeq O(1)$ is one-half the opening angle of the crystal when using non-dimensional units. In this case we solve

$$\left.\begin{aligned} 0 = \Theta_{0,\eta\eta} + \frac{2}{\eta}(\Theta_{0,\eta} - \Theta_0), &\qquad R_{\text{seed}} < \alpha^2\eta < R_{\text{seed}} + \alpha S_0,\ t > 0,\\ \Theta_{0,\eta} = \alpha\delta(\Theta_0 - \Theta_{ch}), &\qquad \alpha^2\eta = R_{\text{seed}},\\ \Theta_0 = 1, &\qquad \alpha^2\eta = R_{\text{seed}} + \alpha S_0, \end{aligned}\right\} \tag{23}$$

where $\alpha^2\eta(z) = R_{\text{seed}} + \alpha z$. The resulting solution takes the form of linear combinations of modified Bessel functions

$$\Theta_0(\eta, t) = \frac{g(\eta)\left[f'(\eta_0) - \alpha\delta f(\eta_0) + \alpha\delta\Theta_{ch} f(\eta_{S_0})\right] - f(\eta)\left[g'(\eta_0) - \alpha\delta g(\eta_0) + \alpha\delta\Theta_{ch} g(\eta_{S_0})\right]}{[f'(\eta_0)g(\eta_{S_0}) - f(\eta_{S_0})g'(\eta_0)] + \alpha\delta[f(\eta_{S_0})g(\eta_0) - f(\eta_0)g(\eta_{S_0})]} \tag{24}$$

where

$$\eta_0 = \eta(0), \quad \eta_{S_0} = \eta(S_0), \quad f(\eta) = \frac{I_1(\sqrt{8\eta})}{\sqrt{\eta}}, \quad g(\eta) = \frac{K_1(\sqrt{8\eta})}{\sqrt{\eta}}$$

and

$$f'(\eta) = \frac{\sqrt{2}}{\eta}I_0(\sqrt{8\eta}) - \frac{I_1(\sqrt{8\eta})}{\eta^{3/2}}, \quad g'(\eta) = -\frac{\sqrt{2}}{\eta}K_0(\sqrt{8\eta}) - \frac{K_1(\sqrt{8\eta})}{\eta^{3/2}}.$$

The corresponding expression for the growth rate is

$$\gamma + S_{0,t} =$$

$$\frac{1}{\alpha}\frac{g'(\eta_{S_0})\left[f'(\eta_0)-\alpha\delta f(\eta_0)+\alpha\delta\Theta_{ch}f(\eta_{S_0})\right]-f'(\eta_{S_0})\left[g'(\eta_0)-\alpha\delta g(\eta_0)+\alpha\delta\Theta_{ch}g(\eta_{S_0})\right]}{[f'(\eta_0)g(\eta_{S_0})-f(\eta_{S_0})g'(\eta_0)]+\alpha\delta[f(\eta_{S_0})g(\eta_0)-f(\eta_0)g(\eta_{S_0})]}$$

Two limiting cases are considered. To compare with the cylindrical case one sets $R_{\text{seed}} = 1$ and expands (24) in a power series of α yielding

$$\Theta_0(z,t) = \frac{\cosh\sqrt{2}z}{\cosh\sqrt{2}S_0}\left\{1 - \frac{\alpha}{8}\left[6(z - S_0) + \sqrt{8}(z^2 - U)\tanh\sqrt{2}z \quad - \right.\right.$$
$$\left.\left.\sqrt{8}(S_0^2 - U)\tanh\sqrt{2}S_0\right]\right\} + \mathcal{O}(\alpha^2)$$

with

$$U = \frac{3}{2} + 2\delta\left(1 - \Theta_{ch}\cosh\sqrt{2}S_0\right)$$

which should be compared to equation (22). Since $R_{\text{seed}} \ll 1$, a simple form of (24) can be obtained by expanding the solution in R_{seed}, as

$$\Theta_0(z,t) = \sqrt{\frac{S_0}{z}}\frac{I_1(\sqrt{8z/\alpha})}{I_1(\sqrt{8S_0/\alpha})}\left[1 + \frac{R_{\text{seed}}}{\alpha}\left(\sqrt{\frac{2}{\alpha z}}\frac{I_0(\sqrt{8z/\alpha})}{I_1(\sqrt{8z/\alpha})} - \right.\right.$$
$$\left.\left.\sqrt{\frac{2}{\alpha S_0}}\frac{I_0(\sqrt{8S_0/\alpha})}{I_1(\sqrt{8S_0/\alpha})} - \frac{1}{z} + \frac{1}{S_0}\right)\right] + \mathcal{O}(R_{\text{seed}}^2).$$

3.1.3. *Comments*

We note that the model for a one-dimensional temperature variation in the axial direction is not new. For example, it has been used in [17] as the model allows for simple analytical solutions. However, the model has not been formally justified in the crystal growth literature.

From the explicit solutions for cylindrical crystals one can observe that the rate of interface growth (therefore the rate of crystal growth) is asymptotically (for large S_0) $\sqrt{2} - \gamma$, when γ is constant. For small S_0, on the other hand, the speed is approximately $2S_0 - \gamma$ for insulated chuck or $\delta + (2 - \delta^2)S_0 - \gamma$ for a cold chuck. This suggests, for both cases, that initially a small γ is necessary to establish the growth. In addition, the value of γ will also affect the total time required for growing crystals of certain sizes.

3.2. *Radial variations: resolution of the first order model*

Having solved the zeroth order model, to give Θ_0 and S_0, we can resolve the radial variations in temperature which occur at first order in Θ_1 and also consider the shape of the crystal/melt interface as it evolves, through S_1. From resolution of the zeroth order model we have that

$$\Theta_{1,r} = \frac{r}{2}\left(\frac{1}{\mathrm{St}}\Theta_{0,t} - \Theta_{0,zz}\right),$$

which integrates with respect to r to give

$$\Theta_1(r,z,t) = \Theta_1(0,z,t) + \frac{r^2}{4}\left(\frac{1}{\mathrm{St}}\Theta_{0,t} - \Theta_{0,zz}\right),$$

or

$$\Theta_1(r,z,t) = \Theta_1^0(z,t) + r^2\Theta_1^1(z,t), \tag{25}$$

where $\Theta_1^0(z,t) = \Theta_1(0,z,t)$ and using equation (18)

$$\Theta_1^1(z,t) = \frac{1}{2R(z)}(R'(z)\Theta_{0,z}(z,t) - \Theta_0(z,t)). \tag{26}$$

Note that the function $\Theta_1^1(z,t)$ is known from the data and the zeroth order solution. By adopting the same procedure as for the zeroth order model we can find $\Theta_1^0(z,t)$, i.e. integrating (13) with respect to r and using the boundary condition at $r = R$ to eliminate $\Theta_{2,r}$. We derive:

$$\begin{aligned}
&\frac{1}{\mathrm{St}}\Theta_{1,t}^0 - \Theta_{1,zz}^0 - \frac{2}{R}(R'\Theta_{1,z}^0 - \Theta_1^0)\\
&= -\frac{R^2}{2}\left(\frac{1}{\mathrm{St}}\Theta_{1,t}^1 - \Theta_{1,zz}^1\right) + 2R(R'\Theta_{1,z}^1 - \Theta_1^1) - \frac{R'^2}{R}\Theta_0,
\end{aligned} \tag{27}$$

$$\Theta_1^0(z,0) = 0, \qquad 0 < z < S_0(t), \tag{28}$$

$$\Theta_{1,z}^0 = \delta\Theta_1^0, \qquad z = 0, \tag{29}$$

$$\Theta_1^0(z,t) = -S_1^0(t)\Theta_{0,z}(z,t), \qquad z = S_0(t) \tag{30}$$

where $S_1^0(t) = S_1(0,t)$ and its value is determined by the Stefan condition (17) at $r = 0$ as

$$S_{1,t}^0 = \left(\Theta_{1,z}^0 + S_1^0\Theta_{0,zz}\right)_{z=S_0(t)}, \qquad S_1^0(0) = 0. \tag{31}$$

The evolution of the interface shape is governed by (17) which reduces to

$$S_{1,t} = \left[\Theta_{1,z}^0 + S_1\Theta_{0,zz} + r^2\left(\Theta_{1,z}^1 + \frac{2(\Theta_1^1)^2}{\Theta_{0,z}}\right)\right]_{z=S_0(t)}, \qquad S_1(r,0) = 0 \tag{32}$$

in which r appears as a parameter.

We note that (27)-(32) has the same structure as the zeroth order problem (16), (18)-(20), but is inhomogeneous, i.e. the zeroth order solution provides the forcing (or heating). A further key difference is in the coupling with the crystal/melt interface position S_1. Equation (30) provides the lower boundary condition for Θ_1 and S_1 advances through (32), which is consequently a first order quasi-linear partial differential equation.

In general, the coupled system (27)-(32) must be solved numerically. For the pseudo-steady case, the formula can be simplified as follows.

From the definition of Θ_1^1 and using the pseudo-steady condition $\Theta_{0,zz} = -4\Theta_1^1$, expressions (27) & (32) reduce to

$$\begin{aligned}\Theta^0_{1,zz} &+ \frac{2}{R}(R'\Theta^0_{1,z} - \Theta^0_1) \\ &= \frac{1}{4}(-RR''' + 5R'R'' + 2R')\Theta_{0,z} - \frac{1}{4R}(5RR'' - 4R'^2 + 2R)\Theta_0 \qquad (33)\end{aligned}$$

and

$$\begin{aligned}S_{1,t} = \Bigg[\Theta^0_{1,z} &+ \frac{2}{R}(1 - R'\Theta_{0,z})S_1 + \\ &\frac{r^2}{2R^2}\left(R' + (RR'' - 2R'^2 - R)\Theta_{0,z} + \frac{1}{\Theta_{0,z}}\right)\Bigg]_{z=S_0(t)} \qquad (34)\end{aligned}$$

3.2.1. *Constant radius crystals*

Here $R(z) = 1$ and $\Theta_{0,zz} = 2\Theta_0$. From the definition of Θ_1^1 we have

$$\Theta_1^1 = \frac{1}{2R}(R'\Theta_{0,z} - \Theta_0) = -\frac{1}{2}\Theta_0$$

giving $\Theta^1_{1,z} = -\Theta_{0,z}/2$ and $\Theta^1_{1,zz} = -\Theta_0$. Consequently (33) becomes

$$\Theta^0_{1,zz} - 2\Theta^0_1 = -\frac{1}{2}\Theta_0.$$

Solving for Θ_1^0 in the case that $\delta = 0$ yields

$$\Theta_1^0 = \frac{1}{4\sqrt{2}}\frac{\cosh\sqrt{2}z}{\cosh\sqrt{2}S_0}\left(S_0\tanh\sqrt{2}S_0 - z\tanh\sqrt{2}z - 8S_1^0\tanh\sqrt{2}S_0\right)$$

from which S_1 can be obtained with equation (32).

3.2.2. *Conical crystals*

Here $R(z) = R_{\text{seed}} + \alpha z$ and we have

$$\Theta_1^1 = \frac{1}{2R}(R'\Theta_{0,z} - \Theta_0) = \frac{1}{2R}(\alpha\Theta_{0,z} - \Theta_0)$$

with the identity $\Theta_1^1 = -\Theta_{0,zz}/4$ from (21). Using these we derive

$$\Theta_{1,z}^1 = -\left(\frac{3\alpha^2}{2R^2} + \frac{1}{2R}\right)\Theta_{0,z} + \frac{3\alpha}{2R^2}\Theta_0,$$

$$\Theta_{1,zz}^1 = \left(\frac{6\alpha^3}{R^3} + \frac{3\alpha}{R^2}\right)\Theta_{0,z} - \left(\frac{6\alpha^2}{R^3} + \frac{1}{R^2}\right)\Theta_0$$

and finally simplify (33) to:

$$\Theta_{1,zz}^0 + \frac{2\alpha}{R}\Theta_{1,z}^0 - \frac{2}{R}\Theta_1^0 = \frac{\alpha}{2}\Theta_{0,z} + \frac{1}{2R}(2\alpha^2 - R)\Theta_0. \tag{35}$$

In general numerical methods will have to be used to find a solution to either case above, even for the pseudo-steady solution. In the following we turn our discussion to thermal stress inside the crystal.

4. Thermal Stress

The thermal stress experienced by the crystal during its growth leads to the generation of structural defects in the crystal [16]. If we want to eliminate these undesirable defects then one must control the thermal stress. Although InSb is anisotropic with respect to its elasticity, we focus on the isotropic case in this paper. Some of the anisotropic effects are discussed in [2].

4.1. *Plain strain assumption*

Assuming that the displacement vector is of the form $\vec{u} = \langle u(r), 0, 0\rangle$ (plain strain) and converting to non-dimensional units reduces the equilibrium expression for $\vec{u}$ to [2]

$$\frac{\partial}{\partial r}\left[\frac{1}{r}\frac{\partial}{\partial r}(ru)\right] = \epsilon\frac{(1+\nu)}{(1-\nu)}\frac{\partial\Theta_1}{\partial r}, \quad u(0) < \infty, \quad \sigma_{rr}(R) = 0.$$

Note that the stress has been non-dimensionalized by $\alpha_0 \Delta T E/(1-\nu)$. We have assumed here that $\partial\Theta/\partial z \simeq 0$ since we want to focus on the sole effect of any radial temperature variations. The solution satisfying the boundary conditions is

$$u(r) = \epsilon\frac{(1+\nu)}{(1-\nu)}\left[\frac{1}{r}\int_0^r \Theta_1(s)s\,ds + (1-2\nu)\frac{r}{R^2}\int_0^R \Theta_1(s)s\,ds\right]$$

and the corresponding nontrivial stresses are

$$\left.\begin{aligned}
\sigma_{rr} &= \epsilon\left[\frac{1}{R^2}\int_0^R \Theta_1(s)s\,ds - \frac{1}{r^2}\int_0^r \Theta_1(s)s\,ds\right], \\
\sigma_{\theta\theta} &= \epsilon\left[\frac{1}{R^2}\int_0^R \Theta_1(s)s\,ds + \frac{1}{r^2}\int_0^r \Theta_1(s)s\,ds - \Theta_1(r)\right], \\
\sigma_{zz} &= \epsilon\left[\frac{2}{R^2}\int_0^R \Theta_1(s)s\,ds - \Theta_1(r)\right]
\end{aligned}\right\} \tag{36}$$

with σ_{zz} modified using St. Venant's principle.

From (25) we obtain

$$\begin{aligned}
&\sigma_{rr} = \tfrac{1}{4}\epsilon\Theta_1^1(z,t)\left(R(z)^2 - r^2\right), \quad \sigma_{\theta\theta} = \tfrac{1}{4}\epsilon\Theta_1^1(z,t)\left(R(z)^2 - 3r^2\right), \\
&\sigma_{zz} = \tfrac{1}{2}\epsilon\Theta_1^1(z,t)\left(R(z)^2 - 2r^2\right).
\end{aligned} \tag{37}$$

Using the expression $2\sigma_{\mathrm{VM}}^2 = (\sigma_{rr} - \sigma_{\theta\theta})^2 + (\sigma_{rr} - \sigma_{zz})^2 + (\sigma_{\theta\theta} - \sigma_{zz})^2$ to compute the von Mises stress gives

$$\sigma_{\mathrm{VM}} = -\frac{1}{4}\epsilon\Theta_1^1(z,t)R(z)^2\left[1 - 4\left(\frac{r}{R(z)}\right)^2 + 7\left(\frac{r}{R(z)}\right)^4\right]^{1/2}. \tag{38}$$

The object in the square brackets is a shape factor which ranges from a value of $\sqrt{3/7}$ at a radius of $r = \sqrt{3/7}R(z)$ to a maximum value of 2 at the outer edge of the crystal. For $\sqrt{4/7}R(z) < r \leq R(z)$ this factor is greater than one.

In the pseudo-steady case Θ_1^1 is given by expression (26) and (21) so that equation (38) becomes

$$\sigma_{\mathrm{VM}} = \frac{1}{16}\epsilon\Theta_{0,zz}R(z)^2\left[1 - 4\left(\frac{r}{R(z)}\right)^2 + 7\left(\frac{r}{R(z)}\right)^4\right]^{1/2} \tag{39}$$

indicating that the stress level is a characteristic of the concavity of the temperature in the axial direction. It should be noted that the stress is also linearly proportional to the Biot number ϵ. This indicates that an increase in the crystal radius will also increase the stress level, other conditions being equal. It also indicates that the increase of radius can be offset by reducing the heat transfer coefficient h_{gs}, suggesting that a possible way for reducing the stress is by changing the local heat flux from the crystal lateral surface.

As other stresses, such as the total resolved stress are often considered more relevant for causing defects, it is important to point out that the same characteristics remain for different representations of the thermal stress, or the crystals being pulled in different directions, which is discussed in [2].

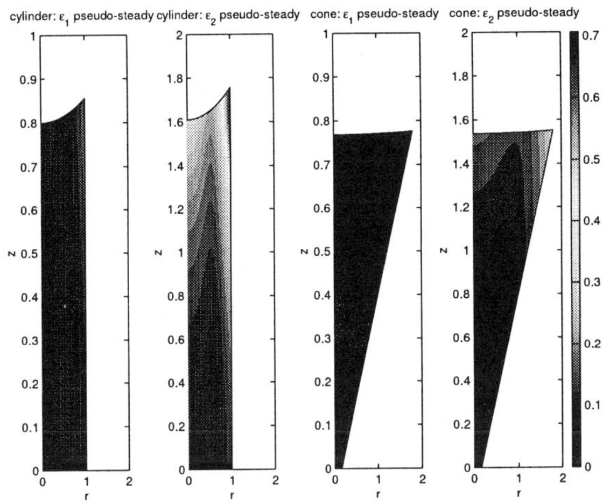

Figure 2. *von Mises stress in MPa for the cylindrical and conical crystal cases.*

4.2. *Results*

Due to page limitation, we only present some typical numerical results on thermal stress obtained using the formula derived earlier. More results are available in [2]. Assuming pulling in the {111} direction, the values of E and ν are $E_{\{111\}} = 6.18 \times 10^4$ MPa and $\nu_{\{111\}} = 0.364$. Thus, $E/(1-\nu)|_{\{111\}} = 9.72 \times 10^4$ MPa. As a result, the dimensional constant for the stress calculations is $\alpha_0 \Delta T E/(1-\nu) \simeq 107$ MPa.

Figure 2 shows the stress contours of the von Mises stress in units of MPa for the cylinder and the cone at the end of the growth. For a fixed value of ϵ the stress in the conical case is about one-half that of the cylindrical case. Also, increasing ϵ increases the stress level dramatically. By growing a conical crystal the stress can be reduced significantly. For a given temperature the amount of stress at which crystal deformation begins to occur is known as the critical resolved shear stress, σ_{crss}. In the case of InSb, σ_{crss} varies from 0.245 MPa [12] to 4.90 MPa [5] as the temperature varies from $T_m = 798.4$ K to 491 K respectively indicating that the conical crystal remains below this critical stress level.

5. Conclusion

In this study, we present a semi-analytical approach for the temperature and thermal stress inside a type III-V (InSb) crystal. An important feature of the approach is that it allows us to derive explicit relationships between the thermal stress and relevant physical and geometrical parameters. This is achieved by using asymptotic expansion of the solution in the Biot number, characterizing the lateral heat flux. The asymptotic solution is obtained by solving essentially one-dimensional problems. The results show that the stress induced by radial temperature variation is related to the size of the crystal (radius) and heat flux through the side surface. On the other hand, the effect of the crystal radius on the stress induced by the axial temperature variation is much weaker. The heat flux through the side surface is an important factor for reducing the overall thermal stress inside the crystal. The other advantages of our semi-analytical approach is that it can be extended to cases with more complicated models for the melt and gas flows. For example, the effect of the gas flow on the lateral heat flux between the crystal surface and the gas can be modelled by a non-constant heat exchange coefficient h_{gs}. The motion of the melt may be modelled by a similar approach, which will be the subject of a subsequent paper.

Acknowledgement. The authors wish to thank Firebird Semiconductors Inc., MITACS, NSERC (Canada) and BC ASI for their financial support.

References

1. Alexander, H. & Hassen, P. (1968). Dislocation and Plastic Flow in the Diamond Structure. *Solid State Physics*, 22, pp. 27-158.
2. Bohun, C.S., Frigaard, I., Huang, H. & Liang, S. (2003). A semi-analytical model for the Cz growth of type III-V compounds, submitted.
3. Brown, R.A. (1988). Theory of transport processes in single crystal growth from the melt. *AIChE Journal*, 34, pp. 881-911.
4. Chan, Y.T., Gilbeling, H.J. & Grubin, H.L. (1988). Numerical simulation of Czochralski growth. *Journal of Applied Physics*, 64(3), pp. 1425-1439.
5. Chaudhuri, A.R., Patel, J.R. & Rubin, L.G. (1962). Velocities and densities of dislocations in germanium and other semiconductor crystals. *Journal of Applied Physics*, 33, pp. 2736-2746.
6. Derby, J.J. & Brown, R.A. (1988). On the quasi-steady-state assumption in modeling Czochralski crystal growth. *Journal of Crystal Growth*, 87(2-3), pp. 251-260.
7. Dupret, F. & van den Bogaert, N. (1994). Modelling Bridgeman and Czochralski Growth. In *Handbook of Crystal Growth*, 2B, Chapter 15, Hurle, D.T.J. ed., North-Holland: Amsterdam.
8. Gulluoglu, A.N. & Tsai, C.T. (1999). Effect of growth parameters on dislocation generation in InP single crystal grown by the vertical gradient freeze process. *Acta Materialia*, 47(8), pp. 2313-2322.
9. Hurle, D.T.J. (1994). Handbook of Crystal Growth, 1 & 2. North-Holland: Amsterdam.

10. Hurle, D.T.J. (1993). Crystal Pulling from the Melt. Springer-Verlag: Berlin.
11. Jordan, A.S., Caruso, R. & von Neida, A.R. (1980). A Thermoelastic Analysis of Dislocation Generation in Pulled GaAs Crystals. *The Bell System Technical Journal*, 59(4), pp. 593-637.
12. Mil'vidskii, M.G. & Bochkarev, E.P. (1978). Creation of defects during the growth of semiconductor single crystals and films. *Journal of Crystal Growth*, 44(1), pp. 61-74.
13. Müller, G. (2002). Experimental analysis and modeling of melt growth processes. *Journal of Crystal Growth*, 237-239, Part 3, pp. 1628-1637.
14. Müller, G. (1988). Convection and Inhomogeneities in Crystal Growth from Melt. In *Crystals, 12*, Freyhardt, F.C. ed., Springer-Verlag: Berlin.
15. Prasad, V., Zhang, H. & Anselmo, A. (1997). Transport Phenomena in Czochralski Crystal Growth Process. *Advances in Heat Transfer*, 30, pp. 313-435.
16. Sinno, T., Dornberg, E., von Ammon, W., Brown, R.A. & Dupret, F. (2002). Defect Engineering of Czochralski Single-Crystal Silicon. *Material Science and Engineering Reports*, 28, pp. 149-198.
17. Tatachenko, Y.A. (1993). Shaped Crystal Growth. Kluwer Academic Publishers: Boston.
18. Tanahasji, K., Kikuchi, M., Higashino, T., Inoue, N. & Mizokawa, Y. (2000). Concentration of point defects changed by thermal stress in growing CZ silicon crystal: effect of the growth rate. *Journal of Crystal Growth*, 210(1), pp. 45-48.
19. Vökl, J. (1994). Stress Computation in Czochralski. In *Handbook of Crystal Growth*, 2B, Chapter 14, Hurle, D.T.J. ed., North-Holland: Amsterdam.

ASYMPTOTIC BEHAVIOUR OF THE TRACE FOR SCHRÖDINGER OPERATOR ON IRREGULAR DOMAINS*

HUA CHEN AND CHUN YU

School of Mathematics and Statistics,
Wuhan University,
Wuhan, 430072, P.R. China

1. Main Result

In 1953, L.M. Gelfand and B.M. Levitan studied the trace of the Sturm-Liouville problem

$$\begin{cases} -y'' + q(x)y = \lambda y, & x \in (0, \pi), \\ y(0) = y(\pi) = 0, \end{cases} \tag{1}$$

where $q(x)$ is bounded and differentiable on $[0, \pi]$. They obtained the following trace identity

$$\sum_{j=1}^{\infty} \left[\lambda_j - j^2 - \frac{1}{\pi} \int_0^{\pi} q(x)dx \right] = -\frac{q(0) + q(\pi)}{4} + \frac{1}{2\pi} \int_0^{\pi} q(x)dx.$$

This result reveals a direct relation between the eigenvalues and the operator quantities. After that, the research focus is on the trace identity of Sturm-Liouville equation with general boundary conditions and the equation with singularities.

The trace identity in k-dimensional case has been studied by Cao Cewen in 1979. He considered the eigenvalue problem

$$\begin{cases} -\Delta u + q(x)u = \lambda u, & \text{in } \Omega, \\ u = 0, & \text{on } \partial\Omega, \end{cases} \tag{2}$$

where $\Omega \subset \mathbb{R}^k$ is a bounded, connected domain with piecewise smooth boundary $\partial\Omega$, Δ is the k-dimensional Dirichlet Laplacian on Ω, $q(x)$ is

*The research is supported by the NNSFC

bounded and differentiable in Ω. Let λ_j be the j-th eigenvalue, μ_j be the j-th eigenvalue of the Dirichlet Laplacian, i.e. the case of $q(x) \equiv 0$, he proved the following average displacement formula

$$\lim_{N\to\infty} \frac{1}{N}\sum_{j=1}^{N}(\lambda_j - \mu_j) = \frac{1}{|\Omega|_k}\int_\Omega q(x)dx$$

which $|\Omega|_k$ represents the k-dimensional Lebesgue measure of Ω. and in a special case, when Ω is k-dimensional open cube, he proved

$$\lim_{R\to+\infty} R^{-\frac{k-1}{2}} \sum_{\mu_j\le R} (\lambda_j - \mu_j - \frac{1}{|\Omega|_k}\int_\Omega q(x)dx)$$
$$= -\frac{|\partial\Omega|_{k-1}}{4\omega_{k-1}}\left(\frac{1}{|\partial\Omega|_{k-1}}\int_{\partial\Omega} qd\sigma - \frac{1}{|\Omega|_k}\int_\Omega q(x)dx\right)$$

where $|\partial\Omega|_{k-1}$ denotes the $(k-1)$-dimensional Lebesgue measure of $\partial\Omega$, $\omega_{k-1} = (2\sqrt{\pi})^{k-1}\Gamma(\frac{k-1}{2}+1)$. The second result is the extension of Gelfand-Levitan's trace identity under the special situation.

In this paper, we consider the eigenvalue problem of Schrödinger operator defined on more general domain

$$\begin{cases} Lu \equiv -\Delta u + q(x)u = \lambda u, & \text{in } \Omega, \\ u = 0, & \text{on } \partial\Omega, \end{cases} \tag{P}$$

where $\Omega = \bigcup\limits_{m=1}^{\infty} \Omega_m$ is an open subset of $\mathbb{R}^k$ with boundary $\partial\Omega$, and the connected regions Ω_m with piecewise smooth boundaries are bounded and $\Omega_j \bigcap \Omega_i = \emptyset$ if $j \neq i$. We assume throughout this paper that $|\Omega_1|_k \ge |\Omega_2|_k \ge \cdots \ge |\Omega_m|_k \ge \cdots$, and $|\Omega_m|_k \to 0$ as $m \to +\infty$, and $|\Omega|_k = \sum\limits_{m=1}^{\infty} |\Omega_m|_k < +\infty$ where $|\cdot|_k$ denotes the k-dimensional Lebesgue measure of Ω. As is well-known that the scalar λ is said to be an eigenvalue of (P) if there exists a nonzero $u \in H_0^1(\Omega)$ which satisfies (P) in the distributional sense. The problem (P) has discrete eigenvalues which can be written as a sequence $(\lambda_j)_{j=1}^{\infty}$ with $\lambda_j \to +\infty$ as $j \to +\infty$.

When $q(x) \equiv 0$, the eigenvalue problem of Dirichlet Laplacian associated with (P) is as follows:

$$\begin{cases} L_0 u \equiv -\Delta u = \lambda u, & \text{in } \Omega, \\ u = 0, & \text{on } \partial\Omega, \end{cases} \tag{D}$$

The spectrum of (D) is discrete and consists of a sequence $(\mu_j)_{j=1}^{\infty}$ and can be written in non-decreasing order according to their finite multiplicity:

$$0 < \mu_1 \leq \mu_2 \leq \cdots \leq \mu_j \leq \cdots \quad \text{with } \mu_j \to +\infty \text{ as } j \to +\infty.$$

Here we denote $v_j \in H_0^1(\Omega)$ as unit eigenfunctions associated with μ_j.

Now, we can state our main results as follows:

Theorem 1.1. *The following formula*

$$\lim_{N\to\infty} \frac{1}{N} \sum_{j=1}^{N} [\lambda_j - \mu_j - (qv_j, v_j)] = 0$$

is true, where $(\cdot,\cdot)$ *denotes the inner product in* $L^2(\Omega)$.

Theorem 1.2. *There exists* $R \to +\infty$ *such that*

$$\frac{1}{A(R)} \sum_{\mu_j \leq R} [\lambda_j - \mu_j - (qv_j, v_j)] = O\left(\frac{\log R}{R}\right),$$

where $A(R) = \#\{k \geq 1; \mu_k \leq R\}$.

Remark 1.1. The results can be extended to the case of that the domain $\Omega \subset \mathbb{R}^k$ is a fractal drum, i.e., Ω is an opened subset with fractal boundary $\partial\Omega$. In this case, if $\delta \in (k-1, k)$ is the Minkowski dimension of the boundary $\partial\Omega$, then we know that $A(R) = (2\pi)^{-k}\omega_k|\Omega|_k R^{k/2} + O(R^{\delta/2})$ as $R \to \infty$.

2. Preparation

For our purpose, we shall first discuss the problem on a single domain Ω_j, i.e.,

$$\begin{cases} Lu \equiv -\Delta u + q(x)u = \lambda u, & \text{in } \Omega_j, \\ u = 0, & \text{on } \partial\Omega_j, \end{cases} \tag{P_j}$$

and

$$\begin{cases} Lu_0 \equiv -\Delta u = \lambda u, & \text{in } \Omega_j, \\ u = 0, & \text{on } \partial\Omega_j, \end{cases} \tag{D_j}$$

Without loss of generality, we assume that $q(x)$ is nonnegative, otherwise, instead of $q(x)$, we shall use $q(x) + E$, where $E > 0$ and $|q(x)| \leq E$. Suppose that Q_j is the upper bound of $q(x)$ in Ω_j and Q the upper bound of $q(x)$ in Ω. So, $0 \leq Q_j \leq Q$.

L_0 and L are unbounded self-adjoint operator in the real Hilbert space $H_0^1(\Omega_j)$. $L = L_0 + q$ is nonnegative since L_0 is nonnegative. We write the eigenvalues of L_0 and L in Q_j as $0 < \mu_1^{(j)} \leq \mu_2^{(j)} \leq \cdots$ and $0 < \lambda_1^{(j)} \leq \lambda_2^{(j)} \leq \cdots$ respectively. $v_n^{(j)}$ and $u_n^{(j)}$ are unit eigenfunctions associated with $\mu_n^{(j)}$ and $\lambda_n^{(j)}$ respectively, and $\{v_n^{(j)}\}$ and $\{u_n^{(j)}\}$ are both complete orthonormal systems in $H_0^1(\Omega_j)$. We put all $\mu_n^{(j)}$ and $\lambda_n^{(j)}$ together, and rearrange them, which are denoted by $\{\sigma_n^{(j)}\}$, in non-decreasing order. For any positive number R, let

$$\begin{aligned} A_j(R) &= \#\{n \geq 1; \mu_n^{(j)} \leq R\}, \\ B_j(R) &= \#\{n \geq 1; \lambda_n^{(j)} \leq R\}, \\ C_j(R) &= \#\{n \geq 1; \sigma_n^{(j)} \leq R\}. \end{aligned}$$

Thus we have

Lemma 2.1. *For any positive number R, we have*
1) $0 \leq \mu_n^{(j)} \leq \lambda_n^{(j)} \leq \mu_n^{(j)} + Q_j$;
2) $A_j(R - Q_j) \leq B_j(R) \leq A_j(R)$;
3) $C_j(R) = A_j(R) + B_j(R)$;
$2A_j(R - Q_j) \leq C_j(R) \leq 2A_j(R)$.

Secondly, we have for $\{f_n\}$ being orthogonal unit vectors,

$$\sum_{m=1}^{N} (Lf_n, f_n) \geq \sum_{n=1}^{N} \lambda_n^{(j)}, \text{ and } \sum_{m=1}^{N} (L_0 f_n, f_n) \geq \sum_{n=1}^{N} \mu_n^{(j)}.$$

Thus

Lemma 2.2. *For any positive number R,*

$$0 \leq - \sum_{\mu_n^{(j)} \leq R} [\lambda_n^{(j)} - \mu_n^{(j)} - (qv_n^{(j)}, v_n^{(j)})] \leq F_j(R),$$

where

$$F_j(R) = \sum_{\mu_n^{(j)} \leq R} (qv_n^{(j)}, v_n^{(j)}) - \sum_{\lambda_n^{(j)} \leq R} (qu_n^{(j)}, u_n^{(j)}).$$

Proof. Since $(Lv_n^{(j)}, v_n^{(j)}) = \mu_n^{(j)}$, $(Lu_n^{(j)}, u_n^{(j)}) = \lambda_n^{(j)}$, then

$$\begin{aligned}\sum_{\mu_n^{(j)}\le R} (qv_n^{(j)}, v_n^{(j)}) &= \sum_{n=1}^{A_j(R)} (Lv_n^{(j)}, v_n^{(j)}) - \sum_{n=1}^{A_j(R)} (L_0 v_n^{(j)}, v_n^{(j)}) \\ &= \sum_{n=1}^{A_j(R)} (\lambda_n^{(j)} - \mu_n^{(j)}) = \sum_{\mu_n^{(j)}\le R} (\lambda_n^{(j)} - \mu_n^{(j)})\end{aligned}$$

and

$$\begin{aligned}\sum_{\lambda_n^{(j)}\le R} (qu_n^{(j)}, u_n^{(j)}) &= \sum_{n=1}^{B_j(R)} (Lu_n^{(j)}, u_n^{(j)}) - \sum_{n=1}^{B_j(R)} (L_0 u_n^{(j)}, u_n^{(j)}) \\ &= \sum_{n=1}^{B_j(R)} (\lambda_n^{(j)} - \mu_n^{(j)}) = \sum_{\lambda_n^{(j)}\le R} (\lambda_n^{(j)} - \mu_n^{(j)})\end{aligned}$$

By Lemma 2.1, $\lambda_n^{(j)} - \mu_n^{(j)} \ge 0$, $B_j(R) \le A_j(R)$, hence

$$\begin{aligned}0 &\le \sum_{\mu_n^{(j)}\le R} (qv_n^{(j)}, v_n^{(j)}) - \sum_{\mu_n^{(j)}\le R} (\lambda_n^{(j)} - \mu_n^{(j)}) \\ &= - \sum_{\mu_n^{(j)}\le R} [(\lambda_n^{(j)} - \mu_n^{(j)}) - (qv_n^{(j)}, v_n^{(j)})] \\ &\le \sum_{\mu_n^{(j)}\le R} (qv_n^{(j)}, v_n^{(j)}) - \sum_{\lambda_n^{(j)}\le R} (\lambda_n^{(j)} - \mu_n^{(j)}) \\ &\le \sum_{\mu_n^{(j)}\le R} (qv_n^{(j)}, v_n^{(j)}) - \sum_{\lambda_n^{(j)}\le R} (qu_n^{(j)}, u_n^{(j)}).\end{aligned}$$

□

Lemma 2.3. *Let S, R be any two real numbers satisfying* $0 < S < R$, *then*

$$0 \le F_j(R) \le \sum_{\sigma_n^{(j)}\le S} \frac{Q_j^2}{R - \sigma_n^{(j)}} + \sum_{S<\sigma_n^{(j)}\le R} Q_j.$$

Proof. Notice that $\{u_n^{(j)}\}$ and $\{v_n^{(j)}\}$ are both complete orthonormal systems in $H_0^1(\Omega_j)$, we deduce from Lemma 2.2 and the Parseval's formula

$$\begin{aligned}
F_j(R) &= \sum_{\mu_n^{(j)}\le R} (qv_n^{(j)}, v_n^{(j)}) - \sum_{\lambda_m^{(j)}\le R} (qu_m^{(j)}, u_m^{(j)}) \\
&= \sum_{\mu_n^{(j)}\le R}\sum_{m=1}^{\infty} (qv_n^{(j)}, u_m^{(j)})(v_n^{(j)}, u_m^{(j)}) - \sum_{\lambda_n^{(j)}\le R}\sum_{n=1}^{\infty} (qu_m^{(j)}, v_n^{(j)})(u_m^{(j)}, v_n^{(j)}) \\
&= \sum_{\mu_n^{(j)}\le R}\sum_{\lambda_m^{(j)}>R}^{\infty} (qv_n^{(j)}, u_m^{(j)})(v_n^{(j)}, u_m^{(j)}) \\
&\qquad - \sum_{\lambda_n^{(j)}\le R}\sum_{\mu_n^{(j)}>R}^{\infty} (qu_m^{(j)}, v_n^{(j)})(u_m^{(j)}, v_n^{(j)}) \\
&= I - II.
\end{aligned}$$

To estimate I, we write $\eta_n = \sum_{\lambda_m^{(j)}>R}^{\infty} (qv_n^{(j)}, u_m^{(j)})(v_n^{(j)}, u_m^{(j)})$, then $|\eta_n| \le \left\|qv_n^{(j)}\right\| \left\|v_n^{(j)}\right\| \le Q_j$. On the other hand

$$\begin{aligned}
(\lambda_m^{(j)} - \mu_n^{(j)})(u_m^{(j)}, v_n^{(j)}) &= \lambda_m^{(j)}(u_m^{(j)}, v_n^{(j)}) - \mu_n^{(j)}(u_m^{(j)}, v_n^{(j)}) \\
&= (Lu_m^{(j)}, v_n^{(j)}) - (u_m^{(j)}, L_0 v_n^{(j)}) \\
&= (qu_m^{(j)}, v_n^{(j)}).
\end{aligned}$$

Thus, for $\mu_m^{(j)} \le R$,

$$|\eta_n| = \sum_{\lambda_m^{(j)}>R} \frac{(qv_n^{(j)}, u_m^{(j)})^2}{\lambda_m^{(j)} - \mu_n^{(j)}} < \sum_{\lambda_m^{(j)}>R} \frac{(qv_n^{(j)}, u_m^{(j)})^2}{R - \mu_n^{(j)}} \le \frac{Q_j^2}{R - \mu_n^{(j)}}.$$

So

$$\begin{aligned}
|I| &= \sum_{\mu_m^{(j)}\le R} |\eta_n| \\
&= \sum_{\mu_m^{(j)}\le S} |\eta_n| + \sum_{S<\mu_m^{(j)}\le R} |\eta_n| \\
&= \sum_{\mu_m^{(j)}\le S} \frac{Q_j^2}{R - \mu_n^{(j)}} + \sum_{S<\mu_m^{(j)}\le R} Q_j.
\end{aligned}$$

As the same way, we have

$$|II| \le \sum_{\lambda_m^{(j)}\le S} \frac{Q_j^2}{R - \lambda_n^{(j)}} + \sum_{S<\lambda_m^{(j)}\le R} Q_j.$$

Finally, $0 \leq F_j(R) \leq |I| + |II|$, Lemma 2.3 follows. □

The following two lemmas are the restrictions uniformly on the increasing rate of $A_j(R)$.

Lemma 2.4. *For arbitrary $\varepsilon > 0$, there exists sufficiently small $\delta > 0$ so that we have for j*

$$\lim_{R\to+\infty} \frac{A_j(R) - A_j(R-\delta R)}{A_j(R)} \leq \varepsilon.$$

Proof. Since $A_j(R) = (2\pi)^{-k}\omega_k|\Omega_j|_k R^{k/2} + o(R^{k/2})$, as $R \to +\infty$, where ω_k denotes the volume of the unit ball in $\mathbb{R}^k$, then

$$\begin{aligned}\frac{A_j(R) - A_j(R-\delta R)}{A_j(R)} &= 1 - \frac{(2\pi)^{-k}\omega_k|\Omega_j|_k (R-\delta R)^{k/2} + 0\left((R-\delta R)^{k/2}\right)}{(2\pi)^{-k}\omega_k|\Omega_j|_k R^{k/2} + o(R^{k/2})} \\ &\sim 1 - (1-\delta)^{k/2}, \text{ as } R \to +\infty.\end{aligned}$$

Thus, for any $\varepsilon > 0$ take $0 < \delta < 1 - (1-\varepsilon)^{2/k}$, Lemma 2.4 holds. □

Lemma 2.5. *For the operator L_0, we have uniformly estimate on j:*

$$\frac{A_j(2R)}{A_j(R)} = O(1), \quad \text{as } R \to +\infty.$$

3. Proof of the Main Results

In section 2, we consider the asymptotic behaviour of the trace for the operators L and L_0 in a single domain Ω_j. Next, we shall consider the same problem in the case of $\Omega = \bigcup_{j=1}^{\infty} \Omega_j$. First, we confirm that the eigenvalues of (P) have the following relation with the eigenvalues of $(P_j)_{j=1}^{\infty}$.

Lemma 3.1. *Suppose that $\lambda_n^{(j)}$ is an eigenvalue of (P_j), then $\lambda_n^{(j)}$ is certainly the eigenvalue of (P); on the other hand, for an eigenvalue λ_n of (P), there exists j such that λ_n is the eigenvalue of (P_j).*

Also we can prove that the eigenvalues between problem (D) and $(D_j)_{j=1}^{\infty}$ have the same relation. Thus to consider the eigenvalue problem (P) and (D) would be equivalent to consider the eigenvalues $(P_j)_{j=1}^{\infty}$ and $(D_j)_{j=1}^{\infty}$ together. Now, we rearrange the all eigenvalues $\bigcup_{j=1}^{\infty}\bigcup_{m=1}^{\infty}\left\{\mu_m^{(j)}\right\}$ of (D) which are rewritten by $\{\mu_n\}$ in non-decreasing order. Denote that

v_n are the unit eigenfunctions associated with μ_n. For any μ_n, there exist j and m such that $\mu_n = \mu_m^{(j)}$. Corresponding to μ_n, we rewrite $\lambda_m^{(j)}$ by λ_n. Then by Lemma 3.1, $\bigcup\limits_{n=1}^{\infty} \{\lambda_n\}$ is the whole eigenvalues $\bigcup\limits_{j=1}^{\infty} \bigcup\limits_{m=1}^{\infty} \left\{\lambda_m^{(j)}\right\}$. Perhaps, λ_n are not arranged in non-decreasing order, but $\lambda_n \to +\infty$ as $n \to \infty$. Denote $A(R) = \#\{n|\mu_n \leq R\}$, then $A(R) = \sum\limits_{j=1}^{\infty} A_j(R)$, and the sum is finite for and fixed R.

Lemma 3.2.

$$\lim_{R\to+\infty} \frac{1}{A(R)} \sum_{j=1}^{\infty} \sum_{\mu_n^{(j)} \leq R} [\lambda_n^{(j)} - \mu_n^{(j)} - (qv_n^{(j)}, v_n^{(j)})] = 0$$

or

$$\lim_{R\to+\infty} \frac{1}{A(R)} \sum_{n=1}^{A(R)} [\lambda_n - \mu_n - (qv_n, v_n)] = 0.$$

Proof. By Lemma 2.2, it is necessary to show that

$$\lim_{R\to+\infty} \frac{1}{A(R)} \sum_{j=1}^{\infty} F_j(R) = 0,$$

i.e.,

$$\lim_{R\to+\infty} \sum_{j=1}^{\infty} \frac{F_j(R)}{A_j(R)} \cdot \frac{A_j(R)}{A(R)} = 0.$$

For any given $\varepsilon > 0$, take $\delta > 0$ which satisfies the condition of Lemma 2.4 and $\theta = \frac{\delta}{2}$, $S = R - \theta R$ as mentioned in Lemma 2.3. Then we have

$$\begin{aligned} 0 &\leq \frac{F_j(R)}{A_j(R)} \\ &\leq \frac{1}{A_j(R)} \sum_{\sigma_n^{(j)} \leq S} \frac{Q_j^2}{R - \sigma_n^{(j)}} + \frac{1}{A_j(R)} \sum_{S < \sigma_n^{(j)} \leq R} Q_j \\ &\leq \frac{Q_j^2}{R\theta} \cdot \frac{C_j(R - \theta R)}{A_j(R)} + Q_j \cdot \frac{C_j(R) - C_j(R - \theta R)}{A_j(R)} \\ &\leq \frac{Q^2}{R\theta} \cdot \frac{C_j(R)}{A_j(R)} + Q \cdot \frac{C_j(R) - C_j(R - \theta R)}{A_j(R)}. \end{aligned}$$

From Lemma 2.1, we obtain

$$\begin{aligned} 0 &\le \frac{F_j(R)}{A_j(R)} \\ &\le \frac{2Q^2}{R\theta} + 2Q \cdot \frac{A_j(R) - A_j(R - \theta R - Q_j)}{A_j(R)} \\ &\le \frac{2Q^2}{R\theta} + 2Q \cdot \frac{A_j(R) - A_j(R - \delta R)}{A_j(R)}, \end{aligned}$$

where we take R sufficiently large such that $Q_j \le Q < \theta R$.

For sufficiently large R, Lemma 2.4 gives that, for j uniformly,

$$\frac{A_j(R) - A_j(R - \delta R)}{A_j(R)} \le \varepsilon.$$

On the other hand, $\frac{2Q^2}{R\theta} \le Q\varepsilon$ for sufficiently large R, thus

$$0 \le \frac{F_j(R)}{A_j(R)} \le 3Q\varepsilon,$$

which implies that, for R large enough,

$$0 \le \frac{1}{A(R)} \sum_{j=1}^{\infty} F_j(R) \le 3Q\varepsilon \sum_{j=1}^{\infty} \frac{A_j(R)}{A(R)} = 3Q\varepsilon.$$

Since ε is arbitrary, Lemma 3.2 holds. □

Now we can prove our main results.

The proof of Theorem 1.1. Let $t_1 < t_2 < \cdots < t_n < \cdots$ ($t_n \to +\infty$ as $n \to +\infty$) be the leap points of the right continuous function $A(R)$. It is trivial to see that $A(t_{k+1}) \le A(t_k)$. For given $\varepsilon > 0$, by Lemma 2.4, there exists $\delta > 0$, such that for j uniformly

$$\lim_{R \to +\infty} \frac{A_j(R)}{A_j(R - \delta R)} \le \frac{1}{1 - \varepsilon}.$$

For such δ, we take k sufficiently large so that $\delta t_{k+1} > 1$, then $A(t_{k+1} - \delta t_{k+1}) \le A(t_{k+1} - 1) \le A(t_k)$, since $A(R)$ and $A_j(R)$ are both non-decreasing. Let $A(t_k) = \alpha_k$, then for k large enough,

$$1 < \frac{\alpha_{k+1}}{\alpha_k} = \sum_{j=1}^{\infty} \frac{A_j(t_{t_k+1})}{A(t_k)}$$
$$\leq \sum_{j=1}^{\infty} \frac{A_j(t_{t_k+1})}{A(t_{k+1} - \delta t_{k+1}))} \frac{A_j(t_k)}{A(t_k)}$$
$$\leq \frac{1}{1-\varepsilon} \sum_{j=1}^{\infty} \frac{A_j(t_k)}{A(t_k)} = \frac{1}{1-\varepsilon},$$

so $\lim_{k\to\infty} \frac{\alpha_{k+1}}{\alpha_k} = 1$ if we take $\varepsilon \to 0^+$.

Notice that $\alpha_1 < \alpha_2 < \cdots < \alpha_k < \cdots$, and $\alpha_k \to +\infty$ as $k \to \infty$. Then for a positive integer N, there exists a natural number k such that $\alpha_k < N \leq \alpha_{k+1}$. Since $\lambda_n - \mu_n \geq 0$, we have

$$-\sum_{n=1}^{N} [\lambda_n - \mu_n - (qv_n, v_n)]$$
$$= -\sum_{n=1}^{\alpha_k} [\lambda_n - \mu_n - (qv_n, v_n)] + \sum_{\alpha_k < n \leq N} (qv_n, v_n)$$
$$= -\sum_{n=1}^{\alpha_k} [\lambda_n - \mu_n - (qv_n, v_n)] + Q(\alpha_{k+1} - \alpha_k),$$

therefore

$$0 \leq -\frac{1}{N} \sum_{n=1}^{N} [\lambda_n - \mu_n - (qv_n, v_n)]$$
$$\leq -\frac{1}{\alpha_k} \sum_{n=1}^{\alpha_k} [\lambda_n - \mu_n - (qv_n, v_n)] + Q\left(\frac{\alpha_{k+1}}{\alpha_k} - \alpha_k\right).$$

Here $N \to \infty$ implies $k \to \infty$, we obtain the following result from Lemma 3.2

$$\lim_{N\to\infty} \frac{1}{N} \sum_{n=1}^{N} [\lambda_n - \mu_n - (qv_n, v_n)] = 0.$$

The proof of Theorem 1.2. Let $C_k^{(j)}$ be the number of $\sigma_k^{(j)}$ in $I_k = (k-1, k]$, then $C^{(j)}(m) = C_1^{(j)} + \cdots + C_m^{(j)}$. Put $R = m$ and $S = m - 1$ as that we have in Lemma 2.3, then

$$\begin{aligned} 0 &\le F_j(m) \\ &\le \sum_{\sigma_n^{(j)} \le m-1} \frac{Q_j^2}{m-\sigma_n^{(j)}} + \sum_{m-1<\sigma_n^{(j)}\le m} Q_j \\ &\le \sum_{k=1}^{m-1} \sum_{\sigma_n^{(j)} \in I_k} \frac{Q^2}{m-\sigma_n^{(j)}} + QC_m^{(j)}, \\ &\le Q^2 \sum_{k=1}^{m-1} \frac{C_k^{(j)}}{m-k} + QC_m^{(j)}, \end{aligned}$$

hence

$$F(m) = \sum_{j=1}^{\infty} F_j(m) \le \sum_{j=1}^{\infty} \left(Q^2 \sum_{k=1}^{m-1} \frac{C_k^{(j)}}{m-k} + QC_m^{(j)} \right),$$

and

$$\begin{aligned} &\sum_{m=N+1}^{2N} F(m) \\ &\le \sum_{m=N+1}^{2N} \left[Q^2 \sum_{k=1}^{m-1} \frac{C_k^{(j)}}{m-k} + QC_m^{(j)} \right] \\ &= \sum_{j=1}^{\infty} \left[Q^2 \sum_{m=N+1}^{2N} \sum_{k=1}^{m-1} \frac{C_k^{(j)}}{m-k} + Q \sum_{m=N+1}^{2N} C_m^{(j)} \right] \\ &\le \sum_{j=1}^{\infty} \left[Q^2 \left(\sum_{k=1}^{N} \sum_{m=N+1}^{2N} + \sum_{k=N+1}^{2N-1} \sum_{m=N+1}^{2N} \right) \frac{C_k^{(j)}}{m-k} + QC^{(j)}(2N) \right] \\ &\le Q^2 \sum_{j=1}^{\infty} C^{(j)}(2N)(1+\log N) + QC(2N) \\ &= C(2N)[Q^2(1+\log N) + Q]. \end{aligned}$$

So

$$\frac{1}{N} \sum_{m=N+1}^{2N} F(m) \le C(2N) \cdot \frac{Q^2(1+\log N)+Q}{N} \equiv \Gamma.$$

Now, we have N nonnegative numbers $F(m)$, in which the arithmetic mean-value does not exceed Γ. It is clear that there exists at least one $m = R$, $N < R \le 2N$ such that $F(R) \le \Gamma$.

Finally, we have by Lemma 2.2, Lemma 2.1 and Lemma 2.5

$$
\begin{aligned}
0 &\le -\frac{1}{A(R)} \sum_{\mu_n \le R} [\lambda_n - \mu_n - (qv_n, v_n)] \\
&\le \frac{1}{A(R)} \sum_{j=1}^{\infty} F_j(R) = \frac{F(R)}{A(R)} \le \frac{2A(2N)}{A(N)} \cdot \frac{Q^2(1+\log N)+Q}{N} \\
&= 2\sum_{j=1}^{\infty} \frac{A_j(2N)}{A_j(N)} \cdot \frac{A_j(2N)}{A(N)} \cdot \frac{Q^2(1+\log N)+Q}{N} \\
&= O\left(\frac{\log N}{N}\right) = O\left(\frac{\log R}{R}\right).
\end{aligned}
$$

Theorem 1.2 is proved.

4. Further Results

By using the asymptotic formula of Theorem 1.1 and Theorem 1.2, in the cases of $n = 1$ or 2, we can deduce that

Corollary 4.1. *We have following trace displacement formula:*

$$\sum_{\mu_k \le R} (\lambda_k - \mu_k) = H\varphi(R) + r(R), \ \text{as } R \to +\infty.$$

where the remainder term $r(R) = o(R^{n/2})$, $H = \frac{1}{|\Omega|_n} \int_\Omega q(x)dx$, *and* $\varphi(R) = (2\pi)^{-n}\omega_n |\Omega|_n R^{n/2}$ *is the Weyl term for the Dirichlet counting function* $A(R)$.

Proof. Let us prove the case of $n = 2$. Without loss of generality, we suppose here $q(x) \ge 0$, thus

$$\lim_{N\to\infty} \frac{1}{A(R)} \sum_{j=1}^{N} (\lambda_j - \mu_j) = \lim_{N\to\infty} \frac{1}{A(R)} \sum_{j=1}^{N} (qv_j, v_j).$$

Let us discuss the problem on a single domain Ω_j, we use the same notations as used above and we have

$$\sum_{k=1}^{\infty} \frac{v_k^{(j)}(x,y)]^2}{(\mu_k^{(j)} + \xi)^2} = \frac{1}{4\pi\xi} - F_\xi^{(j)}(x,y)$$

where $\mu_k^{(j)}$ (or $v_k^{(j)}$) is the Dirichlet eigenvalue (or eigenfunction) of Ω_j, and

$$\int_{\Omega_j} \left| F_\xi^{(j)}(x,y) \right| dxdy = O(\xi^{-\frac{3}{2}}), \ \text{as } \xi \to +\infty.$$

So we get, as $\xi \to +\infty$,

$$\sum_{k=1}^{\infty} \frac{(qv_k^{(j)}, v_k^{(j)})}{(\mu_k^{(j)} + \xi)^2} = \frac{I_j}{4\pi\xi} + O(\xi^{-\frac{3}{2}}),$$

where

$$I_j = \int_{\Omega_j} q(x,y)dxdy.$$

Also we know

$$A_j(R) = \frac{1}{4\pi}|\Omega_j|_2 R + o(R), \text{ as } R \to +\infty,$$

and

$$A(R) = \frac{1}{4\pi}|\Omega|_2 R + o(R), \text{ as } R \to +\infty.$$

Let $\alpha_j(R) = \sum_{\mu_k^{(j)} \le R} (qv_k^{(j)}, v_k^{(j)})$, then $\alpha_j(R)$ is nonnegative and non-decreasing function, we have

$$\alpha_j(R) \le Q_j A_j(R) = O(R), \text{ as } R \to +\infty.$$

We notice that $\alpha_j(R) = 0$ when $0 < R < \mu_1^{(j)}$, and $\alpha_j(\mu_{k+1}^{(j)}) - \alpha_j(\mu_k^{(j)}) = (qv_{k+1}^{(j)}, v_{k+1}^{(j)})$. Therefore, we have

$$\int_0^{\mu_{N+1}^{(j)}} \frac{\alpha_j(R)}{(R+\xi)^3} dR = \sum_{k=1}^{N} \int_{\mu_k^{(j)}}^{\mu_{k+1}^{(j)}} \frac{\alpha_j(\mu_k^{(j)})}{(R+\xi)^3} dR$$

$$= \frac{1}{2}\sum_{k=1}^{N}\left[\frac{\alpha_j(\mu_k^{(j)})}{(\mu_k^{(j)}+\xi)^2} - \frac{\alpha_j(\mu_k^{(j)})}{(\mu_{k+1}^{(j)}+\xi)^2}\right]$$

$$= \frac{1}{2}\left[\sum_{k=1}^{N}\frac{\alpha_j(\mu_k^{(j)})}{(\mu_k^{(j)}+\xi)^2} - \sum_{k=2}^{N+1}\frac{\alpha_j(\mu_{k-1}^{(j)})}{(\mu_k^{(j)}+\xi)^2}\right]$$

$$= \frac{1}{2}\frac{\alpha_j(\mu_1^{(j)})}{(\mu_1^{(j)}+\xi)^2} + \frac{1}{2}\sum_{k=2}^{N+1}\frac{\alpha_j(\mu_j^{(j)}) - \alpha_j(\mu_{k-1}^{(j)})}{(\mu_k^{(j)}+\xi)^2} - \frac{1}{2}\frac{\alpha_j(\mu_{N+1}^{(j)})}{(\mu_{N+1}^{(j)}+\xi)^2}$$

$$= \frac{1}{2}\sum_{k=1}^{N+1}\frac{(qv_k^{(j)}, v_k^{(j)})}{(\mu_k^{(j)}+\mu)^2} + O\left(\frac{1}{\mu_{N+1}^{(j)}}\right).$$

Let $N \to \infty$, we have

$$\int_0^{+\infty} \frac{\alpha_j(R)}{(R+\xi)^3} dR = \frac{I_j}{8\pi\xi} + O(\xi^{-\frac{3}{2}}), \text{ as } \xi \to +\infty.$$

From Tauberian Theorem, we have

$$\lim_{R\to+\infty}\frac{\alpha_j(R)}{R}=\frac{\Gamma(3)}{\Gamma(2)\Gamma(1)}\cdot\frac{I_j}{8\pi}=\frac{I_j}{4\pi}.$$

So,

$$\begin{aligned}&\lim_{R\to+\infty}\frac{1}{R}\sum_{\mu_k\le R}(qv_k,v_k)\\ =&\lim_{R\to+\infty}\frac{1}{R}\sum_{j=1}^{\infty}\sum_{\mu_k^{(j)}\le R}(qv_k^{(j)},v_k^{(j)})\\ =&\lim_{R\to+\infty}\sum_{j=1}^{\infty}\frac{\alpha_j(R)}{R}=\sum_{j=1}^{\infty}\frac{I_j}{4\pi}=\frac{I}{4\pi},\end{aligned}$$

where $I=\sum\limits_{j=1}^{\infty}I_j=\int_\Omega q(x,y)dxdy$, which implies

$$\begin{aligned}&\lim_{R\to+\infty}\frac{1}{A(R)}\sum_{\mu_k\le R}(qv_k,v_k)\\ =&\lim_{R\to+\infty}\frac{R}{A(R)}\frac{1}{R}\sum_{\mu_k\le R}(qv_k,v_k)\\ =&\frac{4\pi}{|\Omega|_2}\cdot\frac{I}{4\pi}=\frac{1}{|\Omega|_2}\int_\Omega q(x,y)dxdy=H.\end{aligned}$$

The Corollary 4.1 is proved. □

Finally, if $\Omega\subset\mathbb{R}^n$, $n=1$ or 2, is a fractal drum, and the Minkowski dimension of $\partial\Omega$ is $\delta\in(n-1,n)$, then we have

Corollary 4.2. *For the remainder term $r(R)$, we have*

$$r(R)=O(R^{\frac{\delta}{2}}),\quad \textit{as } R\to+\infty.$$

References

1. Gelfand I.M., Levitan B.M., On a simple identity for the characteristic values of a differential operator of the second order (Russian) D.A.N. **88**(1953), 593-596.
2. Cao Cewen, On the asymptotic estimation of the trace of a partial differential operator, *Scientia Sinica,* Special Issue **II**(1979), 56-68.
3. Cao Cewen, On the asymptotic trace of the second order self-adjoint elliptic operator, Proceedings 1980 Beijing symposium Diff. Geo. and Equ., **vol 3**, 11707-1114.

4. Chen Hua, Sleeman B.D., Fractal drums and n-dimensional modified Weyl-Berry conjecture, Comm. Math. Phys., **168**(1995), 581-607.
5. E.C. Titchmarsh, Eigenfunction expansions associated with second order differential equations, Part II, Oxford, 1958.

LIMITATIONS AND MODIFICATIONS OF BLACK-SCHOLES MODEL *

LISHANG JIANG AND XUEMIN REN
Institute of Math., Tongji University, Shanghai, China

In this paper we investigate the limitations of Black-Scholes model and provide some modifications and some applications in pricing of real options and option-like contracts also. Our modifications are based on the following assumptions: jump-diffusion process, stochastic interest rate, variable volatility and implied volatility.

1. Introduction

In this paper, we start by a brief review of Black-Scholes model, discuss its limitations in detail, and show some modifications.

1.1. *Motivation of Black-Scholes model*

The call(put) option gives the holder the right to buy(sell) a fixed amount of an asset S at a specified time T in future for an agreed price K. The key problem is to find a way to determine the fair price of the option in its whole life, such that

$$V(S_T, T) = \begin{cases} (S_T - K)^+, \text{ call;} \\ (K - S_T)^+, \text{ put.} \end{cases}$$

There are two different style options: European option can be exercised only on maturity while American option can be exercised at any time before maturity.

1.2. *Foundations of Black-Scholes model*

To derive Black-Scholes formula, we need following assumptions and arguments

*supported by the national natural science foundation of china (no. 10171078)

- Asset price s_t is governed by a geometric Brownian motion

$$\frac{dS_t}{S_t} = \mu dt + \sigma dW_t \tag{1.1}$$

where $E(dW_t) = 0, Var(dW_t) = dt;$

- Interest rate r never changes and is same for borrowing or lending;
- Asset's volatility σ is known and never changes for all maturities;
- There are no transaction costs and taxes for either asset or option;
- Unlimited short selling is allowed;
- The underlying asset can be traded continuously and in infinitesimally small numbers of units;

Upon the assumptions above, for European put option, there is an unique martingale measure Q such that

$$V_t = e^{-r(T-t)} E^q((K - S_t|F_t)), \tag{1.2}$$

From the point of view of PDE, there exists a function $V = (S,t)$ which satisfies

$$\begin{cases} \frac{\partial V}{\partial t} + \frac{\sigma^2}{2} S^2 \frac{\partial^2 V}{\partial S^2} + (r-q) S \frac{\partial V}{\partial S} - rV, \\ V|_{t=T} = (K-S)^+, \end{cases} \tag{1.3}$$

where q is the dividend rate of S_t. From the equation (1.3), we can obtain the famous Black-Scholes formula :

$$V(S,t) = Ke^{-r(T-t)} N(-d_2) - Se^{-q(T-t)} N(-d_1), \tag{1.4}$$

where $d_1 = \frac{\ln \frac{S}{K} + (r-q+\frac{\sigma^2}{2})(T-t)}{\sigma\sqrt{T-t}},$

$$d_2 = d_1 - \sigma\sqrt{T-t}, N(x) = \frac{1}{\sqrt{2\pi}} \int_{-\infty}^{x} e^{-\frac{t^2}{2} dt}.$$

As for American put option, we can find a function $V = V(S,t)$ which satisfies

$$V, \frac{\partial V}{\partial S} \in C[R_+ \times [0,T]],$$

and

$$\begin{cases} min\{-LV, V-(K-S)^+\} = 0, \\ V|_{t=T} = (K-S)^+, \end{cases} \tag{1.5}$$

$$LV = \frac{\partial V}{\partial t} + \frac{\sigma^2}{2} S^2 \frac{\partial^2 V}{\partial S^2} + (r-q) S \frac{\partial V}{\partial S} - rV.$$

where Γ is the optimal exercise boundary.

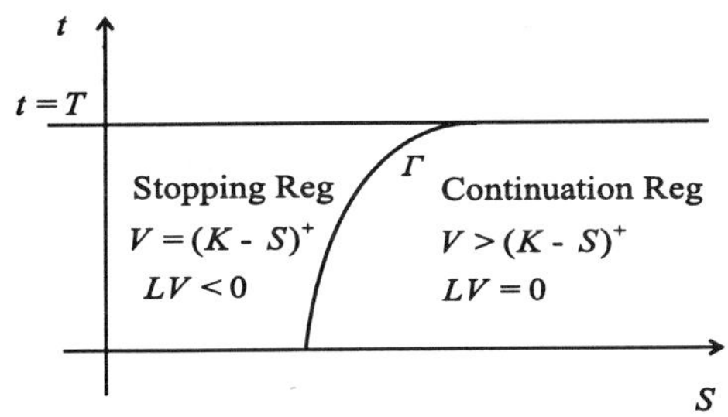

Figure 1

1.3. *Modifications of the model*

F. Black: "when we calculate option values using Black-Scholes model, and compare them with option prices, there is usually a difference".

"There are three reasons for a difference between value and price: we may have the correct value, and the option price may be out of line; we may have used the wrong outputs to the Black-Scholes formula; or the Black-Scholes formula may be wrong. Normally, all three reasons play a part in explaining the difference".

"The Black-Scholes formula may be wrong......, undoubtedly there will be a series of models developed over time that are better than the original Black-Scholes model"[2].

It is clear that each of assumptions mentioned above is unrealistic to some degree: for example,

- Asset prices can jump. This invalidates the assumption that the price of asset follows a geometric Brownian motion since it has continuous sample paths. It also makes the market incomplete, though hedging strategies can still be used to reduce the level of risk.

- The risk-free interest rate does vary and usually in an unpredictable way. We can adapt some models to allow for a stochastic risk-free rate, especially when we price the option with relatively long maturity.
- Among the inputs for Black-Scholes formula, only the volatility σ can not be observed directly from the market. If we derive it from Black-Scholes formula using quoted option prices, we will find the assumption that the volatility is a constant is not true. The reason may be that the distributions of asset returns tend to have fatter tails than suggested by the log-normal model.

Despite all of the potential flaws in the model assumptions, analysis of market derivative prices show that the Black-Scholes model does provide a quite good approximation to the market and an insight into the usefulness of dynamic hedging. The fact that the assumptions not hold in practice does not mean that the model has no use. In our opinion, the best way is to consider some more realistic assumptions.

In the following sections, we will modify Black-Scholes model under the assumption of jump diffusion process and stochastic interest rate, and find a robust method to derive the implied volatility.

2. Jump-diffusion process

If the market gets some good or bad news, the prices of assets can jump. So the jump-diffusion process sounds more realistic.

2.1. *Movement of asset price*

We assume that the asset price follow a jump-diffusion process

$$\frac{dS_t}{S_-} = \mu dt + \sigma dW_t + U dq, \tag{2.1}$$

where a Poisson process dq is defined by

$$dq = \begin{cases} 0, & (\varpi_1) \\ 1, & (\varpi_2) \end{cases} \tag{2.2}$$

$Prob(\varpi_1) = 1 - \lambda dt$, $Prob(\varpi_2) = \lambda dt$, that is, there are no jump in case (ϖ_1) , and a jump in case (ϖ_2) with intensity λ. In case (ϖ_2) , the price will move from S_- to $JS_-(J > 0)$, and $U = (J-1)S_-, -1 < U < +\infty$, is independently equal distribution r.v.

2.2. *Black-Scholes model with jump-diffusion*

It can be showed that under martingale measure, the European put option price $V(S,t)$ satisfies

$$\begin{cases} L_{JD}V = \dfrac{\partial V}{\partial t} + \dfrac{\sigma^2}{2}S^2\dfrac{\partial^2 V}{\partial S^2} + (r - q - \lambda k)S\dfrac{\partial V}{\partial S} - rV + \\ \qquad\qquad \lambda \displaystyle\int_{-1}^{+\infty} [V(S(1+y),t) - V(S,t)]dN(y) = 0, \\ V|_{t=T} = (K - S)^+, \end{cases}$$

where

$$k = E(U) = \int_{-1}^{+\infty} y dN(y).$$

$E[\cdot]$ is the expectation taken over the jump size. For American put option, it turns into

$$\begin{cases} \min\{-L_{JD}V, V - (K-S)^+\} = 0, \\ V|_{t=T} = (K - S)^+. \end{cases} \tag{2.3}$$

$\Gamma : S = S(t)$ is the optimal exercise boundary.

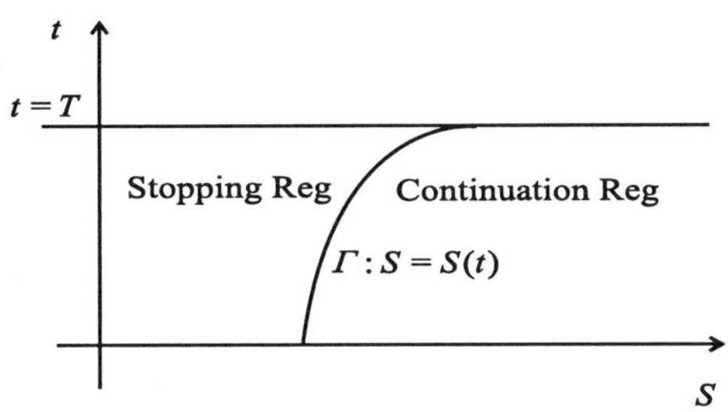

Figure 2

2.3. *Some results*

As we know, there exists a closed form solution for European option. Amin gives the C-R-R Scheme for jump-diffusion as a discrete model to value the European option and proves the convergence in weak sense[1]. Zhang uses a FEM to value American option price and prove the convergence in H^1 sense[13]. For European option, we prove that there are estimates

$$\left|V^{\Delta}(S,t) - V(S,t)\right| \leq C\sqrt{\Delta t},$$

$V^{\Delta}(S,t)$ is given by C-R-R schemes, and

$$\left|\widehat{V}^{\Delta}(S,t) - V(S,t)\right| \le C\Delta t,$$

$\widehat{V}^{\Delta}(S,t)$ is given by modified C-R-R schemes[11]. For American put option, we claim that the optimal exercise boundary $S = S(t)$ satisfies:

$S(t) \in C^{\infty}[0,T)$, $S(t)$ is a monotonic increasing function on the interval $[0,t)$ and

$$S(T) = min\{K, \widehat{S}\},$$

where $\widehat{S}$ is the root of transcendental equation

$$-rK + q\widehat{S} + \lambda \int_{-1}^{+\infty} [(1+y)\widehat{S} - K]^{+} dN(y) = 0, and S_{\infty} \le S(t) \le S(T),$$

where S_{∞} is the optimal exercise point for perpetual American put option[9,?,2].

For American put option, in local uniform sense, we have

$$\Delta t \xrightarrow{lim} \infty\ V^{\Delta}(S,t) = V(S,t), \Delta t \xrightarrow{lim} \infty\ S^{\Delta}t = S(t),$$

where $V(S,t)$ is viscosity solution , $V^{\Delta}(S,t)$ and $S^{\Delta}t$ are given by C-R-R scheme.

3. Stochastic interest rate

3.1. *The models of stochastic interest rate*

We assume the interest rate is governed by following stochastic process:

$$dr_t = \widehat{\mu}(r_t,t)dt + \widehat{\sigma}(r_t,t)d\widehat{W_t}, \tag{3.1}$$

where

$$\widehat{\mu}(r_t,t) = \eta - \theta r_t.$$

It incorporates the mean reverting feature of the interest rate.

$$\widehat{\sigma}(r_t,t) = \begin{cases} \widehat{\sigma_0}, & \text{Vasicek;} \\ \widehat{\sigma_0}\sqrt{r}, & \text{C-I-R.} \end{cases}$$

and

$$E(d\widehat{W_t}) = 0, Var(\widehat{W_t}) = dt$$

$$Cov(dW_t, d\widehat{W_t}) = \rho dt, (|\rho| < 1)$$

Remark 3.1: the range of r_t is

$$Vasicek\ \ model: -\infty < r_t < +\infty$$

$$C-I-R\ \ model: if 2\eta \geq \widehat{\sigma_0}^2, 0 \leq r_t < +\infty$$

3.2 Black-Scholes model with stochastic interest rate

Under a martingale measure, European put option price $V = V(S, r, t)$ satisfies

$$\begin{cases} L_{SI}V = \frac{\partial V}{\partial t} + \frac{\sigma^2}{2}S^2\frac{\partial^2 V}{\partial S^2} + \rho\sigma S\widehat{\sigma}(r,t)\frac{\partial^2 V}{\partial s \partial r} + \frac{1}{2}\widehat{\sigma}^2(r,t)\frac{\partial^2 V}{\partial r^2} + rs\frac{\partial V}{\partial S} \\ +(\eta - \theta r)\frac{\partial V}{\partial r} - rV = 0 (0 \leq S < +\infty, r \in D, 0 \leq t < t), \\ V(S,r,T) = (K-S)^+ (0 \leq S < +\infty, r \in D), \end{cases} \tag{3.2}$$

where

$$D = \begin{cases} R, & \text{Vasicek;} \\ R^+, & \text{C-I-R.} \end{cases}$$

Remark: From Fichera theory for second order partial differential equations with nonnegative characteristic form, we know that European put option problem is well posed, if $2\eta \geq \widehat{\sigma_0}^2$ in C-I-R model, that is, we should not impose any boundary condition on $r = 0$.

We introduce a zero coupon bond with maturity T as a numeraire. $P(r, t)$ denotes the value of this zero coupon bond. Under this relative price system, there exists a martingale measure Q such that $\frac{V_t}{P_t} = E^Q[\frac{V_T}{P_T}|F_t]$. $P(r, t)$ satisfies

$$\begin{cases} \frac{\partial P}{\partial t} + \frac{1}{2}\widehat{\sigma}^2(r,t)\frac{\partial^2 P}{\partial r^2} + (\eta - \theta r)\frac{\partial P}{\partial r} - rP = 0 \ \ , (r \in D, 0 \leq t < T) \\ P(r,T) = 1, \quad (r \in D). \end{cases} \tag{3.3}$$

It has affine structure solution

$$P(r,t) = B(t)e^{-A(t)r}, \tag{3.4}$$

$$A(t) = \begin{cases} \frac{1-e^{\theta(T-t)}}{\theta}, & Vasicek; \\ \frac{1+k}{\mu} \cdot \frac{1-e^{-\mu(T-t)}}{1+ke^{-\mu(T-t)}}, & C-I-R. \end{cases}$$

$$\ln B(t) = \begin{cases} \frac{2\theta\eta - \widehat{\sigma}_0}{2\theta^2}[(T-t) - A(t)] + \frac{\widehat{\sigma}_0}{4\theta}A^2(t), & Vasicek; \\ \frac{2\eta}{\widehat{\sigma}_0^2}\ln(\frac{1+ke^{-\mu(T-t)}}{1+k}) + \frac{\eta(\mu-\theta)}{\widehat{\sigma}_0^2}(T-t), & C-I-R. \end{cases}$$

where

$$\mu = \theta\sqrt{2 + 2(\frac{\widehat{\sigma}_0}{\theta})^2}, \quad k = \frac{\mu - \theta}{\mu + \theta}.$$

Set

$$\widehat{V} = \frac{V}{P(r,t)}, Z = \frac{S}{P(r,t)},$$

So we introduce a relative price system, then the Cauchy problem (3.2) can be reduced as follows:

$$\begin{cases} \frac{\partial \widehat{V}}{\partial t} + \frac{1}{2}\widetilde{\sigma}^2(t) Z^2 \frac{\partial^2 \widehat{V}}{\partial Z^2} = 0, \\ \widehat{V}\mid_{t=T} = (K - Z)^+, \end{cases} \tag{3.5}$$

where

$$\widetilde{\sigma}^2(t) = [A^2(t)\widehat{\sigma}_0^2 - 2\rho\widehat{\sigma}_0\sigma A(t) + \sigma^2].$$

So, we reduce the dimension of the problem fortunately (including both equation and boundary) From (3.5) it yields

$$\widehat{V}(Z,t) = KN(-\widehat{d}_2) - ZN(-\widehat{d}_1),$$

where

$$\widehat{d}_1 = \frac{\ln\frac{Z}{K} + \frac{1}{2}\int_t^T \widetilde{\sigma}_0^2(\tau)d\tau}{\sqrt{\int_t^T \widetilde{\sigma}_0^2(\tau)d\tau}}, \widehat{d}_2 = \widehat{d}_1 - \sqrt{\int_t^T \widetilde{\sigma}_0^2(\tau)d\tau}$$

Thus, the value of the European put option is

$$V(S,r,t) = P(r,t)\widehat{V}(\frac{S}{P(r,t)}, t).$$

We can use this method in pricing interest rate derivatives especially for different kinds of European style options.

(1). pricing convertible bond with default which can exercise only at maturity[8].

(2). pricing cross currency derivatives [10].

3.2. *American style option—valuation of mortgage*

Mortgage is a derivative of risk asset S_t and stochastic interest rater_t. S_t is the price of real estate asset. It has two features: prepayment and default.We can divide the domain into three regions: Σ_1 is continuation region; Σ_2 is prepayment region and Σ_3 is default region. $M(t)$ denotes

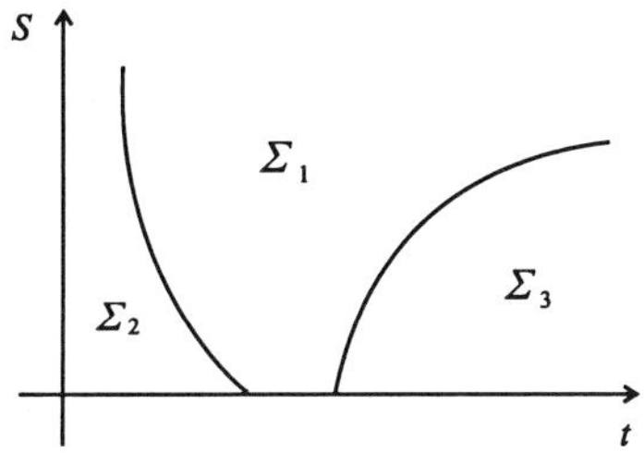

Figure 3

the remain principle and m denotes the loan payment rate. Then, $M(t)$ satisfies an ODE

$$\begin{cases} \frac{dM(t)}{dt} = cM(t) - m, \\ M(0) = M_0, M(T) = 0, \end{cases} \tag{3.6}$$

where c is interest rate for loan, M_0 is the loan amount and T is the maturity of the loan. From (3.6), we obtain

$$M(t) = \frac{1 - e^{-c(T-t)}}{1 - e^{-cT}} M_0, m = \frac{cM_0}{1 - e^{-cT}}.$$

The value of the mortgage $V_t = V(S_t, r_t, t)$ satisfies[12]

$$\begin{cases} V < \min\{M(t), S\} \\ LV = 0 \end{cases} in\ \Sigma_1,$$
$$\begin{cases} V = M(t) < S \\ LV = L(M(t)) > 0 \end{cases} in\ \Sigma_2$$
$$\begin{cases} V = S < M(t) \\ LV = L(S) > 0 \end{cases} in\ \Sigma_3.$$

where

$$LV = \frac{\partial V}{\partial t} + \frac{\sigma^2}{2} S^2 \frac{\partial^2 V}{\partial S^2} + \rho\sigma\widehat{\sigma}(r,t) S \frac{\partial^2 V}{\partial s \partial r} + \frac{1}{2}\widehat{\sigma}^2(r,t)\frac{\partial^2 V}{\partial r^2} + rS\frac{\partial V}{\partial S} + (\eta - \theta r)\frac{\partial V}{\partial r} - rV + m$$

and

$$\widehat{\sigma}(r_t, t) = \begin{cases} \widehat{\sigma}_0, & Vasicek; \\ \widehat{\sigma}_0\sqrt{r}, & C - I - R. \end{cases}$$

For the fixed rate mortgage model [12], the problem is to find $V, \nabla V \in C(R_+ \times D \times [0, T))$, s.t.

$$\begin{cases} \max\{-LV, V - \min\{M(t), S\}\} = 0 \\ V\mid_{t=T} = 0 \end{cases}. \tag{3.7}$$

And for the fixed rate mortgage model without default risk, the problem is to find $V, \nabla V \in C(D \times [0,T))$, s.t.

$$\begin{cases} \max\{-L_0 V, V - M(t)\} = 0, \\ V(r,T) = 0, \end{cases} \tag{3.8}$$

where

$$L_0 V = \frac{\partial V}{\partial t} + \frac{1}{2}\widehat{\sigma}^2(r,t)\frac{\partial^2 V}{\partial r^2} + (\eta - \theta r)\frac{\partial V}{\partial r} - rV + m.$$

In [12] we give a characteristic-difference scheme to solve the variational inequality (3.7) and prove the convergence of the numerical result for Bermudan-Style mortgage problem. The mortgage problem (3.8) as a free boundary problem is concerned with the optimal prepay boundary$r(t)$. We prove that [7]:

$$r(t) \in C^\infty, r(t) is\ a\ monotonic\ increasing\ function$$

and

$$C_0 < r(0) \leq r(t) \leq r(T) < C_1$$

4. Variable volatility & implied volatility

Volatility σ is the only parameter in Black-Scholes formula that is not directly observable. But, we can find some information from the option market, because the option price in the market is the market's view of volatility over the life of the option.

4.1. *Implied volatility surface*

The implied volatility is the volatility derived from Black-Scholes formula using the real option price. For one underlying asset, there are a series of option prices with different strike prices and maturities.

It is clear that the implied volatility is not a constant even for same maturity. This means that it strongly depends on K and T. It also exhibits the phenomena of 'smile' and 'skew'. So, the constant volatility assumption made in Black-Scholes model is not valid in real market. One of the reasonable ways is to assume σ is a determinate function of S_t and t. Following

$T \backslash K$	K_1	K_2	$\cdots$	K_n
T_1	V_{11}	V_{12}	$\cdots$	V_{1n}
T_2	V_{21}	V_{22}	$\cdots$	V_{2n}
$\cdots$	$\cdots$	$\cdots$	$\cdots$	$\cdots$
T_m	V_{m1}	V_{m2}	$\cdots$	V_{mn}

Dupire's idea, $\sigma(S,t)$ can be determined through quoted option prices[4]. If at time $t = t_0$,the asset price is $S = S_0$, a series of option prices with different strike price and maturities are known and can be given by a form of continuous function $V(S_0, t_0, K, T)$, where $0 \leq K < \infty$ and $T_1 \leq T \leq T_2$, then from Dupire's formula, we have

$$\sigma(K,T) = \sqrt{\frac{2\dfrac{\partial V}{\partial T} + (r-q)K\dfrac{\partial V}{\partial K} - qV}{K^2\dfrac{\partial^2 V}{\partial K^2}}}.$$

Unfortunately, the option pricing is typically limited to a finite set of strike prices and maturities, a naive interpolation yields to instability of volatility. So, it is an ill-posed problem.

4.2. *Related Optimal Control Problem*

Problem I (to determine implied volatility): Find $\sigma(S)$ such that

$$\begin{cases} \dfrac{\partial V}{\partial t} + \dfrac{\sigma^2(S)}{2}S^2\dfrac{\partial^2 V}{\partial S^2} + (r-q)S\dfrac{\partial V}{\partial S} - rV = 0 \\ V\big|_{t=T} = (S-K)^+ \end{cases} \tag{4.1}$$

and

$$V(S_0, t_0, K, T) = C(K), \quad (0 \leq K < \infty)$$

where $t_0 \in [0,T)$, T are constants. According to Dupire's idea, $V = V(S,t;K,T)$ of a function of K, T is a solution of Cauchy problem (4.1). Thus, follow Dupire's idea, the problem I can be rewritten in the following form: find $\sigma(K)$ such that

$$\begin{cases} \dfrac{\partial V}{\partial \tau} - \dfrac{\sigma^2(K)}{2} K^2 \dfrac{\partial^2 V}{\partial K^2} + (r-q)K\dfrac{\partial V}{\partial K} - qV = 0 \\ V|_{\tau=0} = (S-K)^+ \quad (\tau = T - t_0) \end{cases} \tag{4.2}$$

and

$$V(S_0, t_0, K, \tau_0) = C(K), \quad (\tau_0 = T - t_0) \tag{4.3}$$

This is a typical terminal-state observation problem!
set

$$y = \ln \frac{K}{S}, \quad v = \frac{1}{S} e^{q\tau} V$$

which leads to Cauchy problem

$$(A_0) \begin{cases} \dfrac{\partial v}{\partial \tau} = a(y)(v_{yy} - v_y) - (r-q)v_y \\ v|_{\tau=0} = (1 - e^y)^+, \end{cases}$$

and

$$v(y, \tau_0) = v_0(y), \quad (\tau_0 > 0),$$

where

$$a(y) = \frac{1}{2}\sigma^2(K) = \frac{1}{2}\sigma(Se^y), \quad v_0(y) = \frac{1}{S_0} e^{q\tau_0} C(S_0 e^y),$$

$$0 \le v_0(y) \le 1,$$

$$|v_0(y) - H(-y)| = O(e^{-|y|}) \text{ as } |y| \to \infty.$$

Problem [6,5] : Find: $\hat{a}(y) \in A$, such that

$$J(\hat{a}) = \inf_{a \in A} J(a)$$

where

$$J(a) = \frac{1}{2} \int_R |v(y, \tau_0, a) - v_0(y)|^2 \, dy + \frac{N}{2} \int_R |\nabla a|^2 dy$$

$$v(y,\tau;a)\text{— solution of Cauchy problem } (A_0),$$

$$A=\left\{a(y)\in C(R), 0<a_0\le a(y)\le a_1,\ \int_R |\nabla a|^2 dy<\infty\right\}$$

Theorem 1 [6] Optimal control problem (A) admits at least a minimizer $\hat{a}(y)\in A$

Let $\widehat{a}(y)$ *be a minimizer and* $\hat{v}(y,\tau)=v(y,\tau;\hat{a})$, define

$$a_\lambda(y)=(1-\lambda)\hat{a}(y)+\lambda h(y)\in A, \forall 0\le\lambda\le 1, h(y)\in A$$

It is easy to check that

$$\xi(y,\tau)=\left.\frac{dv(y,\tau;a_\lambda}{d\lambda}\right|_{\lambda=0}$$

is a solution of Cauchy problem

$$\begin{cases} \dfrac{\partial\xi}{\partial\tau}-\hat{a}(y)(\dfrac{\partial^2\xi}{\partial y^2}-\dfrac{\partial\xi}{\partial y})-(r-q)\dfrac{\partial\xi}{\partial y}=(h(y)-\hat{a}(y))(\dfrac{\partial^2\hat{v}}{\partial y^2}-\dfrac{\partial\hat{v}}{\partial y}) \\ \xi|_{\tau=0}=0 \end{cases}$$

As $\hat{a}(y)\in A$ is a minimizer, the function $j(\lambda)=J(a_\lambda(y))$ $(\lambda\in[0,1])$, for any $h(y)\in A$ attains its minimum at $\lambda=0$, so $j'(0)\ge 0$, i.e.

$$\frac{d}{d\lambda}\left.\left(\int_R |v(y,\tau;a_\lambda)-v_0(y)|^2 dy+\frac{N}{2}\int_R |\nabla a_\lambda|^2 dy\right)\right|_{\lambda=0}\ge 0 \qquad (4.4)$$

By a straight calculation, we have

$$\int_R [(\hat{v}(y,\tau_0)-v_0(y))\,\xi(y,\tau_0)]\,dy+N\int_R \nabla\hat{a}\nabla(h-\hat{a})dy\ge 0, \forall h\in A$$

By duality technique, we prove

Theorem 2 (Necessary condition) [6,5]: There exists a triple $\{\widehat{a},\widehat{v},\widehat{\phi}\}$ which satisfies the following system:

$$(A_1)\begin{cases} \dfrac{\partial\hat{v}}{\partial\tau}-\hat{a}(y)\left(\dfrac{\partial^2\hat{v}}{\partial y^2}-\dfrac{\partial\hat{v}}{\partial y}\right)-(r-q)\dfrac{\partial\hat{v}}{\partial y}=0 & (y\in R, 0<\tau\le\tau_0) \\ \hat{v}|_{\tau=0}=0 \end{cases}$$

$$(A_2)\begin{cases}\dfrac{\partial\hat{\phi}}{\partial\tau}+\dfrac{\partial^2}{\partial y^2}(\hat{a}(y)\hat{\phi})+\dfrac{\partial}{\partial y}(\hat{a}(y)\hat{\phi})+(r-q)\dfrac{\partial\hat{\phi}}{\partial y}=0 & (y\in R, 0<\tau\leq\tau_0)\\ \hat{\phi}\Big|_{\tau=\tau_0}=\hat{v}(y,\tau_0)-v_0(y)\end{cases}$$

$$(A_3)N\int_R \nabla\hat{a}\nabla(h-\hat{a})dy+\int_0^{\tau_0}d\tau\int_R\hat{\phi}\left(\frac{\partial^2\hat{v}}{\partial y^2}-\frac{\partial\hat{v}}{\partial y}\right)(h-\hat{a})dy\geq 0, \forall h\in A$$

Remark: The optimal control problem is nonconvex. Therefore one cannot expect uniqueness of the minimizer in the global. However in some special situations the uniqueness can be expected to prove only in the local time, i.e. $|tau_0| = |T - t_0| \ll 1$ [6].

4.3. *Numerical results* [5]

Given $\tau_0 = T - t_0, V_0 = e^{q\tau_0}C(K)/S_0$, $\hat{a}(y)$ can be obtained from system $(A_1) - (A_3)$ by iteration procedure. If $a_n(y)$ is known, then

$$a_n(y) \to v_n(y) \to \phi_n y \to a_{n+1}(y), \ (n = 0, 1, \cdots, N-1)$$

and we define

$$\sigma(S) = 2\sqrt{a_N(\ln\frac{S}{S_0})}.$$

References

1. Amin.K.I.: Jump diffusion option valuation in discrete time, J. Finance Vol 48(1993) 1833-1863.
2. Black F.: The holes in Black-Scholes, <From Black-Scholes to black holes>, Risk publishing,51-56.
3. Yang C., Biao B. : Free boundary and American options in a jump-diffusion model , working paper,Tongji University (2003).
4. Dupire B.: Pricing with smile, Risk, Vol 7, 18-20.
5. Jiang L., Chen Q., Wang L. and Zhang J.: A new well-posed algorithm to recover implied local volatility, Quantitative Finance, (to appear).
6. Jiang L., Tao Y: Identifying the volatility of underlying assets from option prices, Inverse Problems, Vol 17(2001) 137-155.
7. Jiang L., Biab B. and Yi F.: A parabolic variational inequality arising from valuation of fixed rate mortgages, working paper, Tongji University, 2003.
8. Li S. and Ren X.: The pricing theory of convertible bonds with the conversion clause at maturity considering the risk of default, working paper, Tongji University,2003.

9. Qian X.,Jiang L. and Xu C.: Convergence of BTM for American options in a jump-diffusion model, SIAM Numer. Anal. (to appear).
10. Xu G.:Analysis of pricing European call foreign currency option under the Vasicek interest model, working paper,. Tongji University,2003.
11. Xu C., Jiang L. and Qian X.: Numerical analysis on BMT for a jump-diffusion model, J. Computational and Applied Math., Vol 56(2003) 23-45.
12. Yuan G., Jiang L. and Luo J.: Pricing the fixed-rate-mortgage contracts—Limiting on payment dates to prepay or default, System Engineering-Theory & Practice, Vol 23 No. 9 (2003) 48-55.
13. Zhang X.: Formules quasi-explicites pour les American options dans un modele de diffusion avec sautés, Math. Comp. Simu.,Vol 38(1998) 51-61..

EXACT BOUNDARY CONTROLLABILITY OF UNSTEADY FLOWS IN A NETWORK OF OPEN CANALS

TATSIEN LI (DAQIAN LI)

Fudan University,
Shanghai 200433, China
dqli@fudan.edu.cn

We will establish the exact boundary controllability of unsteady flows in a tree-like network of open canals with general topology (joint work with B. P. Rao).

1. Introduction

The mathematical model of unsteady flows in an open canal was established by de Saint-Venant [1] in 1871. As an one-dimensional model, Saint-Venant system is a first order quasilinear hyperbolic system and has been frequently used by hydraulic engineers in their practice.

In recent years, the control problem for Saint-Venant system was studied by many authors (see [2]–[7]). In particular, from a unified point of view, G. Leugering and E. G. Schmidt [2] obtained the corresponding model of Saint-Venant system for a network of open canals, in which the interface conditions at any given joint point of open canals are given.

In this talk we will establish the exact boundary controllability of unsteady flows in a tree-like network of open canals with general topology.

2. General consideration

First of all, we will give a general consideration on the exact boundary controllability for hyperbolic equations(systems),

For a given hyperbolic equation (system), for any given initial data φ and final data ψ, if we can find a time $T > 0$ and suitable boundary controls on the boundary $\partial\Omega$ of the domain Ω, such that the corresponding mixed initial-boundary value problem with the initial data φ admits a unique classical solution $u = u(t, x)$ on the whole domain $[0,\ T] \times \bar{\Omega}$, which verifies exactly the final condition

$$t = T: \quad u = \psi(x), \qquad x \in \Omega, \tag{1}$$

namely, if, by means of boundary controls, the system can drive any given initial state φ to any given final state ψ at $t = T$, then, we say that this system possesses the exact boundary controllability.

More precisely, if the exact boundary controllability can be realized only for initial and final states small enough in a certain sense, we say that the system possesses the local exact boundary controllability; Otherwise, we say the system possesses the global exact boundary controllability.

Since the hyperbolic wave has a finite speed of propagation, the exact boundary controllability of a hyperbolic equation (system) requires that the controllability time T must be suitably large so that two maximum determinate domains associated with the initial state and the final state respectively are separated. Then, in order to have a classical solution to the corresponding mixed initial boundary value problem on the domain $[0, T] \times \bar{\Omega}$, we should first prove the existence and uniqueness of the semi-global classical solution, namely, the classical solution on the time interval $0 \le t \le T$, where $T > 0$ is a preassigned and possibly quite large number. The exact boundary controllability will be based on the existence and uniqueness of semi-global classical solution to the mixed initial-boundary value problem for quasilinear hyperbolic equations (systems).

There are a number of publications concerning the exact controllability for linear hyperbolic equations (systems) (see J. L. Lions [8], D. L. Russell [9] etc.). For the semilinear case, using the HUM method suggested by J. L. Lions and Schauder's fixed point theorem, E. Zuazua [10] proved the global (resp. local) exact boundary controllability for semilinear wave equations in the asymptotically linear case (resp. the super-linear case with suitable growth conditions). Furthermore, using a global inversion theorem, I. Lasiecka and R. Triggiani [11] established an abstract result on the exact controllability for semilinear equations. As applications, they gave the global exact boundary controllability for wave and plate equations in the asymptotically linear case. However, only a few results are known for quasilinear hyperbolic systems. In an earlier work, M. Cirinà [12]–[13] considered the zero exact boundary controllability for quasilinear hyperbolic systems with linear boundary controls, but the author needed some very strong conditions on the coefficients of the system(globally bounded and globally Lipschitz continuous) and his results are essentially valid only for the system of diagonal form.

In order to get the exact boundary controllability in the quasilinear hyperbolic case, we present a constructive method which works for general quasilinear hyperbolic systems with general nonlinear boundary conditions at least in one-dimensional case. The main idea can be shown as follows.

In order to realize the exact boundary controllability, it is only necessary to find a time $T > 0$ such that the given hyperbolic equation (system) admits a classical solution $u = u(t,x)$ on the domain $[0,\ T] \times \bar{\Omega}$, which verifies simultaneously the initial condition

$$t = 0: \quad u = \varphi(x), \quad x \in \Omega \tag{2}$$

and the final condition (1). In fact, putting $u = u(t,x)$ into the boundary conditions, we get immediately the boundary controls. By uniqueness, the classical solution to the corresponding mixed initial-boundary value problem with the initial data φ must be $u = u(t,x)$, which automatically satisfies the given final data ψ.

Moreover, if the solution $u = u(t,x)$ constructed above also satisfies a part of boundary conditions, then we need only to put $u = u(t,x)$ into the other part of boundary conditions to get the corresponding boundary controls, and, as a result, the number of boundary controls will be reduced and the boundary controls can be asked to act only on a part of boundaries, however, the controllability time will be enlarged.

Of course, for the purpose of application, the controllability time T will be asked to be as small as possible.

3. Statement on the exact boundary controllability of unsteady flows in a network of open canals

We now consider unsteady flows in a network composed of N horizontal and cylindrical canals. Let L_a be the length of the a-th canal, a_1 and a_0 the x-coordinates of two ends of the a-th canal: $L_a = a_1 - a_0$ $(a = 1, \cdots, N)$.

Suppose that there is no friction, the corresponding Saint-Venant system (cf. [2], [14]) is

$$\begin{cases} \dfrac{\partial A^{(a)}}{\partial t} + \dfrac{\partial (A^{(a)} V^{(a)})}{\partial x} = 0, \\ \dfrac{\partial V^{(a)}}{\partial t} + \dfrac{\partial S^{(a)}}{\partial x} = 0, \end{cases} \qquad t \geq 0,\ a_0 \leq x \leq a_1 \quad (a = 1, \cdots, N), \tag{3}$$

where for the a-th canal, $A^{(a)} = A^{(a)}(t,x)$ stands for the area of the cross section at x occupied by the water at time t, $V^{(a)} = V^{(a)}(t,x)$ the average velocity over the cross section and

$$S^{(a)} = \frac{1}{2}(V^{(a)})^2 + gh_a(A^{(a)}) + gY_b^{(a)} \quad (a = 1, \cdots, N), \tag{4}$$

where g is the gravity constant, constant $Y_b^{(a)}$ denotes the altitude of the bed of the a-th canal and

$$h_a = h_a(A^{(a)}) \tag{5}$$

is the depth of the water in the a-th canal, $h_a(A^{(a)})$ being a suitably smooth function of $A^{(a)}$ such that

$$h'_a(A^{(a)}) > 0. \tag{6}$$

The initial condition is

$$t = 0: \quad (A^{(a)}, V^{(a)}) = (A_0^{(a)}(x), V_0^{(a)}(x)), \quad a_0 \le x \le a_1 \ (a = 1, \cdots, N). \tag{7}$$

When $x = a_0$ (resp. $x = a_1$) is a simple node,the boundary condition on $x = a_0$ (resp. $x = a_1$) is given by

$$x = a_0 \ (resp. \ x = a_1): \quad A^{(a)}V^{(a)} = q_0^{(a)}(t) \ (resp. \ q_1^{(a)}(t)). \tag{8}$$

When , for instance, $x = a_0$ is a multiple node and the open canals jointed at $x = a_0$ are indexed by $\bar{a}, \ \bar{b}, \ \bar{c}, \ \cdots$, the interface conditions at $x = a_0$ are given by the total flux-type interface condition

$$x = a_0: \quad \sum_{i=\bar{a},\bar{b},\bar{c},\cdots} \pm A^{(i)}V^{(i)} = Q_{a_0}(t) \tag{9}$$

and the energy-type interface conditions

$$\begin{aligned} x = a_0: \quad & \frac{1}{2}(V^{(i)})^2 + gh_i(A^{(i)}) + gY_b^{(i)} \\ & = \frac{1}{2}(V^{(\bar{a})})^2 + gh_a(A^{(\bar{a})}) + gY_b^{(\bar{a})} \quad (i = \bar{b}, \ \bar{c}, \ \cdots) \end{aligned} \tag{10}$$

(cf. [2], [14]).

Consider an equilibrium state $(A^{(a)}, V^{(a)}) = (A_0^{(a)}, V_0^{(a)})$ with $A_0^{(a)} > 0 \ (a = 1, \cdots, N)$ for system (3), which belongs to a subcritical case, i.e.,

$$|V_0^{(a)}| < \sqrt{gA_0^{(a)}h'_a(A_0^{(a)})} \quad (a = 1, \cdots, N), \tag{11}$$

and satisfies the energy-type interface conditions (10).

In a neighbourhood of any given subcritical equilibrium state $(A^{(a)}, V^{(a)}) = (A_0^{(a)}, V_0^{(a)})$ $(a = 1, \cdots, N)$, (3) is a hyperbolic system with real eigenvalues

$$\begin{cases} \lambda_1^{(a)} = V^{(a)} - \sqrt{gA^{(a)}h_a'(A^{(a)})} < 0, \\ \lambda_2^{(a)} = V^{(a)} + \sqrt{gA^{(a)}h_a'(A^{(a)})} > 0, \end{cases} \quad (a = 1, \cdots, N). \tag{12}$$

For $a = 1, \cdots, N$, introducing the Riemann invariants r_a and s_a as follows:

$$\begin{cases} 2r_a = V^{(a)} - V_0^{(a)} - G_a(A^{(a)}), \\ 2s_a = V^{(a)} - V_0^{(a)} + G_a(A^{(a)}), \end{cases} \tag{13}$$

where

$$G_a(A^{(a)}) = \int_{A_0^{(a)}}^{A^{(a)}} \sqrt{\frac{gh_a'(A^{(a)})}{A^{(a)}}} dA^{(a)}, \tag{14}$$

we have

$$V^{(a)} = r_a + s_a + V_0^{(a)}, \tag{15}$$

$$A^{(a)} = H_a(s_a - r_a) > 0, \tag{16}$$

where H_a is the inverse function of $G_a(A^{(a)})$,

$$H_a(0) = A_0^{(a)} \tag{17}$$

and

$$H_a'(0) = \sqrt{\frac{A_0^{(a)}}{gh_a'(A_0^{(a)})}} > 0. \tag{18}$$

Taking (r_a, s_a) $(a = 1, \cdots, N)$ as new unknown variables, the equilibrium state $(A^{(a)}, V^{(a)}) = (A_0^{(a)}, V_0^{(a)})(a = 1, \cdots, N)$ corresponds to $(r_a, s_a) = (0, 0)$ $(a = 1, \cdots, N)$ and system (3) can be reduced to the following system of diagonal form:

$$\begin{cases} \dfrac{\partial r_a}{\partial t} + \lambda_1^{(a)}(r_a, s_a)\dfrac{\partial r_a}{\partial x} = 0, \\ \dfrac{\partial s_a}{\partial t} + \lambda_2^{(a)}(r_a, s_a)\dfrac{\partial s_a}{\partial x} = 0, \end{cases} \quad (a = 1, \cdots, N), \tag{19}$$

where

$$\begin{cases} \lambda_1^{(a)}(r_a, s_a) &= r_a + s_a + V_0^{(a)} - \sqrt{gH_a(s_a - r_a)h_a'(H_a(s_a - r_a))} < 0, \\ \lambda_2^{(a)}(r_a, s_a) &= r_a + s_a + V_0^{(a)} + \sqrt{gH_a(s_a - r_a)h_a'(H_a(s_a - r_a))} > 0, \\ & \qquad (a = 1, \cdots, N). \end{cases} \tag{20}$$

The boundary condition (8) on $x = a_0$ is now written as

$$x = a_0: \quad (r_a + s_a + V_0^{(a)})H_a(s_a - r_a) = q_0^{(a)}(t). \tag{21}$$

It is easy to see that in a neighbourhood of $(r_a, s_a) = (0, 0)$, (21) can be equivalently rewritten as

$$x = a_0: \quad s_a = g_a(t, r_a) + h_a(t) \tag{22}$$

or

$$x = a_0: \quad r_a = \tilde{g}_a(t, s_a) + \tilde{h}_a(t), \tag{23}$$

in which

$$g_a(t, 0) \equiv \tilde{g}_a(t, 0) \equiv 0 \tag{24}$$

and

$$\begin{aligned} &C^1 \text{ norm of } h_a(\cdot) \ (resp.\ \tilde{h}_a(\cdot)) \text{ small enough} \\ \Longleftrightarrow\ &C^1 \text{ norm of } (q_{(a)}(\cdot) - A_0^{(a)}V_0^{(a)}) \text{ small enough.} \end{aligned} \tag{25}$$

The interface conditions (9)–(10) on $x = a_0$ are now written as

$$x = a_0: \quad \sum_{i=\bar{a},\bar{b},\bar{c},\cdots} \pm(r_i + s_i + V_0^{(i)})H_i(s_i - r_i) = Q_{a_0}(t) \tag{26}$$

and

$$\begin{aligned} x = a_0: \ &\frac{1}{2}\left(r_i + s_i + V_0^{(i)}\right)^2 + gh_i(H_i(s_i - r_i)) + gY_b^{(i)} \\ &= \left(\frac{1}{2}(r_{\bar{a}} + s_{\bar{a}} + V_0^{(\bar{a})})^2 + gh_{\bar{a}}(H_{\bar{a}}(s_{\bar{a}} - r_{\bar{a}})) + gY_b^{(\bar{a})}\right) \\ &\qquad (i = \bar{b},\ \bar{c},\ \cdots). \end{aligned} \tag{27}$$

For the purpose of the exact boundary controllability of unsteady flows in a network of open canals, we want to find a time $T > 0$ such that for any given initial data

$$t = 0: \ (A^{(a)},\ V^{(a)}) = (A_0^{(a)}(x),\ V_0^{(a)}(x)), \quad a_0 \leq x \leq a_1 \ (a = 1, \cdots, N) \tag{28}$$

and any given final data

$$t = T: \ (A^{(a)},\ V^{(a)}) = (A_T^{(a)}(x),\ V_T^{(a)}(x)), \quad a_0 \leq x \leq a_1 \ (a = 1, \cdots, N) \tag{29}$$

in a C^1 neighbourhood of the subcritical equilibrium state $(A_0^{(a)}, V_0^{(a)})$ $(a = 1, \cdots, N)$, there exist suitable boundary flux controls at some simple nodes and multiple nodes, such that the corresponding mixed initial-boundary value problem for Saint-Venant systems (3) with the initial data and the corresponding boundary conditions and interface conditions admits a unique semi-global piecewise C^1 solution on the time interval $0 \le t \le T$, which verifies exactly the final data.

To this end, it suffices to construct a piecewise C^1 solution to Saint-Venant system (3) on the time interval $0 \le t \le T$, which satisfies the initial condition on $t = 0$, the final condition on $t = T$, all the energy-type interface conditions and a part of flux boundary conditions and total flux-type interface conditions. If we can do so, putting this solution to the rest of flux boundary conditions and total flux-type interface conditions, we get the desired flux boundary controls and total flux interface controls which realize the exact boundary controllability.

In what follows we will see that there are still many possibilities to get this kind of semi-global piecewise C^1 solution. The main principle is to make the number of controls or the optimal controllability time as small as possible. However, for a smaller number of controls, we need a larger optimal controllability time, and vise versa.

4. Preliminaries for quasilinear hyperbolic systems of diagonal form

Since Saint-Venant system can be reduced to a quasilinear hyperbolic system of diagonal form, in this section we first give some results for quasilinear hyperbolic systems of diagonal form

$$\frac{\partial u_i}{\partial t} + \lambda_i(u)\frac{\partial u_i}{\partial x} = F_i(u) \qquad (i = 1, \cdots, n), \tag{30}$$

where $u = (u_1, \cdots, u_n)^T$ is the unknown vector function of (t, x), $\lambda_i(u)$ and $F_i(u)$ are C^1 functions of u and

$$F_i(0) = 0 \qquad (i = 1, \cdots, n). \tag{31}$$

Suppose that on the domain under consideration

$$\lambda_r(u) < 0 < \lambda_s(u) \qquad (r = 1, \cdots, m;\ s = m+1, \cdots, n). \tag{32}$$

Consider the mixed initial-boundary value problem for system (30) with the following initial condition

$$t = 0: \quad u = \varphi(x) \qquad 0 \le x \le L \tag{33}$$

and boundary conditions

$$x = 0: \quad u_s = G_s(t, u_1, \cdots, u_m) + H_s(t) \quad (s = m+1, \cdots, n), \tag{34}$$

$$x = L: \quad u_r = G_r(t, u_{m+1}, \cdots, u_n) + H_r(t) \quad (r = 1, \cdots, m), \tag{35}$$

where φ, G_i and $H_i (i = 1, \cdots, n)$ are C^1 functions with respect to their arguments, and, without loss of generality, we assume that

$$G_i(t, 0, \cdots, 0) \equiv 0 \qquad (i = 1, \cdots, n). \tag{36}$$

Moreover, the conditions of C^1 compatibility are assumed to be satisfied at points $(t, x) = (0, 0)$ and (0,L) respectively.

Lemma 4.1 ([15]): Under the assumptions mentioned above, for any preassigned and possibly quite large $T > 0$, the mixed initial-boundary value problem (30) and (33)–(35) admits a unique C^1 solution (called the **semi-global C^1 solution**) $u = u(t, x)$ with small C^1 norm on the domain

$$R(T) = \{(t, x) |\ 0 \le t \le T, \quad 0 \le x \le L\}, \tag{37}$$

provided that the C^1 norms $\|\varphi\|_{C^1[0,L]}$ and $\|H_i\|_{C^1[0,T]}$ $(i = 1, \cdots, n)$ are small enough (depending on T).

We have the following two theorems which lead to the local exact boundary controllability directly (cf. [16]–[18]).

Theorem 4.1: Let

$$T > L \max_{i=1,\cdots,n} \frac{1}{|\lambda_i(0)|}. \tag{38}$$

Under the assumptions mentioned at the beginning of this section, for any given initial state $\varphi(x)$ and final state $\psi(x)$ with small C^1 norms $\|\varphi\|_{C^1[0,L]}$ and $\|\psi\|_{C^1[0,L]}$, there exists a C^1 solution $u = u(t, x)$ with small C^1 norm on the domain

$$R(T) = \{(t, x) |\ 0 \le t \le T, \quad 0 \le x \le L\}, \tag{39}$$

which satisfies simultaneously the initial condition (33) and the final condition

$$t = T: \quad u = \psi(x), \quad 0 \le x \le L. \tag{40}$$

To construct this kind of solution is a non-standard problem. Roughly speaking, the procedure contains the following steps:

1) By means of the initial data we solve a forward mixed problem to get a semi-global C^1 solution on the time interval $0 \le t \le T_1$, where $\frac{T}{2} > T_1 \approx \frac{T}{2}$.

2) By means of the final data we solve a backward mixed problem to get a semi-global C^1 solution on the time interval $T - T_1 \le t \le T$.

3) By suitably supplying the value of u for $T_1 \le T \le T - T_1$ on $x = \frac{L}{2}$ and changing the order of t and x, we solve the rightward and leftward mixed problem respectively to get the desired solution.

From Theorem 4.1, there exist boundary controls $H_i \in C^1[0,T]$ $(i = 1, \cdots, n)$ with small C^1 norm, which realize the exact boundary controllability. Here, the number of boundary controls is equal to n, the number of unknown variables, and the boundary controls act on two ends.

This procedure is denoted by

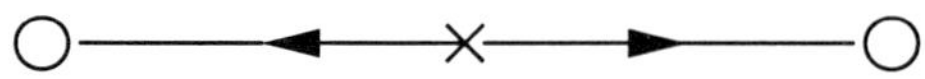

Figure 1.

Here, "×" stands for the starting point, the middle point $x = \frac{L}{2}$, on which the value of solution is given, "o" the arriving end on which the value of solution can be determined and then there are boundary controls, and "$\longrightarrow$" dedicates the direction pointing to the control end, along which we solved our rightward or leftward problem respectively.

Theorem 4.2: Let

$$T > 2L \max_{i=1,\cdots,n} \frac{1}{|\lambda_i(0)|}. \tag{41}$$

Under the assumptions mentioned at the beginning of this section, suppose furthermore that

$$n = 2m \tag{42}$$

and in a neighbourhood of $u = 0$ the boundary condition (34) can be equivalently rewritten as

$$x = 0: \quad u_r = \bar{G}_r(t, u_{m+1}, \cdots, u_n) + \bar{H}_r(t) \qquad (r = 1, \cdots, m), \tag{43}$$

where $\bar{G}_r$ and $\bar{H}_r (r = 1, \cdots, m)$ are C^1 functions with respect to their arguments,

$$\bar{G}_r(t, 0, \cdots, 0) \equiv 0 \qquad (r = 1, \cdots, m) \tag{44}$$

and, consequently,

$$\begin{aligned} &C^1 \textit{ norms of } H_s(\cdot) \ (s = m+1, \cdots, n) \textit{ small enough} \\ &\Leftrightarrow C^1 \textit{ norms of } \bar{H}_r(\cdot) \ (r = 1, \cdots, m) \textit{ small enough.} \end{aligned} \tag{45}$$

Then, for any given initial state φ, final state ψ and boundary functions $H_s(t)$ $(s = m+1,\cdots,n)$ at $x = 0$ with small C^1 norms $\|\varphi\|_{C^1[0,L]}$, $\|\psi\|_{C^1[0,L]}$ and $\|H_s\|_{C^1[0,T]}$ $(s = m+1,\cdots,n)$, such that the conditions of C^1 compatibility are satisfied at points $(t,x) = (0,0)$ and $(T,0)$ respectively, there exists a C^1 solution $u = u(t,x)$ with small C^1 norm on the domain $R(T)$, which satisfies the initial condition (33), the final condition (40) and the boundary condition (34) on $x = 0$.

The procedure of constructing this solution contains the following steps:

1) By means of the initial data and the boundary conditions (34) on $x = 0$, we solve a forward mixed problem to get a semi-global C^1 solution on the time interval $0 \leq t \leq T_1$, where$\frac{T}{2} > T_1 \approx \frac{T}{2}$.

2) By means of the final data and the boundary conditions (43) on $x = 0$, we solve a backward mixed problem to get a semi-global C^1 solution on the time interval $T - T_1 \leq t \leq T$.

3) Suitably supplying the value of u for $T_1 \leq T \leq T - T_1$ on $x = 0$, such that the boundary conditions (34) on $x = 0$ are satisfied for the whole interval $0 \leq t \leq T$, and changing the order of t and x,we solve the rightward mixed problem to get the desired solution.

From Theorem 4.2, there exist boundary controls $H_r(t)$ $(r = 1,\cdots,m)$ at $x = 1$ with small C^1 norms $\|H_r\|_{C^1[0,T]}$ $(r = 1,\cdots,m)$, which realize the exact boundary controllability. Here, the number of boundary controls is equal to $m = \frac{n}{2}$ and the boundary controls act only on one end, however, the controllability time is doubled.

This procedure is denoted by

Figure 2.

Here "$\bullet$" stands for the starting end on which there are no boundary controls, "o" the arriving end on which there are boundary controls, and "$\longrightarrow$"dedicates the direction of solving the rightward problem and points to the control end.

5. Corresponding consideration for Saint-Venant system

5.1. For a single canal with flux boundary conditions on both ends, Theorems 4.1 and 4.2 can be applied directly to get the results denoted by (cf. [14])

Figure 3.

and

Figure 4.

respectively.

5.2. For a string-like network of open canals , based on a generalization of Theorems 4.1 and 4.2 to the mixed problem with interface conditions (see [19]), we can get the corresponding results denoted by

Figure 5.

and

Figure 6.

or simply denoted by

Figure 7.

and

Figure 8.

Here, the piecewise C^1 solution is still constructed along the direction $\rightarrow$ from the node "$\bullet$" to the end "o". In the first case, the node "$\bullet$"is anyone of double nodes and both two simple nodes are the control ends "o". There are two flux controls acting on both simple nodes respectively; while, in the second case, the end "$\bullet$" is a simple node and another simple node is the control end "o". There is only one flux control acting on a simple node. Both cases are completely similar to the cases for a single canal, however, the controllability time should be much larger: it is a corresponding sum of controllability time for each individual canal (for the detail, see [19]) .

5.3. By rescaling and folding all corresponding x-intervals to an interval $0 \leq x \leq 1$, Theorem 4.2 can be almost directly applied to a star-like network of open canals. Then we can get the desired piecewise C^1 solution by solving the problem from the joint end (multiple node) to all simple nodes, as denoted by

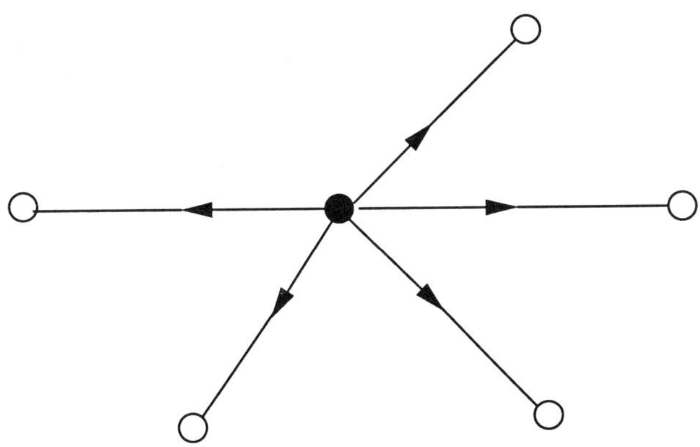

Figure 9.

Here, all simple nodes are the control ends "o", while there is no control on the joint end "$\bullet$". The number of boundary flux controls is equal to the number of canals.

5.4. We now give another way to construct the desired piecewise C^1 solution for a star-like network by a procedure of solving the problem from the simple end of one canal to the joint end, and then from the joint end to other simple ends respectively, denoted by

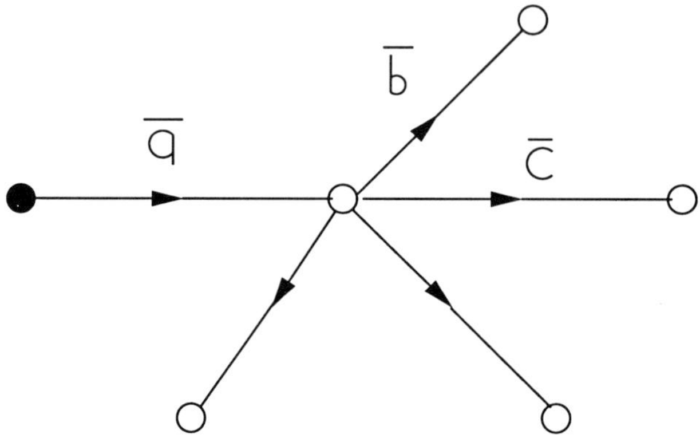

Figure 10.

To show this, we first apply Theorem 4.2 to get a C^1 solution $(A^{(\bar{a})}(t,x), V^{(\bar{a})}(t,x))$ or $(r_{\bar{a}}(t,x), s_{\bar{a}}(t,x))$ for the $\bar{a}$-th canal, which satisfies the initial data, the final data and the flux boundary condition (21) on the simple end of the $\bar{a}$-th canal. Thus, we have the value of $(r_{\bar{a}}, s_{\bar{a}})$ at the joint end as a vector function of t; then, the energy-like interface condition (27) for $i = \bar{b}$ gives a boundary condition on this joint end for the $\bar{b}$-th canal, which satisfies the assumption given in Theorem 4.2. Therefore, by Theorem 4.2 we can get a C^1 solution for the $\bar{b}$-th canal, which satisfies the initial data, the final data and, together with $(r_{\bar{a}}(t,x), s_{\bar{a}}(t,x))$, verifies the energy-like interface condition (27) for $i = \bar{b}$. For the other canals, we can do the same thing. As a result, we get a piecewise C^1 solution for this star-like network, which exactly satisfies the initial data, the final data, the flux boundary condition (21) on the simple end of the $\bar{a}$-th canal and the energy-like interface conditions (27). Putting this solution to the flux boundary condition on the simple end of other canals and to the total flux-like interface condition (26), we get the flux boundary controls and the total flux interface control.

As shown in the figure, the nodes where there are boundary or interface controls are denoted by "o", the node "•" as a starting point is a control-free one, "$\rightarrow$" dedicates the direction along which the problem is solved. In this situation, the number of controls is still equal to the number of canals, however, there is a total flux control acting on the joint end.

5.5. Combining the results for a string-like network and for a star-like network, we can get a piecewise C^1 solution for a star-like network of string-like sub-networks, which satisfies the initial data, the final data, the energy-like interface condition (27) on each multiple node, the total flux-like interface condition (26) on each multiple node of string-like sub-network as well as the total flux-like interface condition (26) on the multiple node of star-like network or one flux boundary condition on a simple node, denoted by

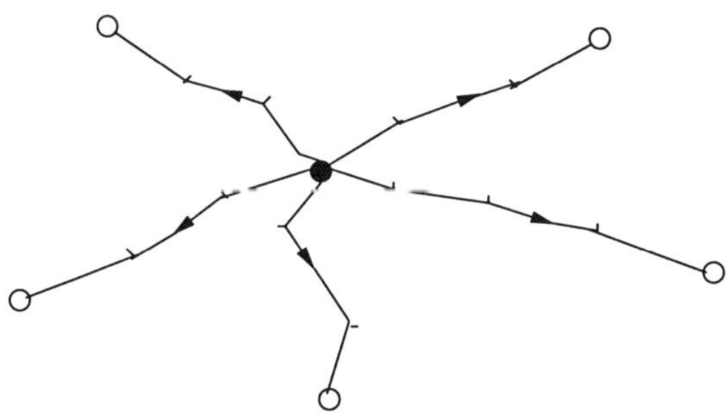

Figure 11.

or

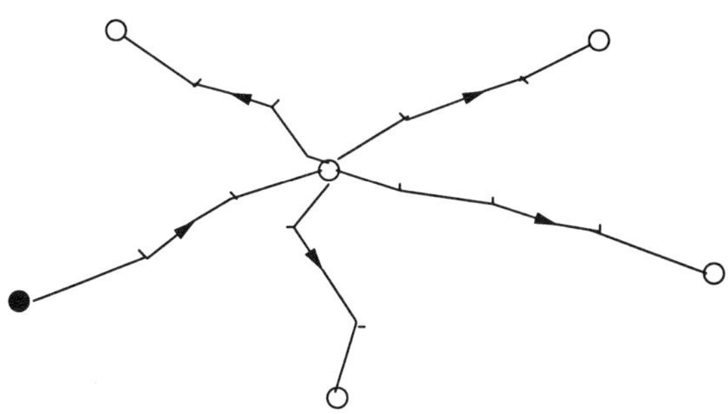

Figure 12.

Both cases are completely similar to the cases for a star-like network, however, the controllability time should be much larger.

5.6. If we wish reduce the controllability time, by the method used above for a star-like network we can construct the solution from one canal to the next canal through the joint end for a string-like canal. As a result, each joint end should be denoted by "o", on which there is a total flux control. Thus, the number of controls increased, however, the controllability time diminishes correspondingly. It can be denoted by

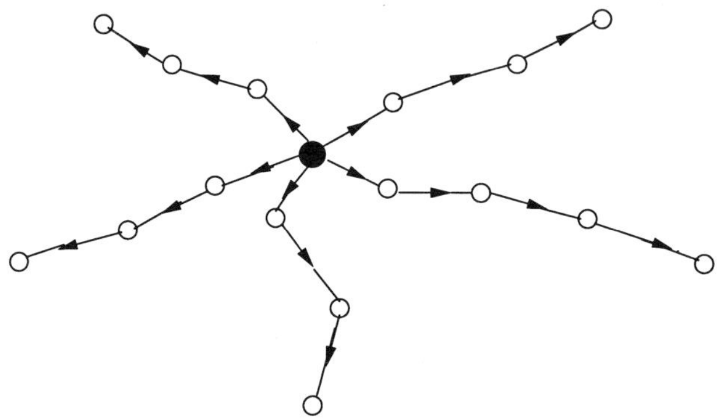

Figure 13.

or

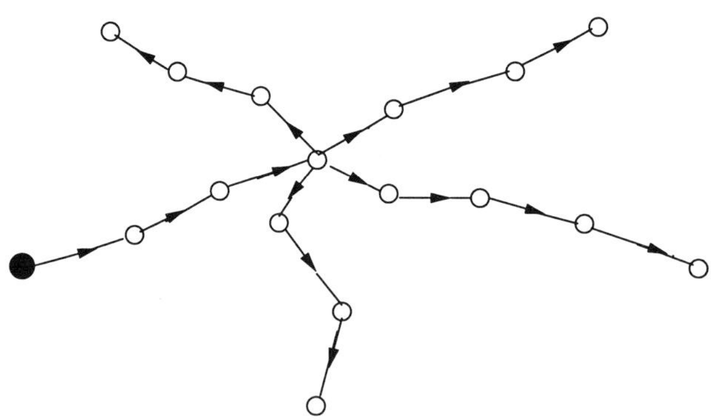

Figure 14.

6. Exact boundary controllability of unsteady flows in a tree-like network of open canals

A tree-like network of open canals means a connected network without loop, as shown in the following figure.

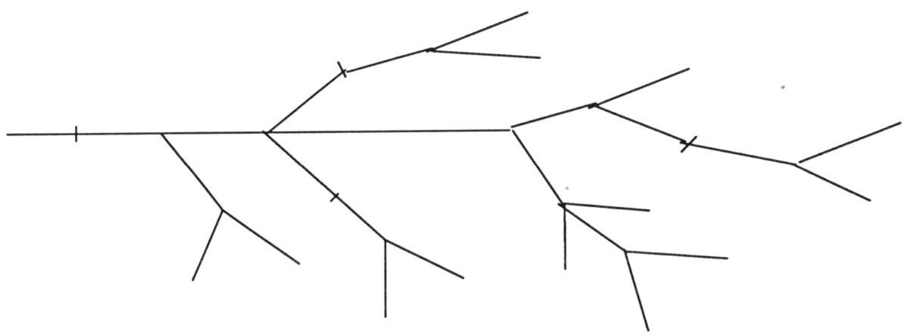

Figure 15.

Arbitrarily taking a node as the starting point, we construct a piecewise C^1 solution satisfying the initial data, the final data, the energy-like interface condition (27) on each multiple node and flux boundary conditions on some simple nodes or total flux-like interface conditions on some multiple nodes as follows:

When the starting point is a simple node, we use the result for a single canal (see Figure 4) or for a string-like network of canals (see Figure 8) to get a C^1 solution for the corresponding canal or a piecewise C^1 solution for the corresponding string-like sub-network, which satisfies the initial data, the final data, the flux boundary condition on this simple node and the interface conditions on the multiple nodes of this string-like sub-network.

Then, we successively use the result for a star-like network (see Figure 10 or Figure 12) to construct a piecewise C^1 solution for the rest of canals. This procedure gives the desired exact boundary controllability.

For example,

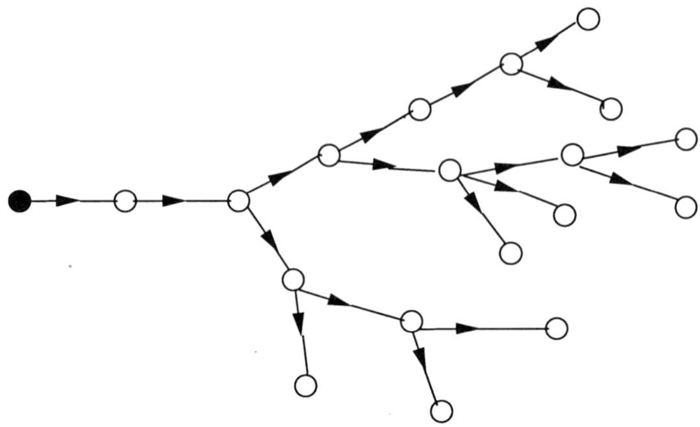

Figure 16. The number of controls = 18.

or

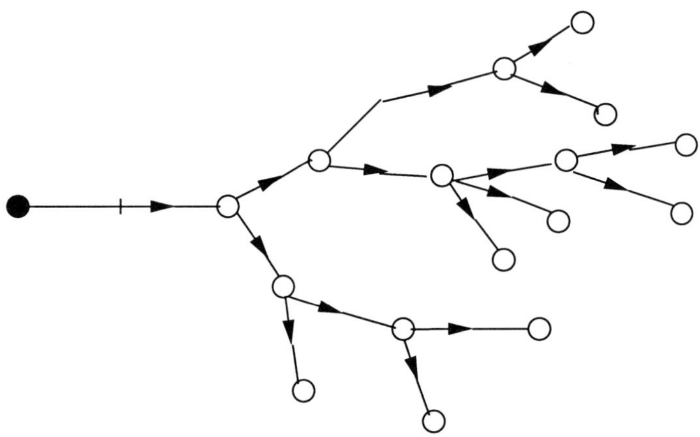

Figure 17. The number of controls = 16.

When the starting point is a multiple node, we use the result for a star-like network (see Figures 9, 11 or 13) to get a piecewise C^1 solution for the corresponding star-like sub-network with this starting point as the joint end, which satisfies the initial data, the final data, the interface conditions on this multiple node. Then, we can repeat the previous procedure as in the first situation.

For example,

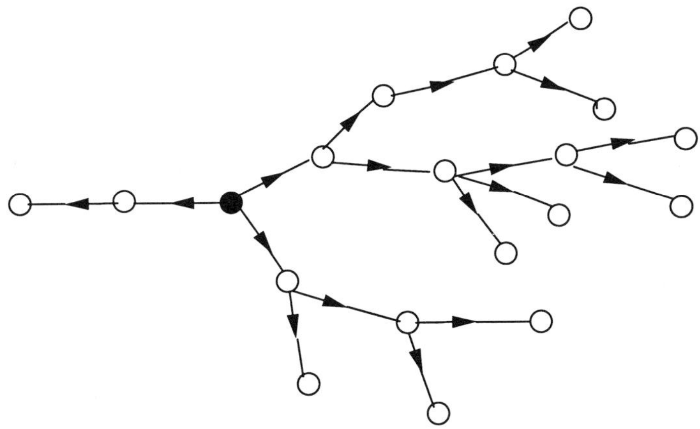

Figure 18. The number of controls = 18.

or

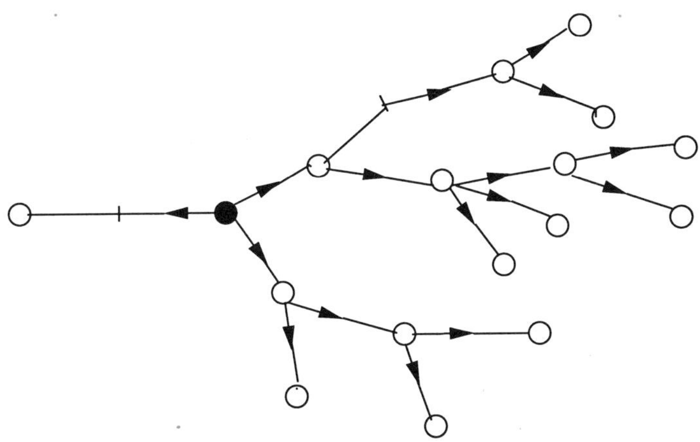

Figure 19. The number of controls = 16.

7. Some remarks

1). The controllability time obtained for each case is optimal.

2). The flux boundary condition (8) can be replaced by the water level boundary condition

$$x = a_0 : \quad A^{(a)} = q_0^{(a)}(t). \tag{46}$$

Moreover, the energy-like interface conditions (10) can be also replaced by the water lever interface conditions

$$h_i(A^{(i)}) + Y_b^{(i)} = h_{\bar{a}}(A^{(\bar{a})}) + Y_b^{(\bar{a})} \quad (i = \bar{b}, \bar{c}, \cdots). \tag{47}$$

3). If one considers the friction, in a network of open canals, Saint-Venant system (3) becomes

$$\begin{cases} \dfrac{\partial A^{(a)}}{\partial t} + \dfrac{\partial (A^{(a)}V^{(a)})}{\partial x} = 0, & t \geq 0, \quad a_0 \leq x \leq a_1 \\ \dfrac{\partial V^{(a)}}{\partial t} + \dfrac{\partial S^{(a)}}{\partial x} + F_a(A^{(a)}, V^{(a)}) = 0, & (a = 1, \cdots, N), \end{cases} \tag{48}$$

where, for $a = 1, \cdots, N$, $S^{(a)}$ is still given by (4),

$$F_i(A_i, 0) \equiv 0 \tag{49}$$

and

$$V_i F_i(A_i, V_i) \geq 0. \tag{50}$$

In this case, for any given equilibrium state $(A^{(a)}, V^{(a)}) = (A_0^{(a)}, 0)$ with $A_{i0} > 0$ $(a = 1, \cdots, N)$, we still have the exact boundary controllability.

4). Since both the initial and final states belong to a neighbourhood of an arbitrarily given subcritical equilibrium state, our results give only the local exact boundary controllability for unsteady flows in canals, however, that is enough to get the corresponding global exact boundary controllability. In fact, in order to drive any given initial state in a neighbourhood of a subcritical equilibrium state $(A_0^{(a)}, V_0^{(a)})$ $(a = 1, \cdots, N)$ to any given final state in a neighbourhood of another subcritical equilibrium state $(A_1^{(a)}, V_1^{(a)})$ $(a = 1, \cdots, N)$, it suffices to use the previous results to first drive the initial state to the equilibrium $(A_0^{(a)}, V_0^{(a)})$ $(a = 1, \cdots, N)$, then successively transfer equilibrium states to $(A_1^{(a)}, V_1^{(a)})$ $(a = 1, \cdots, N)$, and finally reach the final state.

Thus, we get the corresponding global exact boundary controllability, however, the controllability time must be enlarged many times.

References

1. B. de Saint-Venant, Théorie du mouvement non permanent des eaux, avec application aux crues des rivières et l'introduction des marées dans leur lit, *C. R. Acad. Sci.*, **73**, 147 237 (1871).
2. G. Leugering and E. G. Schmidt, On the modelling and stabilization of flows in networks of open canals, *SIAM J. Control Optim.* **41**, 164 (2002).

3. M. Gugat and G. Leugering, Global boundary controllability of the de St.Venant equations between steady states, *Ann. I. H. Poincaré, Analyse Non Linéaire* **20**, 1 (2003).
4. J.-M. Coron, B. d'Andréa-Novel and G. Bastin, A Lyapunov approach to control irrigation canals modeled by Saint-Venant equations, in CD-Rom Proceedings, Paper F1008-5, ECC99, Karlsruhe, Germany, (1999).
5. J. de Halleux, B. d'Andréa-Novel, J.-M. Coron and G. Bastin, A Lyapunov approach for the control of multi reach channels modelled by Saint-Venant equations, in CD-Rom Proceedings, NOLCOS'01, St-Petersburg, Russia, June 2001, 1515 (2001).
6. J. de Halleux and G. Bastin, Stabilization of Saint-Venant equations using Riemann invariants: Application to waterways with mobile spillways, in CD-Rom Proceedings, Barcelona, Spain, July (2002).
7. J. de Halleux, C. Prieur, J.-M. Coron, B. d'Andréa-Novel and G. Bastin Boundary feedback control in networks of open channels, Preprint, (2002).
8. J.-L. Lions, Contrôlabilité Exacte, Perturbations et Stabilisation de Systèmes Distribués, Vol. **I**, Masson, (1988).
9. D. L. Russell, Controllability and stabilizability theory for linear partial differential equations, Recent progress and open questions, *SIAM Rev.* **20**, 639 (1978).
10. E. Zuazua, Exact controllability for the semilinear wave equation, *J. Math. Pures et Appl.* **69**, 1 (1990).
11. I. Lasiecka and R. Triggiani, Exact controllability of semilinear abstract systems with applications to waves and plates boundary control problems, *Appl. Math. Optim.* **23**, 109 (1991).
12. M. Cirinà, Boundary controllability of nonlinear hyperbolic systems, *SIAM J. Control Opt.* **7**, 198 (1969).
13. M. Cirinà, Nonlinear hyperbolic problems with solutions on preassigned sets, *Michigan Math. J.* **17**, 193 (1970).
14. T. T. LI, Exact boundary controllability of unsteady flows in a network of open canals, to appear in Mathematische Nachrichten.
15. T. T. LI and Y. JIN, Semi-glabal C1 solution to the mixed initial-boundary value problem for quasilinear hyperbolic systems, *Chin. Ann. Math.* **22B**, 325 (2001).
16. T. T. LI, B. P. Rao and Y. JIN, Solution C^1 semi-globale et contrôlabilité exacte frontière de systèmes hyperboliques quasi linéaires, *C. R. Acad. Sci. Paris, t. 333* Série **I**, 219 (2001).
17. T. T. LI and B. P. Rao, Exact boundary controllability for quasilinear hyperbolic systems, *SIAM J. Control Optim.* **41**, 1748 (2003).
18. T. T. LI and B. P. Rao, Local exact boundary controllability for a class of quasilinear hyperbolic systems, *Chin. Ann. Math.* **23B**, 209 (2002).
19. T. T. LI, Exact controllability for quasilinear hyperbolic systems and its application to unsteady flows in a network of open canals, to appear in Mathematical Methods in the Applied Sciences.

HIERARCHY OF PARTIAL DIFFERENTIAL EQUATIONS AND FUNDAMENTAL SOLUTIONS ASSOCIATED WITH SUMMABLE FORMAL SOLUTIONS OF A PARTIAL DIFFERENTIAL EQUATION OF NON KOWALEVSKI TYPE

MASATAKE MIYAKE AND KUNIO ICHINOBE
GRADUATE SCHOOL OF MATHEMATICS, NAGOYA UNIVERSITY

1. Introduction

Let p and q be natural numbers such that $q > p$, and consider the following Cauchy problem in two dimensional complex planes:

$$\text{(CP)} \qquad \begin{cases} \partial_t^p u(t,x) = \partial_x^q u(t,x), \quad t, x \in \mathbf{C}, \\ u(0,x) = \varphi(x) \in \mathcal{O}_x, \ \partial_t^j u(0,x) = 0 \ (1 \le j < p), \end{cases}$$

where $\mathcal{O}_x (= \mathbf{C}\{x\})$ denotes the set of holomorphic functions in a neighborhood of the origin $x = 0$ or the set of convergent series in x. The Cauchy problem (CP) has a unique formal solution $\hat{u}(t,x)$ given by

$$\hat{u}(t,x) = \sum_{n=0}^{\infty} \varphi^{(qn)}(x) \frac{t^{pn}}{(pn)!} \in \mathcal{O}_x[[t]]_{\sigma(0)}, \quad \sigma(0) := \frac{q-p}{p}, \tag{1}$$

which is divergent in general by the assumption that $q > p$. Here the symbol $\mathcal{O}_x[[t]]_\sigma$ $(\sigma > 0)$ denotes the set of formal power series $\hat{f}(t,x) = \sum_{n=0}^{\infty} f_n(x) t^n$ $(f_n(x) \in \mathcal{O}_x)$ such that

$$\left(\hat{\mathcal{B}}_\sigma \hat{f}\right)(s,x) := \sum_{n=0}^{\infty} f_n(x) \frac{s^n}{\Gamma(1+\sigma n)} \in \mathcal{O}_{s,x}, \tag{2}$$

where $\Gamma(\cdot)$ denotes the Gamma function, and $\hat{\mathcal{B}}_\sigma$ is called the formal Borel transformation of order σ (in the variable t). A formal power series $\hat{f}(t,x) \in \mathcal{O}_x[[t]]_\sigma$ is called a formal power series of Gevrey order σ.

The following questions are natural ones for divergent formal solution $\hat{u}(t,x)$ of Gevrey type:

[Q1] When is $\frac{q-p}{p}$-Gevrey summable for the formal solution $\hat{u}(t,x)$? The definition of Gevrey summability will be given in the following section.

[Q2] When [Q1] is established, we want to know the integral representation of the $\frac{q-p}{p}$-Gevrey sum by using a kernel function which we call the fundamental solution.

In the case of $(p,q) = (1,2)$, i.e., the case of the complex heat equation, the answers to these questions are obtained completely in a paper by D. Lutz, M. Miyake and R. Schäfke[9] in 1999. For our Cauchy problem (CP) in this paper, [Q1] was solved by M. Miyake[10], and [Q2] was solved by K. Ichinobe[6]. Moreover, K. Ichinobe[7] extended those results for equations with quasi homogeneity. We have to mention that W. Balser obtained sufficient conditions to the question [Q1] for general equations in two dimensional complex planes with constant coefficients[1,2,3,4,5].

In this sense, there are no problems to the above questions, but we shall consider a further question as follows:

[Q3] When the Cauchy data $\varphi(x)$ is an entire function of finite exponential order, the Gevrey order of the formal solution $\hat{u}(t,x)$ becomes smaller than that in (1). Now a question arises that how about the answers to the above questions in this case.

In order to explain our problem, let us define a class $\mathrm{Exp}(r;\mathbf{C})$ $(r>0)$ of entire functions by

$$\mathrm{Exp}(r;\mathbf{C}) = \{f(x) \in \mathcal{O}(\mathbf{C})\,;\, |f(x)| \le C\exp(\delta|x|^r) \text{ for } {}^{\exists}C, {}^{\exists}\delta > 0\}, \quad (3)$$

where $\mathcal{O}(\mathbf{C})$ denotes the set of entire functions. By an easy calculation, we see that

$$u(x) \in \mathrm{Exp}(r;\mathbf{C}) \Longleftrightarrow \left|u^{(n)}(0)\right| \le CK^n (n!)^{1-1/r}, \quad (4)$$

for all $n \in \mathbf{N} := \{0,1,2,\cdots\}$ by some positive constants C and K. From this, we easily see that if we choose $\varphi(x) \in \mathrm{Exp}(r;\mathbf{C})$ for the Cauchy data, then we have

$$\hat{u}(t,x) \in \mathcal{O}_{t,x} \quad \Longleftrightarrow \quad 0 < r \le \frac{q}{q-p}. \quad (5)$$

Since we are interested in the divergent formal solution, we restrict our study to the case

$$r > \frac{q}{q-p}.$$

We further assume that the Cauchy data $\varphi(x)$ belongs to a class of entire functions scaled by exponentially growth as follows.

$$\varphi(x) \in \mathrm{Exp}\left(\frac{q}{\ell};\mathbf{C}\right), \quad 0 \le \ell \le q-p-1, \quad (\ell \in \mathbf{N}). \quad (6)$$

When $\ell = 0$, we understand that $\mathrm{Exp}(q/0; \mathbf{C}) = \mathcal{O}_x$ which does not make any contradiction in the results.

By the above choice of the Cauchy data $\varphi(x)$ we have for the Gevrey order of the formal solution

$$\hat{u}(t,x) \in \mathcal{O}_x[[t]]_{\sigma(\ell)}, \quad \sigma(\ell) = \frac{q-p-\ell}{p} < \sigma(0). \tag{7}$$

Remark 1. Our Cauchy problem (CP) is only for a simple equation, but the generalization is possible for an equation in two variables which has a quasi homogeneity in derivatives ∂_t and ∂_x. In fact, it is enough to follow the arguments[7] consulting the arguments developed in this paper.

2. $\sigma(0)$-Gevrey summability (Review)

We, first, give a short review of the definition of the Gevrey summability and the Gevrey sum.

Let $\theta \in \mathbf{R}$, $\alpha > 0$ and $0 < \rho \leq +\infty$. Then a sector $S(\theta, \alpha, \rho)$ in the complex t plane is defined by

$$S(\theta,\alpha,\rho) = \left\{t \in \mathbf{C}\,;\, 0 < |t| < \rho,\ |\arg(t) - \theta| < \frac{\alpha}{2}\right\}.$$

We refer θ, α and ρ as the direction, opening angle and the radius of the sector, respectively.

Let $\hat{f}(t,x) = \sum_{n\geq 0} f_n(x)t^n \in \mathcal{O}_x[[t]]_\sigma$ with $\sigma > 0$. Then, by the definition, we say that $\hat{f}(t,x)$ is σ-Gevrey summable in θ direction (in t-plane) if there exists an analytic function $f(t,x)$ on $S(\theta, \sigma\pi + \varepsilon, \rho) \times \{|x| \leq r\}$ for some positive constants ε, ρ and r such that for any closed sector S' with a vertex at the origin such that $(S'\backslash\{0\}) \subset S(\theta, \sigma\pi+\varepsilon, \rho)$ there exist positive constants r', C and K such that the following asymptotic estimates hold

$$\max_{|x|\leq r'} \left| f(t,x) - \sum_{n=0}^{N-1} f_n(x)t^n \right| \leq CK^N(N!)^\sigma |t|^N, \quad \forall t \in S'\backslash\{0\}, \tag{8}$$

for all $N \geq 1$. This kind of asymptotic expansion is called σ-Gevrey asymptotic expansion, and it is written by

$$f(t,x) \sim_\sigma \hat{f}(t,x) \quad \text{in } S(\theta, \sigma\pi + \varepsilon). \tag{9}$$

In this expression, we omit to write the radius ρ of the sector $S(\theta, \sigma\pi+\varepsilon, \rho)$, since the radius is not an essential problem for the Gevrey summability. We note that thanks of the condition that the opening angle of the sector is larger than $\sigma\pi$ for the definite domain of $f(t,x)$, one can show that such an analytic function $f(t,x)$ is unique if such analytic functions exist, and it is called the σ-Gevrey sum of $\hat{f}(t,x)$ in θ direction. For the detail of Gevrey summability, one can consult the book by W. Balser[2].

We denote by $\mathcal{O}_x\{t\}_{\sigma,\theta}$ the set of formal power series $\hat{f}(t,x) \in \mathcal{O}_x[[t]]_\sigma$ which is σ-Gevrey summable in θ direction.

We remark that if $\hat{u}(t,x) \in \mathcal{O}_x\{t\}_{\sigma(0),\theta}$ for the formal solution, then its $\sigma(0)$-Gevrey sum $u(t,x)$ is an actual analytic solution of the equation $\partial_t^p u(t,x) = \partial_x^q u(t,x)$ satisfying the Cauchy data $\varphi(x)$ in the asymptotic meaning as $t \to 0$ along the sector $S(\theta, \sigma(0)\pi + \varepsilon, \rho)$.

Now the characterization of $\sigma(0)$-Gevrey summability of the formal solution $\hat{u}(t,x)$ is obtained as follows;

THEOREM 1[10] *Let $\varphi(x) \in \mathcal{O}_x$ for the Cauchy data. Then, $\hat{u}(t,x) \in \mathcal{O}_x\{t\}_{\sigma(0),\theta}$ if and only if the following conditions are satisfied;*

$$\text{(i)} \quad \varphi(x) \in \mathcal{O}\left(\bigcup_{m=0}^{q-1} S\left(\frac{p\theta + 2\pi m}{q}, \varepsilon, \infty\right)\right) \quad \textit{for} \quad \exists \varepsilon > 0.$$

$$\text{(ii)} \quad |\varphi(x)| \le C \exp(\delta |x|^{q/(q-p)}) \quad \textit{for } x \in \bigcup_{m=0}^{q-1} S\left(\frac{p\theta + 2\pi m}{q}, \varepsilon, \infty\right),$$

by some positive constants C and δ. Here, $\mathcal{O}(\Omega)$ denotes the set of holomorphic functions on the domain Ω.

The conditions (i) and (ii) can be expressed in a simplified form as follows. Let us define

$$\Phi(x,\zeta) := \sum_{m=0}^{q-1} \varphi(x + \omega_q^m \zeta) \quad (\omega_q = e^{2\pi i/q}).$$

Then the conditions (i) and (ii) are equivalent to that

$$\Phi(x,\zeta) \in \mathrm{Exp}_\zeta\left(\frac{q}{q-p}; S\left(\frac{p\theta}{q}, \varepsilon, \infty\right)\right) \tag{10}$$

uniformly in x in a neighborhood of the origin $x = 0$, by some positive constant ε. This means that $\Phi(x,\zeta)$ is defined on $\{|x| \le r\} \times S(p\theta/q, \varepsilon, \infty)$ for some $r > 0$ and satisfies the following growth condition

$$\max_{|x|\le r} |\Phi(x,\zeta)| \le C \exp(\delta |\zeta|^{q/(q-p)}) \quad \text{for } \zeta \in S\left(\frac{p\theta}{q}, \varepsilon, \infty\right),$$

by some positive constants C and δ.

About the integral representation of the $\sigma(0)$-Gevrey sum, we have

THEOREM 2[6,7] *Under the conditions in* Theorem 1, *the $\sigma(0)$-Gevrey sum $u(t,x)$ of $\hat{u}(t,x)$ in θ direction is given by the following integral formula*

$$u(t,x) = \int_0^{\infty(p\theta/q)} \Phi(x,\zeta) E_0(t,\zeta)\, d\zeta, \tag{11}$$

where the integration is taken from the origin to the infinity along the half line of argument $p\theta/q$, i.e., $e^{(p\theta/q)i}\mathbf{R}_{\geq 0}$, and the kernel function $E_0(t,\zeta)$ is given by

$$E_0(t,\zeta) = \frac{\boldsymbol{\Gamma}(\mathbf{p}/p)}{\boldsymbol{\Gamma}(\mathbf{q}/q)} \frac{1}{\zeta} G^{q,0}_{p,q}\left(c_{pq} \frac{\zeta^q}{t^p} \middle| \begin{matrix} \mathbf{p}/p \\ \mathbf{q}/q \end{matrix} \right) \quad \text{(Fundamental solution).} \tag{12}$$

Here, we use the following notations and definitions.

$$\mathbf{q} := (1,2,\cdots,q),\ \frac{\mathbf{q}}{q} := \left(\frac{1}{q},\frac{2}{q},\cdots,\frac{q-1}{q},\frac{q}{q}\right),\ \boldsymbol{\Gamma}\left(\frac{\mathbf{q}}{q}\right) := \prod_{j=1}^{q} \Gamma\left(\frac{j}{q}\right),$$

$c_{pq} = \frac{p^p}{q^q}$, *and* $G^{q,0}_{p,q}\left(Z \middle| \begin{matrix} \mathbf{p}/p \\ \mathbf{q}/q \end{matrix} \right)$ $(Z \in \mathbf{C}\backslash\{0\})$ *denotes the Meijer G-function defined by*

$$G^{q,0}_{p,q}\left(Z \middle| \begin{matrix} \mathbf{p}/p \\ \mathbf{q}/q \end{matrix} \right) = \frac{1}{2\pi i} \int_I \frac{\boldsymbol{\Gamma}(\mathbf{q}/q+\tau)}{\boldsymbol{\Gamma}(\mathbf{p}/q+\tau)} Z^{-\tau}\, d\tau, \tag{13}$$

by an integral path $I = \{\tau = k+iy; y \in \mathbf{R}\}$ with a fixed $-1/q < k < 0$.

3. The case of entire function Cauchy data

PROPOSITION 1. *Let $\varphi(x) \in \mathrm{Exp}(q/\ell;C)$ $(1 \leq \ell \leq q-p-1)$ for the Cauchy data. Then $\hat{u}(t,x) \in \mathcal{O}_x\{t\}_{\sigma(\ell),\theta}$ $(\sigma(\ell) = (q-p-\ell)/p)$ if $\Phi(x,\zeta)$ satisfies the condition* (10). *Moreover, the $\sigma(\ell)$-Gevrey sum $u(t,x)$ is given by the formula* (11).

Remark 2. In the above proposition, the condition is only a sufficient condition which is just the same with the $\sigma(0)$-Gevrey summability and the integral expression of the $\sigma(\ell)$-Gevrey sum is also nothing but the $\sigma(0)$-Gevrey sum. As a natural question, the condition will be expected to be a necessary condition for the $\sigma(\ell)$-Gevrey summability. But we do not know whether it is true or not. On the other hand, one can find a necessary and sufficient condition for the $\sigma(\ell)$-Gevrey summability in different form by following Balser's argument[1,2]: Let

$$\hat{u}(t,x) = \sum_{n=0}^{\infty} \varphi^{(qn)}(x) \frac{t^{pn}}{(pn)!} \in \mathcal{O}_x[[t]]_\sigma \quad (\sigma > 0) \tag{14}$$

be a formal solution of the Cauchy problem (CP). (The restriction for σ such as $\sigma(\ell)$ is not necessary.) Let us define

$$\hat{\psi}_j(x)(=(\partial_x^j \hat{u})(x,0)) = \sum_{n=0}^{\infty} \varphi^{(qn+j)}(0)\frac{x^{pn}}{(pn)!} \quad (0 \le j \le q-1). \tag{15}$$

Then one can prove that

$$\hat{u}(t,x) \in \mathcal{O}_x\{t\}_{\sigma,\theta} \iff \hat{\psi}_j(x) \in \mathbf{C}\{x\}_{\sigma,\theta} \quad (0 \le j \le q-1). \tag{16}$$

In this case, the necessity of the condition is easily proved. For the proof of the sufficiency we take the σ-Gevrey sum $\psi_j(x)$ of $\hat{\psi}_j(x)$. Then one can prove that

$$u(t,x) := \sum_{j=0}^{q-1}\sum_{n=0}^{\infty} \partial_t^{pn}\psi_j(t)\frac{x^{qn+j}}{(qn+j)!} \tag{17}$$

is the σ-Gevrey sum of $\hat{u}(t,x)$. Here we have to mention that it is not an easy task to realize the conditions in (16) as those for the Cauchy data $\varphi(x)$ itself as in Proposition 1.

The proof of Proposition 1 is done by making step by step formal Borel transformations which sends the divergent series to a convergent series, and after checking the analytic continuation property and exponential growth condition along a sector of infinite radius for the obtained convergent series, we make again step by step Laplace transformations to obtain the Gevrey sum. The procedure is already done in the previous papers[10,6,7], therefore the proof does not include or need anything new, but at each step of making the formal Borel transformation we get a partial differential equation of non-Kowalevski type which is Fuchsian in Baouendi-Goulaouic's sense for which the kernel function or the fundamental solution for the Gevrey sum is obtained by a special function which is given by Meijer's G-functions[8]. Such a procedure will be given Section 5.

4. Proof of Proposition 1

The proof of Proposition 1 follows immediately from those for Theorems 1 and 2, and essentially we will repeat the proofs of them which are given in the papers[10,6,7].

The fundamental criterion of the Gevrey summability is given by

LEMMA 1[10] *Let $\hat{f}(t,x) \in \mathcal{O}_x[[t]]_\sigma$. Then the following two statements are equivalent:*

(i) $\hat{f}(t,x) \in \mathcal{O}_x\{t\}_{\sigma,\theta}$.

(ii) *Let* $\{\sigma_j\}_{j=1}^k$ $(\sigma_j > 0,\ k \geq 1)$ *satisfy* $\sigma = \sigma_1 + \cdots + \sigma_k$, *and define* $g(s,x) \in \mathcal{O}_{s,x}$ *by iterated formal Borel transformations*

$$g(s,x) := (\hat{\mathcal{B}}_{\sigma_k} \circ \cdots \circ \hat{\mathcal{B}}_{\sigma_1} \hat{f})(s,x). \tag{18}$$

Then $g(s,x) \in \mathrm{Exp}_s(1/\sigma; S(\theta,\varepsilon,\infty))$ *uniformly in* x *in a neighborhood of* $x=0$, *by a positive constant* ε.

Moreover, under the condition (ii), *the* σ*-Gevrey sum* $f(t,x)$ *of* $\hat{f}(t,x)$ *is given by the following iterated Laplace transformations*

$$f(t,x) = (\mathcal{L}_{1/\sigma_1,\theta} \circ \cdots \circ \mathcal{L}_{1/\sigma_k,\theta}\, g)(t,x), \tag{19}$$

where

$$(\mathcal{L}_{1/\sigma,\theta}\, g)(t,x) := \frac{1}{t^{1/\sigma}} \int_0^{\infty(\theta)} e^{(-s/t)^{1/\sigma}} g(s,x)\, d(s^{1/\sigma}). \tag{20}$$

Exactly speaking, the analytic continuation of $f(t,x)$ in t variable into a sector $S(\theta, \sigma\pi + \varepsilon, \rho)$ for some positive constant ε is done by rotating the argument of the path of integration $e^{i\theta}\,\mathbf{R}_{\geq 0}$ ($\mathbf{R}_{\geq 0} := [0,+\infty)$), which is permitted by the condition that $g(s,x) \in \mathrm{Exp}_s(1/\sigma; S(\theta,\varepsilon,\infty))$ uniformly in x in a neighborhood of the origin.

For the proof of Proposition 1, we define

$$v(s,x) = (\{\hat{B}_{1/p}\}^{q-p-\ell}\hat{u})(s,x) \in \mathcal{O}_{s,x}. \tag{21}$$

Then for the proof of $\sigma(\ell)$-Gevrey summability in θ direction, it is enough to prove the following exponential growth property:

$$v(s,x) \in \mathrm{Exp}_s\left(\frac{p}{q-p-\ell};, S(\theta,\varepsilon,\infty)\right) \tag{22}$$

uniformly in x in a neighborhood of $x=0$. For that purpose we further take ℓ times iterated formal Borel transformations and put

$$\begin{aligned} w(\eta,x) &= (\{\hat{B}_{1/p}\}^{\ell} v)(\eta,x) \\ &= (\{\hat{B}_{1/p}\}^{q-p}\hat{u})(\eta,x) = \sum_{n=0}^{\infty} \varphi^{(qn)}(x) \frac{\eta^{qn}}{(pn)!(n!)^{q-p}}. \end{aligned} \tag{23}$$

It is obvious that

$$v(s,x) = (\mathcal{L}_{p,\theta}^{\ell}\, w)(s,x). \tag{24}$$

This suggests that the desired property for $v(s,x)$ will be reduced from that for $w(\eta,x)$.

In order to establish an integral expression of $w(\eta, x)$ by using the Cauchy data $\Phi(x,\zeta) = \sum_{m=0}^{q-1} \varphi(x+\zeta\omega_q^m)$ and a kernel function, we need to use the generalized hypergeometric function (or series). We notice that

$$\sum_{n=0}^{\infty} \frac{(qn)!}{(pn)!(n!)^{q-p}} Z^n = \sum_{n=0}^{\infty} \frac{(\mathbf{q}/q)_n}{(\mathbf{p}/p)_n (n!)^{q-p}} \left(\frac{q^q}{p^p} Z\right)^n,$$

where $(\mathbf{q}/q)_n = \prod_{j=1}^{q} (j/q)_n$ $((c)_n := c(c+1)\cdots(c+n-1))$. The series on the right hand side is a generalized hypergeometric series which is written by

$$_qF_{q-1}\left(\begin{matrix}\mathbf{q}/q \\ \mathbf{p}/p, \mathbf{1}_{q-p-1}\end{matrix}; \frac{q^q}{p^p} Z\right), \quad \frac{q^q}{p^p}|Z| < 1,$$

where $\mathbf{1}_{q-p-1} := (1, \cdots, 1)$ ($(q-p-1)$-ple of 1's). By using this generalized hypergeometric series or function, we get the following integral expression of $w(\eta, x)$.

$$w(\eta, x) = \frac{1}{2\pi i} \oint_{|\zeta|=r} \frac{\varphi(x+\zeta)}{\zeta} {}_qF_{q-1}\left(\begin{matrix}\mathbf{q}/q \\ \mathbf{p}/p, \mathbf{1}_{q-p-1}\end{matrix}; \frac{q^q}{p^p}\frac{\eta^p}{\zeta^q}\right) d\zeta, \tag{25}$$

by taking small $r > 0$, for x near the origin and η depending on the choice of r such that $|\eta| < (p^p/q^q)^{1/p} r^{q/p}$.

Here we notice that $_qF_{q-1}(\cdots; z)$ is a holomorphic solution at the origin of a Fuchsian ordinary differential equation with three regular singular points at $z = 0, 1, \infty$. Therefore, $_qF_{q-1}(\cdots; z)$ is holomorphic on $\mathbf{C}\backslash[1, +\infty)$, and has a moderate growth when z tends to the singular point $z = 1$ or $z = \infty$.

By this observation we see that the condition (10) for $\Phi(x,\zeta)$ (or see the conditions (i) and (ii) in Theorem 1) implies that

$$w(\eta, x) \in \mathrm{Exp}_\eta\left(\frac{p}{q-p}; S(\theta, \varepsilon, \infty)\right) \tag{26}$$

uniformly in x in a neighborhood of $x = 0$. This means that when $\ell = 0$, the condition (10) implies that $\hat{u}(t,x) \in \mathcal{O}_x\{t\}_{\sigma(0),\theta}$.

In order to prove the $\sigma(\ell)$-Gevrey summability for general ℓ and to get the integral expression (11) of the Gevrey sum, we rewrite the integral representation of $w(\eta, x)$ by deforming the path of integration by considering that the part of hypergeometric function in the integrand of (25) has q singular points in ζ-plane at roots of $\zeta^q = (q^q/p^p)\eta^p$ for any fixed $\eta \in S(\theta, \varepsilon, \infty)$. (We want to move $\eta \to \infty$ along $S(\theta, \varepsilon, \infty)$). Let $\zeta(\eta)$ be the root of $\zeta^q = (q^q/p^p)\eta^p$ of argument $\arg(\eta)p/q$ ($= p\theta/q$ when $\eta \in e^{i\theta}\mathbf{R}_{>0}$) . For the convenience or the simplification of notations we put

$$f(\eta,\zeta) = {}_qF_{q-1}\left(\begin{matrix}\mathbf{q}/q \\ \mathbf{p}/p, \mathbf{1}_{q-p-1}\end{matrix}; \frac{q^q}{p^p}\frac{\eta^p}{\zeta^q}\right).$$

We restrict $\eta \in e^{i\theta}\mathbf{R}_{>0}$ on the half line of argument θ for the simplicity of notation, since it is complicated and tedious to specify the dependence on η of path at each integration. We get the following expression of $w(\eta, x)$.

$$\begin{aligned} w(\eta,x) &= \frac{1}{2\pi i}\int_0^{\zeta(\eta)} \frac{\Phi(x,\zeta)}{\zeta}\left\{f(\eta,\zeta) - f(\eta,\zeta\omega_q^{-1})\right\} d\zeta \\ &= \frac{1}{2\pi i}\int_0^{\infty(p\theta/q)} \frac{\Phi(x,\zeta)}{\zeta}\left\{f(\eta,\zeta) - f(\eta,\zeta\omega_q^{-1})\right\} d\zeta. \end{aligned} \tag{27}$$

By recalling that $v(s,x) = (\mathcal{L}^{\ell}_{p,\theta}w)(s,x)$, we have

$$v(s,x) = \frac{1}{2\pi i}\int_0^{\infty(p\theta/q)} \frac{\Phi(x,\zeta)}{\zeta}\left\{(\mathcal{L}^{\ell}_{p,\theta})(f(\cdot,\zeta) - f(\cdot,\zeta\omega_q^{-1}))\right\}(s)\, d\zeta. \tag{28}$$

Here, we need the integral expression of $f(\eta,\zeta)$ and their difference appeared in the integrand.

$$f(\eta,\zeta) = \frac{C_{pq}}{2\pi i}\int_I \frac{\mathbf{\Gamma}(\mathbf{q}/q+\tau)\Gamma(-\tau)}{\mathbf{\Gamma}(\mathbf{p}/p+\tau)\Gamma(1+\tau)^{q-p-1}}\,(-X_\eta^{-1})^\tau\, d\tau, \quad X_\eta^{-1} = \frac{q^q}{p^p}\frac{\eta^p}{\zeta^q},$$

where $|\arg(-X_\eta^{-1})| \le \pi$, $I = \{\kappa + iy; y \in \mathbf{R}\}$ by a fixed κ such that $-1/q < \kappa < 0$ and $C_{pq} = \mathbf{\Gamma}(\mathbf{p}/p)/\mathbf{\Gamma}(\mathbf{q}/q)$. (See the Appendix[7] for delicate points.) From this expression, we have

$$\begin{aligned} f(\eta,\zeta) - f(\eta,\zeta\omega_q^{-1}) &= C_{pq}\int_I \frac{\mathbf{\Gamma}(\mathbf{q}/q+\tau)}{\mathbf{\Gamma}(\mathbf{p}/p+\tau)\Gamma(1+\tau)^{q-p}}\,X_\eta^{-\tau}\, d\tau \\ &= 2\pi i\, C_{pq}\, G^{q,0}_{q,q}\left(X_\eta \middle| \begin{matrix}\mathbf{p}/p, \mathbf{1}_{q-p} \\ \mathbf{q}/q\end{matrix}\right). \end{aligned}$$

Moreover by taking ℓ times iterated Laplace transformations, we have

$$\begin{aligned} (\mathcal{L}^{\ell}_{p,\theta})\left\{f(\cdot,\zeta) - f(\cdot,\zeta\omega_q^{-1})\right\}(s) &= C_{pq}\int_I \frac{\mathbf{\Gamma}(\mathbf{q}/q+\tau)}{\mathbf{\Gamma}(\mathbf{p}/p+\tau)\Gamma(1+\tau)^{q-p-\ell}}X_s^{-\tau}d\tau \\ &= 2\pi i\, C_{pq}\, G^{q,0}_{q-\ell,q}\left(X_s \middle| \begin{matrix}\mathbf{p}/p, \mathbf{1}_{q-p-\ell} \\ \mathbf{q}/q\end{matrix}\right). \end{aligned}$$

By inserting this into the formula (28), we get the following integral expression of $v(s,x)$.

$$v(s,x) = C_{pq}\int_0^{\infty(p\theta/q)} \frac{\Phi(x,\zeta)}{\zeta} G^{q,0}_{q-\ell,q}\left(X_s \middle| \begin{matrix}\mathbf{p}/p, \mathbf{1}_{q-p-\ell} \\ \mathbf{q}/q\end{matrix}\right) d\zeta. \tag{29}$$

In order to estimate the exponential growth of $v(s,x)$ as $s \to \infty$ along $S(\theta,\varepsilon,\infty)$, we recall the following asymptotic expansion for the Meijer G-function[8]:

$$
\begin{aligned}
G^{q,0}_{q-\ell,q}&\left(z\,\middle|\,\begin{matrix}\mathbf{p}/p,\mathbf{1}_{q-p-\ell}\\ \mathbf{q}/q\end{matrix}\right)\\
&= \frac{(2\pi)^{(\ell-1)/2}}{\ell^{1/2}}\exp\left(-\ell z^{1/\ell}\right) z^{-(q-p-\ell)/2\ell}\left[1+O(z^{-1/\ell})\right],\\
&\quad \text{as } z\to\infty,\ |\arg(z)|\le \ell\pi.
\end{aligned}
\tag{30}
$$

Combining with this estimate and that

$$
\Phi(x,\zeta)\in \mathrm{Exp}_\zeta\left(\frac{q}{q-p-\ell};S\left(\frac{p\theta}{q},\varepsilon,\infty\right)\right)
$$

uniformly in x in a neighborhood of the origin, one can easily prove that $v(s,x)\in \mathrm{Exp}_s(p/(q-p-\ell);S(\theta,\varepsilon,\infty))$ uniformly in x in a neighborhood of the origin for some $\varepsilon>0$ which assures the $(q-p-\ell)/p$-Gevrey summability of the formal solution $\hat{u}(t,x)$ in θ direction.

The integral representation of the $(q-p-\ell)/p$-Gevrey sum is obtained by taking $(q-p-\ell)$ times iterated p Laplace transformations to $v(s,x)$, that is,

$$
\begin{aligned}
u(t,x) &= \{\mathcal{L}^{q-p-\ell}_{p,\theta}v\}(t,x)\\
&= C_{pq}\int_0^{\infty(p\theta/q)}\frac{\Phi(x,\zeta)}{\zeta}\left[\{\mathcal{L}_{p,\theta}\}^{q-p-\ell}\,G^{q,0}_{q-\ell,q}\left(X_s\,\middle|\,\begin{matrix}\mathbf{p}/p,\mathbf{1}_{q-p-\ell}\\ \mathbf{q}/q\end{matrix}\right)\right](t)\,d\zeta\\
&= C_{pq}\int_0^{\infty(p\theta/q)}\frac{\Phi(x,\zeta)}{\zeta}G^{q,0}_{p,q}\left(X_t\,\middle|\,\begin{matrix}\mathbf{p}/p\\ \mathbf{q}/q\end{matrix}\right)d\zeta.
\end{aligned}
\tag{31}
$$

This is nothing but the integral expression (11) which we want to prove.

Remark 3. The above proof shows that under the condition (10) for the Cauchy data $\varphi(x)\in\mathcal{O}_x$ we have $(\hat{\mathcal{B}}^{\ell}_{1/p}\hat{u})(t,x)\in\mathcal{O}_x\{t\}_{(q-p-\ell)/p,\theta}$ and its $(q-p-\ell)/p$-Gevrey sum is given by

$$
(\mathcal{L}^{q-p-\ell}_{p,\theta}w)(t,x)=\int_0^{\infty(p\theta/q)}\Phi(x,\zeta)E_\ell(t,\zeta)\,d\zeta,
$$

where

$$
E_\ell(t,\zeta)=\frac{\mathbf{\Gamma}(\mathbf{p}/p)}{\mathbf{\Gamma}(\mathbf{q}/q)}\frac{1}{\zeta}G^{q,0}_{p+\ell,q}\left(\frac{p^p}{q^q}\frac{\zeta^q}{t^p}\,\middle|\,\begin{matrix}\mathbf{p}/p,\mathbf{1}_\ell\\ \mathbf{q}/q\end{matrix}\right).
$$

5. Hierarchy of differential equations and fundamental solutions

Let $\hat{u}(t,x)=\sum_{n=0}^{\infty}\varphi^{(qn)}(x)t^{pn}/(pn)!$ be the formal solution of the Cauchy problem (CP), and put $(\hat{\mathcal{B}}_{1/p}^{\ell}\hat{u})(s,x)=\hat{w}(s,x)$ $(1\le\ell\le q-p-1)$. Then $\hat{w}(s,x)$ satisfies the following Cauchy problem for a singular partial differential equation of non-Kowalevski type which is Fuchsian in Baouendi and Goulaouic's sense:

$$\begin{cases} \partial_s^p\left(\dfrac{1}{p}s\partial_s\right)^{\ell} w(s,x)=\partial_x^q w(s,x), \\ w(0,x)=\varphi(x), \quad \partial_s^j w(0,x)=0\ (1\le j\le p-1). \end{cases} \tag{32}$$

In fact, it is sufficient to look at the following commutative diagram.

$$\begin{array}{ccc} \dfrac{t^{pn}}{(pn)!} & \xrightarrow{\ (\hat{\mathcal{B}}_{1/p})^{\ell}\ } & \dfrac{s^{pn}}{(pn)!(n!)^{\ell}} \\ \Big\downarrow \partial_t^p & & \Big\downarrow \partial_s^p\left(\dfrac{1}{p}s\partial_s\right)^{\ell} \\ \dfrac{t^{p(n-1)}}{\{p(n-1)\}!} & \xrightarrow{\ (\hat{\mathcal{B}}_{1/p})^{\ell}\ } & \dfrac{s^{p(n-1)}}{\{p(n-1)\}!\{(n-1)!\}^{q-p}} \end{array}$$

Now we consider the above Cauchy problem (32) for a Cauchy data $\varphi(x)\in\mathcal{O}_x$. Then the formal solution $\hat{w}(s,x)$ is given by

$$\hat{w}(s,x)=\sum_{n=0}^{\infty}\varphi^{(qn)}(x)\frac{s^{pn}}{(pn)!(n!)^{\ell}}\in\mathcal{O}_x[[s]]_{(q-p-\ell)/p}.$$

Then the proof in the previous section implies the following,

PROPOSITION 2. *Let assume $\varphi(x)\in\mathcal{O}_x$ for the Cauchy data. Then $\hat{w}(s,x)\in\mathcal{O}_x\{s\}_{(q-p-\ell)/p,\theta}$ if and only if the Cauchy data $\varphi(x)$ satisfies*

$$\Phi(x,\zeta)\in\mathrm{Exp}_{\zeta}\left(\frac{q}{q-p-\ell};S\left(\frac{p\theta}{q},\varepsilon,\infty\right)\right), \tag{33}$$

uniformly in x in a neighborhood of the origin. Moreover, the $(q-p-\ell)/p$-Gevrey sum $w(s,x)$ is given by the following integral expression

$$w(s,x)=\int_0^{\infty(p\theta/q)}\Phi(x,\zeta)E_{\ell}(s,\zeta)\,d\zeta. \tag{34}$$

We omit the proof, since the sufficiency follows directly from the arguments in the previous section and the necessity follows from the argument[10].

6. Singular perturbation

The idea developed in the previous sections can be extended directly to the study of Gevrey summability of divergent solutions to singular perturbed ordinary differential equations. In this section, we shall present the idea. Let p and q be two natural numbers, and consider the following equation

$$(I - t^p \partial_x^q) u(t,x) = \varphi(x) \in \mathcal{O}_x, \tag{35}$$

where I denotes the identity operator. Then a unique formal solution is obtained by

$$\hat{u}(t,x) = \sum_{n=0}^{\infty} \varphi^{(qn)}(x) t^{pn} \in \mathcal{O}_x[[t]]_{q/p}. \tag{36}$$

From this expression of formal solution, we easily see that $\hat{u}(t,x) \in \mathcal{O}_{t,x}$ if and only if $\varphi(x) \in \mathrm{Exp}(1; \mathbf{C})$.

Now as a fundamental result on the Gevrey summability and the integral expression of the Gevrey sum, we can prove the following proposition which corresponds to Theorems 1 and 2.

PROPOSITION 3. *Let $\varphi(x) \in \mathcal{O}_x$. Then $\hat{u}(t,x) \in \mathcal{O}_x\{t\}_{q/p,\theta}$ if and only if*

$$\Phi(x,\zeta) = \sum_{m=0}^{q-1} \varphi(x + \zeta\omega_q^m) \in \mathrm{Exp}_\zeta\left(1; S\left(\frac{p\theta}{q}, \varepsilon, \infty\right)\right), \tag{37}$$

uniformly in x in a neighborhood of the origin for some constant $\varepsilon > 0$. Moreover, in this case, the q/p-Gevrey sum $u(t,x)$ is given by the following integral formula

$$u(t,x) = \int_0^{\infty(p\theta/q)} \Phi(x,\zeta) F_0(t,\zeta) d\zeta, \tag{38}$$

where the kernel function $F_0(t,\zeta)$ is given by

$$F_0(t,\zeta) = \frac{1}{\boldsymbol{\Gamma}(\mathbf{q}/q)} \frac{1}{\zeta} G_{0,q}^{q,0}\left(\frac{1}{q^q}\frac{\zeta^q}{t^p}\,\bigg|\, \mathbf{q}/q\right). \tag{39}$$

The proof of this proposition is reduced to those of Theorems 1 and 2 for the Cauchy problem (CP). In fact, by taking formal Borel transformation of order 1 ($\hat{\mathcal{B}}_1$) to the equation (35), we obtain an integro-differential equation $\{I - \partial_s^{-p}\partial_x^q\}v(s,x) = \varphi(x)$, since the multiplier by t^p is send to an integral operator ∂_s^{-p}. Therefore by operating ∂_s^p to the obtained equation, we get the following Cauchy problem.

$$\partial_s^p v(s,x) = \partial_x^q v(s,x), \quad v(0,x) = \varphi(x),\ \partial_s^j v(0,x) = 0\ (1 \le j \le p-1). \quad (40)$$

Thus, when $q > p$, the problem is reduced to the Cauchy problem (CP) and the previous results are applicable. On the other hand, when $p \ge q$, by putting $\tau = t^p$ we change the problem to the following one, $\{I - \tau\partial_x^q\}w(\tau,x) = \varphi(x)$ by which the problem is reduced to the above case.

We can present the corresponding results to Propositions 1 and 2, but we omit to write them down in explicit form, since they will be easily recognized.

References

1. W. Balser, Divergent solutions of the heat equations: on the article of Lutz, Miyake and Schäfke , Pacific J. Math., Vol. 188 (1999), 53 - 63.
2. W. Balser, Formal power series and linear systems of meromorphic ordinary differential equations, Springer-Verlag, New York, 2000.
3. W. Balser, Summability of formal power series solutions of partial differential equations with constant coefficients, to appear in Proceedings of the International Conference on Differential and Functional Differential Equations, Moscow (2002).
4. W. Balser, Multisummability of formal power series solutions of partial differential equations with constant coefficients, manuscript (2003).
5. W. Balser and M. Miyake, Summability of formal solutions of certain partial differential equations , Acta Sci. Math. (Szeged), Vol. 65 (1999), 543 - 551.
6. K. Ichinobe, The Borel sum of divergent Barnes hypergeometric series and its application to a partial differential equation, Publ. RIMS, Kyoto Univ., Vol. 37 (2001), 91 - 117.
7. K. Ichinobe, Integral representation for Borel sum of divergent solution to a certain non-Kowalevski type equation, to appear in Publ. RIMS, Kyoto Univ., Vol. 39, No. 4 (2003).
8. Y. L. Luke, The special functions and Their Approximations, Vol. I, Academic Press, 1969.
9. D. A. Lutz, M. Miyake and R. Schäfke, On the Borel summability of divergent solutions of the heat equation , Nagoya Math. J., Vol. 154 (1999), 1 - 29.
10. M. Miyake, Borel summability of divergent solutions of the Cauchy problem to non-Kowalevskian equations, Partial differential equations and their applications, (Wuhan, 1999), 225 - 239, World Sci. Publishing River Edge, NJ, 1999.

ON THE SINGULARITIES OF SOLUTIONS OF NONLINEAR PARTIAL DIFFERENTIAL EQUATIONS IN THE COMPLEX DOMAIN, II

HIDETOSHI TAHARA*

Department of Mathematics,
Sophia University,
Kioicho, Chiyoda-ku, Tokyo 102-8554, JAPAN
E-mail: h-tahara@hoffman.cc.sophia.ac.jp

The paper deals with the nonlinear first order partial differential equation (E) $\partial u/\partial t = F(t, x, u, \partial u/\partial x)$ in the complex domain $\mathbb{C}_t \times \mathbb{C}_x^n$, and consider the following problem: does (E) has a solution which possesses singularities only on the hypersurface $\{t = 0\}$? In a generic case, the main result is stated as follow: we can find such a negative rational number σ that (E) has no solutions with singularities only on $\{t = 0\}$ of order $u = o(|t|^{\sigma})$ (as $t \longrightarrow 0$) but that (E) has a solution with singularities only on $\{t = 0\}$ of order $u = O(|t|^{\sigma})$ (as $t \longrightarrow 0$).

1. Introduction

Let $\mathbb{C}$ be the complex plane or the set of all complex numbers, let $(t, x) = (t, x_1, \dots, x_n) \in \mathbb{C} \times \mathbb{C}^n$, $y \in \mathbb{C}$, $z = (z_1, \dots, z_n) \in \mathbb{C}^n$, and let $F(t, x, y, z)$ be a holomorphic function with respect to the variables $(t, x, y, z) \in \mathbb{C}_t \times \mathbb{C}_x^n \times \mathbb{C}_y \times \mathbb{C}_z^n$.

In this paper, as in [7] we will consider the first order nonlinear partial differential equation

$$\frac{\partial u}{\partial t} = F\Big(t, x, u, \frac{\partial u}{\partial x}\Big), \tag{1.1}$$

where $u = u(t, x)$ is the unknown function and $\partial u/\partial x = (\partial u/\partial x_1, \dots, \partial u/\partial x_n)$. Throughout this paper, we will assume: *$F(t, x, y, z)$ is a holomorphic function on $\Omega \times \mathbb{C}_y \times \mathbb{C}_z^n$*, where Ω is an open neighborhood of the origin $(0, 0) \in \mathbb{C}_t \times \mathbb{C}_x^n$.

*This work was partially supported by the Grant-in-Aid for Scientific Research No. 14540185 of Japan Society for the Promotion of Science.

The following theorem is one of the most fundamental results in the theory of partial differential equations in the complex domain:

Theorem 1.1 (Cauchy-Kowalewski). *For any holomorphic function* $\varphi(x)$ *the equation (1.1) has a unique holomorphic solution* $u(t,x)$ *in a neighborhood of* $(0,0) \in \mathbb{C}_t \times \mathbb{C}_x^n$ *satisfying* $u(0,x) = \varphi(x)$ *near* $x = 0$.

By this theorem the structure of local holomorphic solutions of (E) is completely known; though, it says nothing about singular solutions (that is, solutions with some singularities). *Does (1.1) has a solution with some singularities? What kinds of singularities appear in non-holomorphic solutions of (1.1)?*

As a first attempt to answer these questions, in this paper we will consider the following problem:

Problem 1.2. Let s be a real number. Does (1.1) have a solution which possesses singularities only on the hypersurface $\{t = 0\}$ with the growth order $u(t,x) = O(|t|^s)$ (as $t \longrightarrow 0$)?

If the equation (1.1) is linear, it is known by Zerner [11] that such singularities do not appear in the solutions of (1.1). In the nonlinear case, Tsuno's result in [9] asserts that such singularities also do not appear in the solutions of (1.1) if s is a non-negative real number; but, if s is a negative real number we have the following example:

Example 1.3. Let $(t,x) \in \mathbb{C}^2$. The equation

$$\frac{\partial u}{\partial t} = u\left(\frac{\partial u}{\partial x}\right)^m \quad \text{with } m \in \mathbb{N}^* (= \{1,2,\ldots\})$$

has a family of solutions $u(t,x) = (-1/m)^{1/m}(x+c)/t^{1/m}$ with an arbitrary constant $c \in \mathbb{C}$. Clearly, this has singularities only on $\{t = 0\}$ of order $u = O(|t|^{-1/m})$ (as $t \longrightarrow 0$).

2. Non-existence of singularities

In this section we will give a brief survey on the non-existence of singularities of order $u(t,x) = O(|t|^s)$ (as $t \longrightarrow 0$).

By the Taylor expansion in (y,z) we can express $F(t,x,y,z)$ in the form

$$F(t,x,y,z) = \sum_{(j,\alpha)\in\Delta} a_{j,\alpha}(t,x)\, y^j\, z^\alpha$$

where $(j,\alpha) \in \mathbb{N} \times \mathbb{N}^n$, Δ is a subset of $\mathbb{N} \times \mathbb{N}^n$, $a_{j,\alpha}(t,x)$ are holomorphic functions on Ω, and $z^\alpha = {z_1}^{\alpha_1} \cdots {z_n}^{\alpha_n}$. Without loss of generality we may suppose that $a_{j,\alpha}(t,x) \not\equiv 0$ for any $(j,\alpha) \in \Delta$; then we can write

$$a_{j,\alpha}(t,x) = t^{k_{j,\alpha}}\, b_{j,\alpha}(t,x)$$

where $k_{j,\alpha}$ is a non-negative integer and $b_{j,\alpha}(0,x) \not\equiv 0$. Using the above, the equation (1.1) may now be written as

$$\frac{\partial u}{\partial t} = \sum_{(j,\alpha)\in\Delta} t^{k_{j,\alpha}} b_{j,\alpha}(t,x)\, u^j \Big(\frac{\partial u}{\partial x}\Big)^\alpha. \tag{2.1}$$

Set $\Delta_2 = \{(j,\alpha) \in \Delta\,;\, j + |\alpha| \geq 2\}$ (where $|\alpha| = \alpha_1 + \cdots + \alpha_n$). We remark that the equation (1.1) is linear if and only if $\Delta_2 = \emptyset$; it is nonlinear otherwise. Since we already have Zerner's result in the linear case, we will assume henceforth that (1.1) is nonlinear, that is, $\Delta_2 \neq \emptyset$. Then we define the index σ by

$$\sigma = \sup_{(j,\alpha)\in\Delta_2} \frac{-k_{j,\alpha} - 1}{j + |\alpha| - 1}, \tag{2.2}$$

which was introduced by Kobayashi [5]. Note that σ is a non-positive real number. Set $\Omega_+ = \{(t,x) \in \Omega;\ \mathrm{Re}\, t > 0\}$. For a function $f(t,x)$ and a neighborhood ω of $x = 0 \in \mathbb{C}^n_x$ we define the norm $\|f(t)\|_\omega = \sup_{x\in\omega} |f(t,x)|$. We have the following result (originally by Kobayashi [5], and improved by Lope-Tahara [6]):

Theorem 2.1. *Suppose $\Delta_2 \neq \emptyset$. Let σ be the index defined by (2.2).*

(1) If a holomorphic solution $u(t,x)$ of (1.1) in Ω_+ satisfies $\|u(t)\|_\omega = o(|t|^\sigma)$ (as $t \longrightarrow 0$), then $u(t,x)$ can be extended holomorphically up to some neighborhood of the origin $(0,0) \in \mathbb{C}_t \times \mathbb{C}^n_x$.

(2) Set

$$\mathcal{M} = \Big\{(j,\alpha) \in \Delta_2\,;\, \frac{-k_{j,\alpha} - 1}{j + |\alpha| - 1} = \sigma\Big\}. \tag{2.3}$$

If $\mathcal{M} = \emptyset$ and if a holomorphic solution $u(t,x)$ of (1.1) in Ω_+ satisfies $\|u(t)\|_\omega = O(|t|^\sigma)$ (as $t \longrightarrow 0$), then $u(t,x)$ can be extended holomorphically up to some neighborhood of the origin $(0,0) \in \mathbb{C}_t \times \mathbb{C}^n_x$.

Hence we can get the following result on the non-existence of the singularities on $\{t = 0\}$.

Corollary 2.2. *Suppose $\Delta_2 \neq \emptyset$. Let σ be the index defined by (2.2).*

(1) There appears no singularities on $\{t=0\}$ with growth order $o(|t|^\sigma)$ (as $t \longrightarrow 0$) in the solutions of (1.1).

(2) If $\mathcal{M}=\emptyset$, there appears no singularities on $\{t=0\}$ with growth order $O(|t|^\sigma)$ (as $t \longrightarrow 0$) in the solutions of (1.1).

3. Criteria for the existence of singularities

Suppose $\Delta_2 \neq \emptyset$, $\mathcal{M} \neq \emptyset$ and set

$$P(x,y,z) = \sum_{(j,\alpha)\in\mathcal{M}} b_{j,\alpha}(0,x)\, y^j\, z^\alpha. \tag{3.1}$$

It is easy to see that $P(x,y,z) \not\equiv 0$ and that $P(x,y,z)$ is a holomorphic function on $\{x \in \mathbb{C}; (0,x) \in \Omega\} \times \mathbb{C}_y \times \mathbb{C}_z^n$; moreover in (3.1) we have $j+|\alpha| \geq 2$. Since $\mathcal{M} \neq \emptyset$, we have $\sigma = (-k_{j,\alpha}-1)/(j+|\alpha|-1)$ for any $(j,\alpha) \in \mathcal{M}$: this implies that σ is a negative rational number.

To prove the existence of singularities of order $|t|^\sigma$ on $\{t=0\}$, we will construct a solution of the form

$$u(t,x) = t^\sigma\big(\varphi(x) + w(t,x)\big), \tag{3.2}$$

where $\varphi(x)$ is a holomorphic function in a neighborhood of $x=0$ with $\varphi(x) \not\equiv 0$, and $w(t,x)$ is a function belonging in the class $\widetilde{\mathcal{O}}_+$ which is defined by the following:

Definition 3.1. A function $w(t,x)$ is said to be in the class $\widetilde{\mathcal{O}}_+$ if it satisfies the following conditions c_1) and c_2): c_1) $w(t,x)$ is a holomorphic function in a domain $\{(t,x) \in \mathcal{R}(\mathbb{C}_t \setminus \{0\}) \times \mathbb{C}_x^n\,;\, 0<|t|<\eta(\arg t), |x|<R\}$ for some positive-valued, continuous function $\eta(s)$ on $\mathbb{R}_s$ and some $R>0$; c_2) there is an $a>0$ such that for any $\theta>0$ we have $\sup_{|x|<R}|w(t,x)| = O(|t|^a)$ (as $t \longrightarrow 0$ under $|\arg t|<\theta$). Here $\mathcal{R}(\mathbb{C}_t \setminus \{0\})$ denotes the universal covering space of $\mathbb{C}_t \setminus \{0\}$.

Since $\sigma<0$, $\varphi(x) \not\equiv 0$ and $w(t,x) \longrightarrow 0$ (as $t \longrightarrow 0$), we easily see that this function (3.2) has really singularities of order $|t|^\sigma$ on $\{t=0\}$. Hence, if we can construct such a solution as in (3.2), we can conclude that singularities of order $|t|^\sigma$ on $\{t=0\}$ really appear in the solutions of (1.1).

Let us recall the result in [7] which guarantees the existence of singularities only on $\{t=0\}$ of the growth order $|t|^\sigma$. For a function $\varphi(x)$ we define a vector field $X(\varphi)$ by

$$X(\varphi) = \sum_{j=1}^{n} \frac{\partial P}{\partial z_j}\Big(x, \varphi, \frac{\partial \varphi}{\partial x}\Big) \frac{\partial}{\partial x_j}.$$

Theorem 3.2 ([7, Theorem 4.1]). *If the condition (C_1), (C_2), or (C_3) stated below is satisfied, we can construct a solution of the form (3.2) with $\varphi(x) \not\equiv 0$ and $w(t,x) \in \widetilde{\mathcal{O}}_+$.*

(C_1) There is a holomorphic function $\varphi(x)$ in a neighborhood of the origin of $\mathbb{C}^n_x$ which satisfies the following conditions 1)~3): 1) $\varphi \not\equiv 0$, 2) $\sigma\varphi \equiv P(x,\varphi,\partial\varphi/\partial x)$, and 3) $X(\varphi) \equiv 0$.

(C_2) There is a holomorphic function $\varphi(x)$ in a neighborhood of the origin of $\mathbb{C}^n_x$ which satisfies the following conditions 1)~3): 1) $\varphi \not\equiv 0$, 2) $\sigma\varphi \equiv P(x,\varphi,\partial\varphi/\partial x)$, and 3) the vector field $X(\varphi)$ is non-singular at $x=0$.

(C_3) There is a holomorphic function $\varphi(x)$ in a neighborhood of the origin of $\mathbb{C}^n_x$ which satisfies the following conditions 1)~4): 1) $\varphi \not\equiv 0$, 2) $\sigma\varphi \equiv P(x,\varphi,\partial\varphi/\partial x)$, 3) the vector field $X(\varphi)$ is singular at $x=0$, and 4) the vector field $t\,\partial/\partial t - X(\varphi)$ satisfies the Poincaré condition.

Remark 3.3. (1) The condition 3) of (C_1) means that

$$\frac{\partial P}{\partial z_j}\Big(x,\varphi,\frac{\partial\varphi}{\partial x}\Big) \equiv 0 \quad \text{for all } j=1,\dots,n.$$

(2) The condition 3) of (C_2) means that

$$\frac{\partial P}{\partial z_j}\Big(0,\varphi(0),\frac{\partial\varphi}{\partial x}(0)\Big) \neq 0 \quad \text{for some } j.$$

(3) The condition 3) of (C_3) means that

$$\frac{\partial P}{\partial z_j}\Big(0,\varphi(0),\frac{\partial\varphi}{\partial x}(0)\Big) = 0 \quad \text{for all } j=1,\dots,n.$$

(4) Let $\lambda_1,\dots,\lambda_n$ be the eigenvalues of the matrix $[a_{i,j}]_{1\le i,j\le n}$ whose (i,j)-component $a_{i,j}$ is given by

$$a_{i,j} = \frac{\partial}{\partial x_i}\Big(\frac{\partial P}{\partial z_j}\Big(x,\varphi(x),\frac{\partial\varphi}{\partial x}(x)\Big)\Big)\Big|_{x=0}.$$

Then the condition 4) of (C_3) is equivalent to the condition that the convex hull of the set $\{1,-\lambda_1,\dots,-\lambda_n\}$ in $\mathbb{C}$ does not contain the origin of $\mathbb{C}$.

The proof of Theorem 3.2 was done by using results of [3],[10] in the case (C_1), by using the Cauchy-Kowalewski theorem in the case (C_2), and by using results of [1],[8] in the case (C_3). For details, see [7].

4. Sufficient conditions for the existence of singularities

Suppose $\Delta_2 \neq \emptyset$, $\mathcal{M} \neq \emptyset$, and let $P(x,y,z)$ be the one in (3.1). We write

$$\frac{\partial P}{\partial x} = \Big(\frac{\partial P}{\partial x_1}, \ldots, \frac{\partial P}{\partial x_n}\Big) \quad \text{and}$$
$$\frac{\partial P}{\partial z} = \Big(\frac{\partial P}{\partial z_1}, \ldots, \frac{\partial P}{\partial z_n}\Big).$$

In this section, using the criteria in section 3 we will present four sufficient conditions for the existence of singularities of the growth order $|t|^\sigma$ only on $\{t=0\}$. The four conditions correspond to the following four cases:

$$\begin{aligned}
&\text{Case (0)} : \frac{\partial P}{\partial x}(x,y,0) \equiv (0,\ldots,0) \text{ and } \frac{\partial P}{\partial z}(x,y,0) \equiv (0,\ldots,0)\,; \\
&\text{Case (1)} : \frac{\partial P}{\partial z}(x,y,z) \equiv (0,\ldots,0)\,; \\
&\text{Case (2)} : \frac{\partial P}{\partial z}(0,y,z) \not\equiv (0,\ldots,0)\,; \\
&\text{Case (3)} : \frac{\partial P}{\partial z}(0,y,z) \equiv (0,\ldots,0) \text{ and } \frac{\partial P}{\partial z}(x,y,z) \not\equiv (0,\ldots,0)\,.
\end{aligned}$$

We call the case (0) *a special case*: this comes from the fact that in the case (0) we can take $\varphi(x)$ as a constant function and we have $\partial\varphi/\partial x \equiv 0$. The classification of the cases (1),(2),(3) will be not so strange; these correspond to the criteria (C_1),(C_2),(C_3) in Theorem 3.2.

Now, let us first consider the case (0). In this case, by the condition $(\partial P/\partial x)(x,y,0) \equiv (0,\ldots,0)$ we see that $P(x,y,0)$ is independent of x and we can write $P(x,y,0) = q(y)$ for some entire function $q(y)$ satisfying $q(0)=0$ and $q'(0)=0$.

Theorem 4.1 (Case (0)). *Suppose $\Delta_2 \neq \emptyset$, $\mathcal{M} \neq \emptyset$ and the conditions in Case (0). Let $q(y)$ be as above and set $\Sigma^* = \{y \in \mathbb{C} \setminus \{0\}\,;\, q(y) = \sigma y\}$. Then, if $\Sigma^* \neq \emptyset$ the equation (1.1) has a solution which possesses singularities only on $\{t=0\}$ with the growth order $|t|^\sigma$.*

Proof. Since $\Sigma^* \neq \emptyset$ is assumed, we can take $b \in \Sigma^*$ which satisfies $q(b) = \sigma b$ and $b \neq 0$. Then, by setting $\varphi(x) = b$ we can verify the criterion (C_1) in Theorem 3.2 in the following way: 1) $\varphi(x) \not\equiv 0$ comes from the fact $b \neq 0$; 2) $P(x,\varphi,\partial\varphi/\partial x) = P(x,b,0) = q(b) = \sigma b = \sigma\varphi$; and 3) $(\partial P/\partial z)(x,\varphi,\partial\varphi/\partial x) = (\partial P/\partial z)(x,b,0) \equiv (0,\ldots,0)$ (since one of the conditions in Case (0) is $(\partial P/\partial z)(x,y,0) \equiv (0,\ldots,0)$). □

Example (0). Let $(t, x) \in \mathbb{C}^2$ and let us consider

$$\frac{\partial u}{\partial t} = u^2 + b(x)\Big(\frac{\partial u}{\partial x}\Big)^2 + c(t, x),$$

where $b(x)$ and $c(t, x)$ are holomorphic functions. Then $\sigma = -1$, $P = y^2 + b(x)z^2$ and so the conditions in Case (0) are satisfied. Since $q(y) = y^2$ we have $\Sigma^* = \{y \in \mathbb{C} \setminus \{0\}\,;\, q(y) = -y\} = \{-1\} \neq \emptyset$. Thus we can apply Theorem 4.1 to this case and obtain the following: this equation has a solution with singularities only on $\{t = 0\}$ of order $|t|^{-1}$.

Next, let us consider the case (1). In this case, by the condition $(\partial P/\partial z)(x, y, z) \equiv (0, \dots, 0)$ we see that $P(x, y, z)$ is independent of z and we can write $P(x, y, z) = p(x, y)$.

Theorem 4.2 (Case (1)). *Suppose $\Delta_2 \neq \emptyset$, $\mathcal{M} \neq \emptyset$ and the condition in Case (1). Let $p(x, y)$ be as above and set $\Sigma^* = \{y \in \mathbb{C} \setminus \{0\}\,;\, p(0, y) = \sigma y\}$. Then, if*

$$\Sigma^* \neq \emptyset \quad \text{and} \quad \frac{\partial p}{\partial y}(0, y) \not\equiv \sigma \quad \text{on } \Sigma^* \tag{4.1}$$

the equation (1.1) has a solution which possesses singularities only on $\{t = 0\}$ with the growth order $|t|^{\sigma}$.

Proof. By the assumption (4.1) we can choose $b \in \mathbb{C}$ such that $b \neq 0$, $p(0, b) = \sigma b$ and $(\partial p/\partial y)(0, b) \neq \sigma$. Applying the implicit function theorem to

$$p(x, y) = \sigma\, y \quad \text{at } (x, y) = (0, b)$$

we have a holomorphic function $\varphi(x)$ in a neighborhood of $x = 0$ which satisfies $p(x, \varphi(x)) = \sigma\varphi(x)$ and $\varphi(0) = b$. Then we can see that this $\varphi(x)$ satisfies the conditions 1)~3) of (C_1) in Theorem 3.2 in the following way: 1) $\varphi(x) \not\equiv 0$ is verified by $\varphi(0) = b \neq 0$; 2) $P(x, \varphi, \partial\varphi/\partial x) = p(x, \varphi) = \sigma\varphi$; and 3) $(\partial P/\partial z)(x, \varphi, \partial\varphi/\partial x) \equiv (0, \dots, 0)$ comes from the condition of Case (1). □

Example (1). Let $(t, x) \in \mathbb{C}^2$ and let us consider

$$\frac{\partial u}{\partial t} = a(x)\, u^2 + t\,\Big(\frac{\partial u}{\partial x}\Big)^2 + c(t, x),$$

where $a(x)$ and $c(t, x)$ are holomorphic functions. Then $\sigma = -1$, $P = a(x)y^2$ and so the condition in Case (1) is satisfied. Since $p(0, y) = a(0)y^2$

we have $\Sigma^* = \{y \in \mathbb{C} \setminus \{0\}\,;\, a(0)y^2 = -y\}$; if $a(0) \neq 0$ we have $\Sigma^* = \{-1/a(0)\} \neq \emptyset$ and $(\partial p/\partial y)(0, -1/a(0)) = -2 \neq \sigma$. Thus, if $a(0) \neq 0$ we can apply Theorem 4.2 to this case and obtain the following: this equation has a solution with singularities only on $\{t = 0\}$ of order $|t|^{-1}$.

Thirdly, let us consider the case (2). This is the most generic case and we have:

Theorem 4.3 (Case (2)). *Suppose $\Delta_2 \neq \emptyset$, $\mathcal{M} \neq \emptyset$ and the condition in Case (2). Set $\Sigma = \{(y, z) \in \mathbb{C} \times \mathbb{C}^n\,;\, P(0, y, z) = \sigma y\}$. Then, if*

$$\frac{\partial P}{\partial z}(0, y, z) \not\equiv (0, \dots, 0) \quad on\ \Sigma \tag{4.2}$$

the equation (1.1) has a solution which possesses singularities only on $\{t = 0\}$ with the growth order $|t|^{\sigma}$.

Proof. By the assumption (4.2) we can choose $(b, c) \in \mathbb{C} \times \mathbb{C}^n$ such that $P(0, b, c) = \sigma b$ and $(\partial P/\partial z)(0, b, c) \neq (0, \dots, 0)$. Since the degree of each term of $P(0, y, z)$ in (3.1) is greater than or equal to 2 with respect to (y, z), we see that the condition $(\partial P/\partial z)(0, b, c) \neq (0, \dots, 0)$ implies $(b, c) \neq (0, 0)$. Without loss of generality we may assume $(\partial P/\partial z_n)(0, b, c) \neq 0$. Set $x' = (x_1, \dots, x_{n-1})$, $z' = (z_1, \dots, z_{n-1})$, $c = (c_1, \dots, c_n)$ and $c' = (c_1, \dots, c_{n-1})$.

Let us apply the implicit function theorem to

$$P(x, y, z) = \sigma\, y \quad \text{at} \quad (x, y, z) = (0, b, c).$$

Since $(\partial P/\partial z_n)(0, b, c) \neq 0$, we have a holomorphic function $H(x, y, z')$ in a neighborhood of $(x, y, z') = (0, b, c')$ which satisfies

$$P(x, y, z', H(x, y, z')) = \sigma\, y \quad \text{and} \quad H(0, b, c') = c_n. \tag{4.3}$$

We now consider the following Cauchy problem:

$$\begin{cases} \dfrac{\partial \varphi}{\partial x_n} = H\Big(x, \varphi, \dfrac{\partial \varphi}{\partial x'}\Big), \\ \varphi(x', 0) = b + c_1 x_1 + \cdots + c_{n-1} x_{n-1}. \end{cases} \tag{4.4}$$

Since $H(x, y, z')$ is holomorphic in a neighborhood of $(x, y, z') = (0, b, c')$, by the Cauchy-Kowalewski theorem we can get a unique holomorphic solution $\varphi(x)$ in a neighborhood of $x = 0$. By (4.4) we have $\varphi(0) = b$, $(\partial \varphi/\partial x')(0) = c'$ and so $(\partial \varphi/\partial x_n)(0) = H(0, \varphi(0), (\partial \varphi/\partial x')(0)) = H(0, b, c') = c_n$; thus we obtain $(\partial \varphi/\partial x)(0) = c$.

Then we can see that this $\varphi(x)$ satisfies the condition 1)$\sim$3) of (C_2) in Theorem 3.2 in the following way: 1) $\varphi(x) \not\equiv 0$ comes from the fact

$(\varphi(0), (\partial\varphi/\partial x)(0)) = (b, c) \neq (0, 0)$; 2) $P(x, \varphi, \partial\varphi/\partial x) = \sigma\varphi$ is verified by (4.3) and the first equation of (4.4); and 3) $(\partial P/\partial z_n)(0, \varphi(0), (\partial\varphi/\partial x)(0)) = (\partial P/\partial z_n)(0, b, c) \neq 0$. □

Example (2). i) Let $(t, x) \in \mathbb{C}^2$ and let us consider

$$\frac{\partial u}{\partial t} = u\left(\frac{\partial u}{\partial x}\right)^m, \quad m \in \mathbb{N}^*.$$

Then $\sigma = -1/m$, $P = yz^m$, $\partial P/\partial z = myz^{m-1}$, and $\Sigma = \{(y, z) \in \mathbb{C} \times \mathbb{C}\,;\, yz^m = (-1/m)y\}$. If we take $(1, (-1/m)^{1/m}) \in \Sigma$ we have $(\partial P/\partial z)(0, 1, (-1/m)^{1/m}) = -(-m)^{1/m} \neq 0$. Thus, we can apply Theorem 4.3 to this case and obtain the following: this equation has a solution with singularities only on $\{t = 0\}$ of order $|t|^{-1/m}$. Compare this with Example 1.3.

ii) Let us consider

$$\frac{\partial u}{\partial t} = a(x)\, u^2 + b(x)\left(\frac{\partial u}{\partial x}\right)^2 + c(t, x),$$

where $a(x)$, $b(x)$ and $c(t, x)$ are holomorphic functions. Then $\sigma = -1$, $P = a(x)y^2 + b(x)z^2$ and so if $b(0) \neq 0$ the condition in Case (2) is satisfied. We have $\Sigma = \{(y, z) \in \mathbb{C} \times \mathbb{C}\,;\, a(0)y^2 + b(0)z^2 = -y\}$. If we take α so that $\alpha + a(0)\alpha^2 \neq 0$ and define β by $\beta^2 = -(\alpha + a(0)\alpha^2)/b(0)$, then we have $\beta \neq 0$, $(\alpha, \beta) \in \Sigma$ and $(\partial P/\partial z)(0, \alpha, \beta) = 2b(0)\beta$. Thus, if $b(0) \neq 0$ we can apply Theorem 4.3 to this case and obtain the following: this equation has a solution with singularities only on $\{t = 0\}$ of order $|t|^{-1}$.

Lastly, let us consider the case (3). We will give a sufficient condition only in the case $n = 1$; in the general case $n \geq 2$ we have no good results. Suppose $n = 1$ and set

$$\Sigma = \left\{y \in \mathbb{C}\,;\, \frac{\partial P}{\partial x}(0, 0, y) = \sigma\, y\right\},$$

$$\frac{\partial^2 P}{\partial z \partial x}(0, 0, \Sigma) = \left\{\frac{\partial^2 P}{\partial z \partial x}(0, 0, y)\,;\, y \in \Sigma\right\}.$$

Theorem 4.4 (Case (3)). *Suppose $n = 1$, $\Delta_2 \neq \emptyset$, $\mathcal{M} \neq \emptyset$ and the conditions in Case (3). If*

$$\frac{\partial^2 P}{\partial z \partial x}(0, 0, \Sigma) \not\subset [0, \infty) \cup \left\{\frac{\sigma}{2}, \frac{\sigma}{3}, \frac{\sigma}{4}, \ldots\right\} \tag{4.5}$$

in $\mathbb{C}$, the equation (1.1) has a solution which possesses singularities only on $\{t = 0\}$ with the growth order $|t|^{\sigma}$.

Proof. By the assumption (4.5) we can choose $b \in \mathbb{C}$ such that

$$\frac{\partial P}{\partial x}(0,0,b) = \sigma b \quad \text{and} \quad \frac{\partial^2 P}{\partial z \partial x}(0,0,b) \notin [0,\infty) \cup \left\{\frac{\sigma}{2}, \frac{\sigma}{3}, \dots\right\}. \tag{4.6}$$

Since $(\partial P/\partial z)(0,y,z) \equiv 0$ is assumed in Case (3), the function $P(0,y,z)$ is independent of z and so we can write $P(0,y,z) = q(y) = y^2 q_0(y)$; this means that $P(x,y,z)$ is written in the form

$$P(x,y,z) = y^2 q_0(y) + xR(x,y,z). \tag{4.7}$$

Under the expression (4.7) we have $(\partial P/\partial x)(0,y,z) = R(0,y,z)$ and so by (4.6) we have

$$R(0,0,b) = \sigma b \quad \text{and} \quad \frac{\partial R}{\partial z}(0,0,b) \notin [0,\infty) \cup \left\{\frac{\sigma}{2}, \frac{\sigma}{3}, \dots\right\}.$$

For simplicity we set $\alpha = (\partial R/\partial z)(0,0,b)$; we have $\alpha \neq 0$, $\alpha \notin [0,\infty)$ and $\alpha \notin \{\sigma/2, \sigma/3, \dots\}$. Since the degree of each term of $R(0,0,z)$ in z is greater than or equal to 2, the condition $(\partial R/\partial z)(0,0,b) \neq 0$ implies $b \neq 0$.

The condition $(\partial R/\partial z)(0,0,b) \neq 0$ is also used to solve the following functional equation with respect to z:

$$xy^2 q_0(xy) + R(x, xy, y+z) = \sigma y \quad \text{at } (x,y,z) = (0,b,0). \tag{4.8}$$

By the implicit function theorem, the equation (4.8) is reduced to the form

$$z = H(x,y), \tag{4.9}$$

where $H(x,y)$ is a holomorphic function in a neighborhood of $(x,y) = (0,b)$ satisfying $H(0,b) = 0$. Moreover we see

$$\frac{\partial H}{\partial y}(0,b) = \frac{\sigma}{(\partial R/\partial z)(0,0,b)} - 1 = \frac{\sigma}{\alpha} - 1;$$

by the assumption $\alpha \notin \{\sigma/2, \sigma/3, \dots\}$ we have $(\partial H/\partial y)(0,b) \notin \{1,2,\dots\}$.

Now we note that $n = 1$ is assumed and then let us consider the following ordinary differential equation

$$P\Big(x, \varphi, \frac{d\varphi}{dx}\Big) = \sigma \varphi.$$

By (4.7) this is equivalent to

$$\varphi^2 q_0(\varphi) + xR\Big(x, \varphi, \frac{d\varphi}{dx}\Big) = \sigma \varphi.$$

If we set $\varphi(x) = x\psi(x)$, this is rewritten into the form

$$x\psi^2 q_0(x\psi) + R\Big(x, x\psi, \psi + x\frac{d\psi}{dx}\Big) = \sigma \psi; \tag{4.10}$$

moreover, if we consider (4.10) under the condition $\psi(0) = b$, by (4.9) we see that (4.10) is reduced to

$$x\frac{d\psi}{dx} = H(x,\psi), \quad \psi(0) = b. \tag{4.11}$$

Since $H(0,b) = 0$ holds, (4.11) is nothing but a Briot-Bouquet's ordinary differential equation and its characteristic exponent is $(\partial H/\partial y)(0,b)$. Since $(\partial H/\partial y)(0,b) \not\in \{1,2,\ldots\}$ is known, we can conclude that (4.11) has a unique holomorphic solution $\psi(x)$ in a neighborhood of $x = 0$.

Thus, to complete the proof of Theorem 4.4 it is sufficient to show that the above $\varphi(x) = x\psi(x)$ satisfies the conditions 1)~4) of (C_3) in Theorem 3.2. We will do this now.

Since $\psi(0) = b \neq 0$, this implies $\varphi(x) \not\equiv 0$; this is 1). 2) is clear from the construction of $\varphi(x)$. Note that the vector field $X(\varphi)$ is given by

$$X(\varphi) = \frac{\partial R}{\partial z}\Big(x,\varphi,\frac{d\varphi}{dx}\Big)\, x\,\frac{\partial}{\partial x};$$

therefore the condition 3) is clear. Since $(\partial R/\partial z)(0,\varphi(0),(d\varphi/dx)(0)) = (\partial R/\partial z)(0,0,b) = \alpha$ holds, the condition 4) is equivalent to the condition that the vector field

$$t\frac{\partial}{\partial t} - \alpha x\frac{\partial}{\partial x}$$

satisfies the Poincaré condition which is also equivalent to $\alpha \not\in [0,\infty)$. Thus, the condition 4) is also verified. □

Example (3). Let $(t,x) \in \mathbb{C}^2$ and let us consider

$$\frac{\partial u}{\partial t} = a(x)\,u^2 + x\left(\frac{\partial u}{\partial x}\right)^2 + c(t,x),$$

where $a(x)$ and $c(t,x)$ are holomorphic functions. Then $\sigma = -1$, $P = a(x)y^2 + xz^2$ and so the conditions in Case (3) are satisfied. We have $\Sigma = \{\alpha \in \mathbb{C}\,;\, \alpha^2 = -\alpha\} = \{0,-1\}$ and $(\partial^2 P/\partial z\partial x)(0,0,\Sigma) = \{0,-2\}$. Since $-2 \not\in [0,\infty)\cup\{-1/2,-1/3,\ldots\}$ we have the condition (4.5). Thus, we can apply Theorem 4.4 to this case and obtain the following: this equation has a solution with singularities only on $\{t = 0\}$ of order $|t|^{-1}$.

References

1. H. Chen and H. Tahara, On the totally characteristic type non-linear partial differential equations in the complex domain, *Publ. RIMS, Kyoto Univ.*, **35** (1999), 621-636.

2. H. Chen, Z. Luo and H. Tahara, Formal solutions of nonlinear first order totally characteristic type PDE with irregular singularities, *Ann. Inst. FourierC Grenoble*, **51** (2001), 1599-1620.
3. R. Gérard and H. Tahara, Holomorphic and singular solutions of nonlinear singular first order partial differential equations, *Publ. RIMS, Kyoto Univ.*, **26** (1990), 979-1000.
4. R. Gérard and H. Tahara, Singular nonlinear partial differential equations, Aspects of Mathematics, **E 28**, Vieweg-Verlag, 1996.
5. T. Kobayashi, Singular solutions and prolongation of holomorphic solutions to nonlinear differential equations, *Publ. RIMS. Kyoto Univ.*, **34** (1998), 43-63.
6. J.E.C. Lope and H. Tahara, On the analytic continuation of solutions to nonlinear partial differential equations, *J. Math. Pures Appl.*, **81** (2002), 811-826.
7. H. Tahara, On the singularities of solutions of nonlinear partial differential equations in the complex domain, "Microlocal Analysis and Complex Fourier Analysis" (edited by T.Kawai and K.Fujita), 273-283, World Sci., 2002.
8. H. Tahara, Solvability of nonlinear totally characteristic type partial differential equations with resonances, *J. Math. Soc. Japan*, **55** (2003), 1095-1113.
9. Y. Tsuno, On the prolongation of local holomorphic solutions of nonlinear partial differential equations, *J. Math. Soc. Japan*, **27** (1975), 454-466.
10. H. Yamazawa, Singular solutions of the Briot-Bouquet type partial differential equations, *J. Math. Soc. Japan*, **55** (2003), 617-632.
11. M. Zerner, Domaines d'holomorphie des fonctions vérifiant une équation aux derivées partielles, *C. R. Acad. Sci. Paris Sér. I. Math.*, **272** (1971), 1646-1648.

IDENTIFYING CORROSION BOUNDARY BY PERTURBATION METHOD

YONGJI TAN*AND XINXING CHEN

Fudan University

E-mail:yjtan@fudan.edu.cn,

012018041@fudan.edu.cn

In this paper we investigate an inverse boundary value problem to determine the boundary corrosion of a tube, which is formulated to a problem to determine a free boundary for a heat initial-boundary value problem . By perturbation method it can be reduced to inverse problems with fixed boundaries. We develops a method to solve these problems and give several numerical examples to show that this method works quite well.

1. Introduction

We consider an insulated tube one end of which is eroded for some physical and chemical reasons. To determine the corrosion, some sensors are pre-installed on the wall of the tube to measure the temperature, since the corrosion can not be measured directly. This problem can be formulated to an inverse problem to determine a free boundary for the initial-boundary problem of an 1-D heat equation which is also a model problem for some complicated situations e.g. the bottom corrosion of the blast furnace[1].

Since the length of corrosion is relatively small, we denote it by $\varepsilon g(t)$ with a small parameter ε . Let $u = u(x,t)$ be the temperature in the tube located x at time t . By some scaling the mathematical model can be written as:

To determine $u(x,t), g(t)$ with $g(0) = 0, g'(t) \geq 0$ such that

$$\begin{cases} \frac{\partial u}{\partial t} - \frac{\partial^2 u}{\partial x^2} = f(x,t) \; t > 0 \quad, \; 0 < x < 1 + \varepsilon g(t) \\ u(x,0) = \varphi(x) \qquad 0 \leq x \leq 1 \\ u(0,t) = 0 \\ \frac{\partial u}{\partial x}|_{x=1+\varepsilon g(t)} = 0 \end{cases} \tag{1}$$

*Project 1017-1020 supported by NSFC

and

$$u(x_l, t) = \psi_l(t) \qquad (1 \le l \le N, 0 < t \le T) ,$$

where x_l and ψ_l are the location and the measuring temperature of the l-th sensor respectively . It is noticed that this free boundary problem is different from Stefen problem [2] . In Stefen problem, one extra Stefen condition is given in the free boundary, nevertheless, in this problem the extra conditions are given inside the domain .

The plan of this paper is as follows . In section 2 we solve direct problem by perturbation method . In section 3 we give the method of reconstruction of the free boundary . In section 4 we show some numerical results . In section 5 we describe some theoretical results.

2. Perturbation Solution for Direct Problem

To solve the inverse problem of identifying corrosion boundary we have to solve many direct problems in which the boundaries are constantly changing. It causes some inconvenience in numerical solution. To overcome this difficulty, we use perturbation method[3]. The problems can be changed into initial-boundary value problems with fixed boundaries. This idea will be more useful for high dimensional problem.

If $g(t)$ is known,the direct problem is:

$$\begin{cases} \frac{\partial u}{\partial t} - \frac{\partial^2 u}{\partial x^2} = f(x,t) \; t > 0 \;\; , \;\; 0 < x < 1 + \varepsilon g(t) \\ u(x,0) = \varphi(x) \qquad 0 \le x \le 1 \\ u(0,t) = 0 \\ \frac{\partial u}{\partial x}|_{x=1+\varepsilon g(t)} = 0 \end{cases} \tag{2}$$

Formally asymptotic expand $u(x,t)$ about ε as

$$u(x,t) = u_0(x,t) + \varepsilon u_1(x,t) + \cdots .$$

Expanding the boundary condition at the right end into Taylor series about x near x=1, we have

$$\frac{\partial u}{\partial x}|_{x=1+\varepsilon g(t)} = \frac{\partial u}{\partial x}(1,t) + \varepsilon g(t)\frac{\partial^2 u}{\partial x^2}(1,t) + \cdots$$

Substituting them into problem(2) and comparing the coefficients of the same power of ε, we obtain two problems satisfied by $u_0(x,t), u_1(x,t)$ as follows:

$$\begin{cases} \frac{\partial u_0}{\partial t} - \frac{\partial^2 u_0}{\partial x^2} = f(x,t) & t>0 \quad , \ 0<x<1 \\ u_0(x,0) = \varphi(x) & 0 \le x \le 1 \\ u_0(0,t) = 0 & \\ \frac{\partial u_0}{\partial x}|_{x=1} = 0 & \end{cases} \tag{3}$$

$$\begin{cases} \frac{\partial u_1}{\partial t} - \frac{\partial^2 u_1}{\partial x^2} = 0 & t>0 \quad , \ 0<x<1 \\ u_1(x,0) = 0 & 0 \le x \le 1 \\ u_1(0,t) = 0 & \\ \frac{\partial u_1}{\partial x}|_{x=1} = -g(t)\frac{\partial^2 u_0}{\partial x^2}(t,1) & \end{cases} \tag{4}$$

It is noticed that they are both initial-boundary value problems with fixed boundaries. However, in (4) the boundary condition at the right end depends on $u_0(x,t)$.

By separation variable technique we solve problem(3) and obtain :

$$\begin{aligned} & u_0(x,t) \\ = & 2\sum_{k=0}^{\infty}\Big[\int_0^t \exp(-\pi^2(k+\frac{1}{2})^2(t-\tau))d\tau \\ & \qquad \int_0^1 f(\tau,\xi)\sin(\pi(k+\frac{1}{2})\xi)d\xi\sin(\pi(k+\frac{1}{2})x)\Big] \\ & +2\sum_{k=0}^{\infty}\int_0^1 \varphi(\xi)\sin(\pi(k+\frac{1}{2})\xi)d\xi\exp(-\pi^2(k+\frac{1}{2})^2t)\sin(\pi(k+\frac{1}{2})x) \end{aligned}$$

Differentiate $u_0(x,t)$ twice about x. Under some smooth and compatible conditions on boundary data, the order of differentiation and integration can be exchanged. Then

$$\begin{aligned} \frac{\partial^2 u_0}{\partial x^2}(t,x) = & -2\sum_{k=0}^{\infty}\Big[\pi^2(k+\frac{1}{2})^2\int_0^t \exp(-\pi^2(k+\frac{1}{2})^2(t-\tau))d\tau \\ & \qquad \int_0^1 f(\tau,\xi)\sin(\pi(k+\frac{1}{2})\xi)d\xi\sin(\pi(k+\frac{1}{2})x)\Big] \\ & -2\sum_{k=0}^{\infty}\Big[\pi^2(k+\frac{1}{2})^2\int_0^1 \varphi(\xi)\sin(\pi(k+\frac{1}{2})\xi)d\xi \times \\ & \qquad \exp(-\pi^2(k+\frac{1}{2})^2t))\sin(\pi(k+\frac{1}{2})x)\Big] \end{aligned}$$

Substituting x=1 into the expression, we obtain

$$\frac{\partial^2 u_0}{\partial x^2}(t,1) = -2\sum_{k=0}^{\infty}\Big[(-1)^k\pi^2(k+\frac{1}{2})^2\int_0^t \exp(-\pi^2(k+\frac{1}{2})^2(t-\tau))d\tau \int_0^1 f(\tau,\xi)\sin(\pi(k+\frac{1}{2})\xi)d\xi\Big] -2\sum_{k=0}^{\infty}(-1)^k\pi^2(k+\frac{1}{2})^2\int_0^1 \varphi(\xi)\sin(\pi(k+\frac{1}{2})\xi)d\xi \exp(-\pi^2(k+\frac{1}{2})^2 t)$$

Now we are going to solve problem (4). Let

$$\mu(t) = -g(t)\frac{\partial^2 u_0}{\partial x^2}(1,t)\ .$$

Transform the unknown function by

$$\omega(x,t) = u_1(x,t) - x\mu(t)\ ,$$

and then problem (4) turns into

$$\begin{cases} \frac{\partial\omega}{\partial t} - \frac{\partial^2\omega}{\partial x^2} = -x\mu'(t)\ t>0\ ,\ 0<x<1 \\ \omega(x,0)=0 \qquad\qquad 0\le x\le 1 \\ \omega(0,t)=0 \\ \frac{\partial\omega}{\partial x}|_{x=1}=0 \end{cases} \tag{5}$$

Similarly, by separation of variable we obtain

$$\omega(t,x) = \sum_{k=0}^{\infty} D_k^1(t)\sin(\pi(k+\frac{1}{2})x)$$

where

$$D_k^1(t) = 2\int_0^t\int_0^1(-x\mu'(\tau))\exp(-\pi^2(k+\frac{1}{2})^2(t-\tau))\sin(\pi(k+\frac{1}{2})x)dxd\tau$$

$$= \frac{(-1)^{k+1}\times 2}{\pi^2(k+\frac{1}{2})^2}[\mu(t)-\pi^2(k+\frac{1}{2})^2\int_0^t\mu(\tau)\exp(-\pi^2(k+\frac{1}{2})^2(t-\tau))d\tau]$$

and

$$\omega(t,x) = \sum_{k=0}^{\infty}\frac{(-1)^{k+1}\times 2}{\pi^2(k+\frac{1}{2})^2}\Big[\mu(t)-\pi^2(k+\frac{1}{2})^2 \int_0^t\mu(\tau)\exp(-\pi^2(k+\frac{1}{2})^2(t-\tau))d\tau\Big]\sin(\pi(k+\frac{1}{2})x)$$

Therefore

$$u_1(t,x) = \sum_{k=0}^{\infty} \frac{(-1)^{k+1}\times 2}{\pi^2(k+\frac{1}{2})^2}\Bigg[\mu(t) - \pi^2(k+\frac{1}{2})^2 \times \int_0^t \mu(\tau)\exp(-\pi^2(k+\frac{1}{2})^2(t-\tau))d\tau\Bigg]\sin(\pi(k+\frac{1}{2})x) + x\mu(t)$$

and the first order formal asymptotic solution for the original problem is

$$\begin{aligned}\bar{u}(x,t) &= u_0(x,t) + \varepsilon u_1(x,t)\\ &= 2\sum_{k=0}^{\infty}\int_0^t \exp(-\pi^2(k+\frac{1}{2})^2(t-\tau))d\tau \times\\ &\qquad \int_0^1 f(\tau,\xi)\sin(\pi(k+\frac{1}{2})\xi)d\xi\,\sin(\pi(k+\frac{1}{2})x)\\ &\quad +2\sum_{k=0}^{\infty}\int_0^1 \varphi(\xi)\sin(\pi(k+\frac{1}{2})\xi)d\xi\,\exp(-\pi^2(k+\frac{1}{2})^2 t)\sin(\pi(k+\frac{1}{2})x)\\ &\quad +\varepsilon\Bigg\{x\mu(t) + \sum_{k=0}^{\infty}\frac{(-1)^{k+1}\times 2}{\pi^2(k+\frac{1}{2})^2}\times\\ &\qquad \Bigg[\mu(t) - \pi^2(k+\frac{1}{2})^2\int_0^t \mu(\tau)\exp(-\pi^2(k+\frac{1}{2})^2(t-\tau))d\tau\Bigg]\times\\ &\qquad \sin\pi(k+\frac{1}{2})x\Bigg\}\end{aligned}$$

where

$$\begin{aligned}\mu(t) &= -g(t)\frac{\partial^2 u_0}{\partial x^2}(t,1)\\ &= 2g(t)\Bigg\{\sum_{k=0}^{\infty}(-1)^k\pi^2(k+\frac{1}{2})^2\times\\ &\qquad \int_0^t \exp(-\pi^2(k+\frac{1}{2})^2(t-\tau))d\tau\int_0^1 f(\tau,\xi)\sin(\pi(k+\frac{1}{2})\xi)d\xi +\\ &\quad \sum_{k=0}^{\infty}(-1)^k\pi^2(k+\frac{1}{2})^2\int_0^1 \varphi(\xi)\sin(\pi(k+\frac{1}{2})\xi)d\xi\,\exp(-\pi^2(k+\frac{1}{2})^2 t)\Bigg\}\end{aligned}$$

It is easy to see that this asymptotic solution depends on g(t) linearly.

3. Reconstruct Free Boundary

The original problem is to find $u(x,t), g(t)$ with $g(0)=0$ such that

$$\begin{cases} \frac{\partial u}{\partial t} - \frac{\partial^2 u}{\partial x^2} = f(x,t)\ t>0\ ,\ 0<x<1+\varepsilon g(t) \\ u(x,0)=\varphi(x) \qquad 0\le x\le 1 \\ u(0,t)=0 \\ \frac{\partial u}{\partial x}|_{x=1+\varepsilon g(t)}=0 \end{cases} \tag{6}$$

and

$$u(x_l,t)=\psi_l(t)\quad,\quad (1\le l\le N, 0<t\le T) \tag{6'}$$

While g(t) is given, we use the asymptotic solution as the approximate solution of problem and denote it by $\bar{u}(x,t;g(t))$. We relax the condition $(6)'$ into

$$\sum_{l=1}^{N}\int_0^T (u(x_l,t)-\psi_l(t))^2dt = min$$

The problem of reconstructing free boundary is reduced to a variational problem: Find g(t) such that

$$\sum_{l=1}^{N}\int_0^T (\bar{u}(x_l,t;g(t))-\psi_l(t))^2dt = min$$

where $\bar{u}(x_l,t;g(t))$ is the first order asymptotic solution of problem (2), provided that g(t) is given.

We discrete the problem by using polynomial to approximate g(t)

$$g(t)\approx\sum_{j=1}^{n} a_j t^j,$$

and taking the measurements at the discrete time points

$$t_s(1\le s\le L)$$

Denoting $\bar{u}(x_l,t;a_1,a_2,\cdots,a_n) = \bar{u}(x_l,t;g(t))$, the variational problem turns into a function minimization :

$$minE(a_1,a_2,\cdots,a_n)=\sum_{j=1}^{L}\sum_{l=1}^{N}(\bar{u}(x_l,t_j;a_1,a_2,\cdots,a_n)-\psi_l(t_j))^2$$

Noticing that the asymptotic solution depends on g(t) linearly, this problem is a typical least square problem. We can easily solve it and obtain $a_1,a_2,\cdots,a_n$, and solve the inverse problem .

4. Numerical Examples

Example 1

$$\begin{cases} \frac{\partial u}{\partial t} - \frac{\partial^2 u}{\partial x^2} = f(x,t) \ t > 0 \ , \ 0 < x < 1 + \frac{t+t^2}{100} \\ u(x,0) = 0 \qquad 0 \le x \le 1 \\ u(0,t) = 0 \\ \frac{\partial u}{\partial x}(1 + \frac{t+t^2}{100}, t) = 0 \end{cases} \tag{7}$$

$$f(x,t) = (1 + \frac{t+t^2}{100})^2 x - \frac{x^3}{3} + (1 + \frac{t+t^2}{100})\frac{(1+2t)tx}{50} + 2tx$$

The exact solution for this problem is

$$u(x,t) = t((1 + \frac{t+t^2}{100})^2 x - \frac{x^3}{3})$$

Taking $\varepsilon = 0.01$, solving this problem by perturbation method and comparing the result with exact solution in 9×25 net points, we find the maximal relative error is about 3%. It is acceptable.

Then by taking the values at x=0.3,0.5,0.7 of the exact solution as measurements and using above method to solve the inverse problem we obtain $\bar{g}(t) = \sum_{j=1}^{L} a_j t^j$ as the approximation of $g(t)$.

The graph for both of exact boundary('$-$') and reconstructed boundary('$*$') is shown in figure 1 and the biggest error happens at the end of time interval where the value of relative error is about 6%.

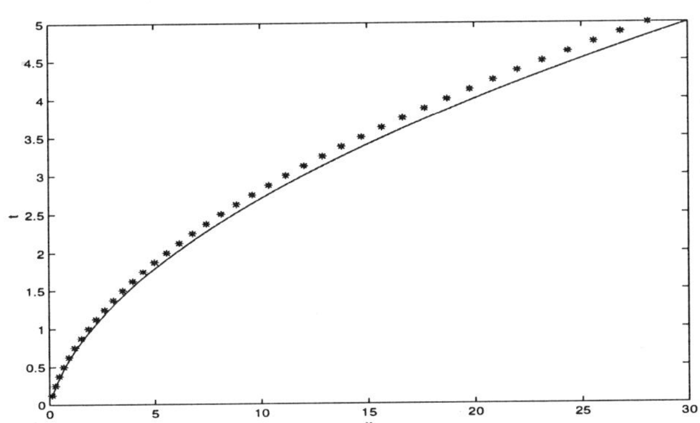

Figure 1. $g(t) = t + t^2$

Example 2 Same problem with

$$f(x,t) = \frac{x}{1+x}, \quad \varphi(x) = \frac{x^3}{6} - \frac{x}{2}, \quad g(t) = \frac{t - 5t^2 + 14t^3 - t^4}{120},$$
$$\varepsilon = 0.01 \quad , \quad 0 < t \leq 10$$

We can not find exact solution for this problem. By finite difference method (FDM) we solve the direct problem and obtain the approximate values of u(x,t) in net points. Using these approximate values at x=0.3,0.5,0.7 as measurements to solve the inverse problem we obtain the 5-th polynomial approximation of the free boundary:

$$\bar{g}(t) = -0.1923t + 0.0767t^2 + 0.0893t^3 - 0.0060t^4 - 0.00005t^5$$

The error of $g(t)$ and $\bar{g}(t)$ is shown in Figure 2 .

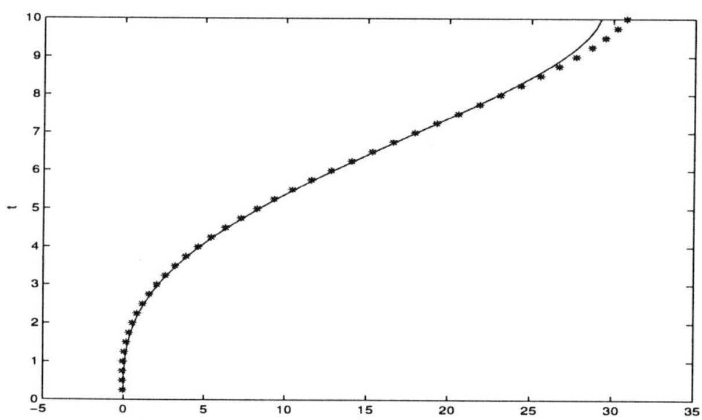

Figure 2. $g(t) = \frac{t-5t^2+14t^3-t^4}{120}$

Example 3 Same problem with

$$f(x,t) = xe^x, \quad \varphi(x) = 0, \quad g(t) = 10(atan(t-5) + atan(5)),$$
$$\varepsilon = 0.01, \quad 0 < t \leq 5$$

Similarly by use of FDM we obtain the approximate solution. Then by taking the values of approximate solution at x=0.3,0.5,0.7 as measurements we reconstruct the unknown free boundary as

$$\bar{g}(t) = -0.1291t + 0.5798t^2 - 0.1459t^3 + 0.0067t^4 + 0.0047t^5$$

The real and the reconstruct free boundaries are shown in Figure 3 .

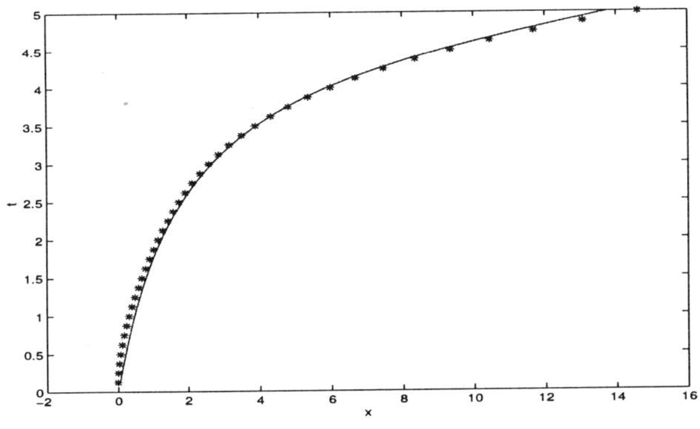

Figure 3. $g(t) = 10(atan(t-5) + atan(5))$

5. Some Theoretical Results

Finally we give some related theoretical results to show that the existence and the uniqueness of the solution for the direct corrosion problem and the formal perturbation solution is really an asymptotic solution .

Theorem 1(Existence and uniqueness). Suppose that $s(t)$ is continuously differentiable in $[0,T]$, vanishes at $t=0$, and has a low bound $c > -1$; suppose that $f(x,t)$ vanishes at x=0 and is *Hölder* continuous about x uniformly in $\bar{\Omega}$ ($\Omega = \{0 < t \leq T, 0 < x < 1+s(t)\}$), and suppose $\varphi(x)$ is continuously differentiable in $[0,1]$ and vanishes at x=0 . Then the corrosion boundary problem

$$\begin{cases} \frac{\partial u}{\partial t} - \frac{\partial^2 u}{\partial x^2} = f(x,t) & in \quad \Omega \\ u(x,0) = \varphi(x) & on \quad [0,1] \\ u(0,t) = 0 \quad \frac{\partial u}{\partial x}|_{x=1+s(t)} = 0 & in \quad (0,T] \end{cases} \tag{8}$$

exists a unique solution .

Based on the result of [4], the proof of this theorem is not difficult.

Theorem 2 . Denote $D_1 = (0,1)$, $\Omega_1 = (0,1) \times (0,T]$, and $\Omega_2 = (0,2) \times (0,T]$. Let $\Omega = \{0 < t \leq T, 0 < x < 1+\varepsilon g(t)\}$ and $\Omega \subseteq \Omega_2$, and ε be small enough . Let $f, f_x \in \bar{C}_\alpha(\bar{\Omega}_2)$ and $f(0,t) = 0$, $\varphi(x) \in C_{3+\alpha}(\bar{D}_1)$ and $\varphi(0) = \varphi_x(1) = \varphi_{xx}(0) = 0$, and $g(t) \in \bar{C}_{1+\alpha}([0,T])$. Then the solution $u(x,t)$ of the corrosion boundary problem (2) has an asymptotic expand about ε as

$$u(x,t) = u_0(x,t) + \varepsilon u_1(x,t) + O(\varepsilon^2) \quad in \quad \bar{\Omega}_1 \cap \bar{\Omega}$$

where $u_0(x,t)$ is the solution of (3) , and $u_1(x,t)$ is the solution of (4) .

References

1. Karstein Sorli, Inge M.Skaar, Monitoring the wear-line of a melting furnace, Inverse problems in Engineering : Theory and Practice 3rd Int.Conference on Inverse Problems in Engineering, June 13-18, 1999, Port Ludlow, WA.USA
2. John Ockendon, Sam Howison, Andrew Lacey, Alexander movchan, Applied Partial Differential Equations(revised edition) , Oxford university press, Oxford, 2003
3. John Hinch, Perturbation methods, Cambridge University Press, Cambridge, 1991
4. Avner Friedman, Partial differential equations of parabolic type, Prentice Hall, Englewood Cliffs, 1964

EXISTENCE AND STABILITY OF LAMELLAR AND WRIGGLED LAMELLAR SOLUTIONS IN THE DIBLOCK COPOLYMER PROBLEM

JUNCHENG WEI

Department of Mathematics
Chinese University of Hong Kong
Shatin, Hong Kong
E-mail:wei@math.cuhk.edu.hk

We consider the lamellar phases in the diblock copolymer system which can be written as a system of elliptic equations. Using Γ-convergence, the existence and stability of K-interface solutions in 1D are characterized. Then these solutions extend trivially to 2D and 3D to become perfect lamellar solutions. The stability of these lamellar solutions is completely characterized by obtaining the asymptotic expansions of their eigenvalues and eigenfunctions. Consequently we find that they are stable, i.e. are local minimizers in space, only if they have sufficiently many interfaces. Interestingly the 1-D global minimizer is near the borderline of 3-D stability. Finally using bifurcation analysis, we find wriggled lamellar solutions of the Euler-Lagrange equation of the total free energy. They bifurcate from the perfect lamellar solutions. The stability of the wriggled lamellar solutions is reduced to a relatively simple finite dimensional problem, which may be solved accurately by a numerical method. Our tests show that most of them are stable. The existence of such stable wriggled lamellar solutions explains why in reality the lamellar phase is fragile and it often exists in distorted forms.

Key words diblock copolymer, Γ-convergence, lamellar solutions, distortion, stability, wriggled lamellar solution, perfect lamellar solution,

2000 Mathematics Subject Classification 58E07, 35J55, 34D15, 45J05, 82D60

1. Introduction

In this article, we review and summarize recent advances on the study of diblock copolymer system. A diblock copolymer is a soft material, characterized by fluid-like disorder on the molecular scale and a high degree of order at longer length scales. A molecule in a diblock copolymer melt is linear sub-chain of A monomers grafted covalently to another sub-chain of B monomers. Because of the repulsion between the unlike monomers, the different type sub-chains tend to segregate below some critical temperature, but as they are chemically bonded in chain molecules, even a complete segregation of sub-chains cannot lead to a macroscopic phase separation. Only a local micro-phase separation occurs: micro-domains rich in A and B are formed. These micro-domains form morphology patterns/phases in a larger scale. The most commonly observed undistorted phases are the spherical, cylindrical and lamellar, depicted in Figure 1. Here we seek undistorted or distorted, lamellar patterns.

We consider a scenario that a diblock copolymer melt is placed in a domain D and maintained at fixed temperature. D is scaled to have unit volume in space. Let $a \in (0,1)$ be the relative number of the A monomers in a chain molecule, and $b = 1-a$ be the relative number of the B monomers in a chain. The relative A monomer density field u is an order parameter. $u \approx 1$ stands for high concentration of A monomers. The melt is incompressible so the relative B monomer density is $1-u$ and $u \approx 0$ stands for high concentration of B monomers.

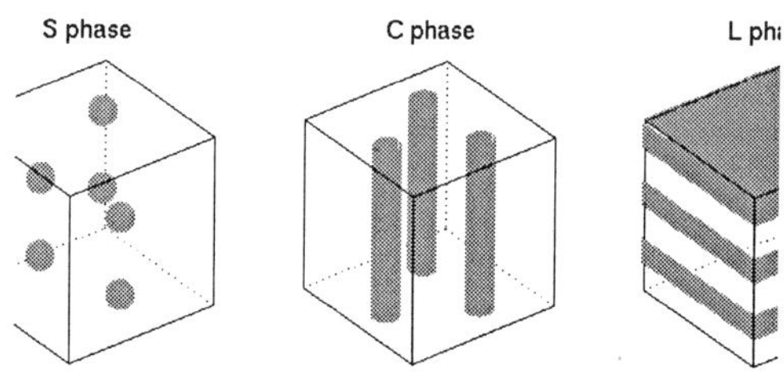

Figure 1. The spherical, cylindrical, and lamellar morphology phases commonly observed in diblock copolymer melts. The dark color indicates the concentration of type A monomer, and the white color indicates the concentration of type B monomer.

Ohta and Kawasaki [14] introduced an equilibrium theory, in which the free energy of the system is a functional of the relative A monomer density:

$$I(u) = \int_D \{\frac{\epsilon^2}{2}|\nabla u|^2 + \frac{\epsilon\gamma}{2}|(-\Delta)^{-1/2}(u-a)|^2 + W(u)\}, \qquad (1.1)$$

defined in $X_a = \{u \in W^{1,2}(D) : \overline{u} = a\}$, where $\overline{u} := \int_D u$ is the average of u on D. The original formula in [14] is given on the entire $\mathbf{R}^3$. The expression here on a bounded domain D first appeared in Nishiura and Ohnishi [12]. The local function W is smooth and has the shape of a double well. It has the global minimum value 0 at two numbers: 0 and 1. To avoid unnecessary technical difficulties we assume that $W(p) = W(1-p)$. The two global minimum points are non-degenerate: $W''(0) = W''(1) \neq 0$.

The most unusual in (1.1) is the nonlocal expression $(-\Delta)^{-1/2}(u-a)$. It reflects the connectivity of polymer chains. $(-\Delta)^{-1/2}$ is the square root of the positive operator $(-\Delta)^{-1}$ from $\{w \in L^2(D) : \overline{w} = 0\}$ to itself. The integral of the nonlocal part in (1.1) may be rewritten as

$$\int_D |(-\Delta)^{-1/2}(u-a)|^2 = \int_D \int_D G_D(x,y)(u(x)-a)(u(y)-a)\,dxdy.$$

G_D is the Green function of $-\Delta$ with the Neumann boundary condition. It splits to a fundamental solution part and a regular part. The fundamental solution in $\mathbf{R}^3$ is $\frac{1}{4\pi|x-y|}$, a long range Coulomb type interaction, which is common in many important physical systems (Muratov [10]).

ϵ and γ are positive dimensionless parameters that depend on various physical quantities. In the strong segregation region where morphology patterns form, ϵ is very small. γ is of order 1 when we choose the size of the sample to be comparable to the size of the microdomains. We develop a particular two parameter perturbation method. We do singular perturbation analysis with respect to ϵ and bifurcation analysis with respect to γ. The challenge is to combine these two techniques to derive fine analytical results. Even though this mathematical method is tailored for the diblock copolymer problem, we believe that it may be applied to other ones with multiple parameters. Examples include the Seul-Andelman membrane problem [22, 26], charged Langmuir monolayers [2, 23], and smectic films [24, 25].

The Euler-Lagrange equation of I is

$$-\epsilon^2\Delta u + f(u) - \overline{f(u)} + \epsilon\gamma(-\Delta)^{-1}(u-a) = 0, \quad \partial_\nu u = 0 \text{ on } \partial D. \tag{1.2}$$

f is the derivative of W. The term $\overline{f(u)}$ is equal to the Lagrange multiplier corresponding to the constraint $\overline{u} = a$. The equation (1.2) may also be written as an elliptic system:

$$\begin{cases} -\epsilon^2\Delta u + f(u) + \epsilon\gamma v = \text{Const.} \\ -\Delta v = u - a \\ \partial_\nu u = \partial_\nu v = 0 \text{ on } \partial D, \quad \overline{u} - a = \overline{v} = 0. \end{cases} \tag{1.3}$$

Here Const. is the Lagrange multiplier.

From now on, we shall concentrate on the study of (1.3). To avoid clumsy notations a quantity's dependence on ϵ is usually suppressed. For example we write u, the lamellar solution, instead of u_ϵ. On the other hand we often emphasize a quantity's independence of ϵ with a superscript 0. For example the limit of a lamellar solution u as $\epsilon \to 0$ is denoted by u^0. In

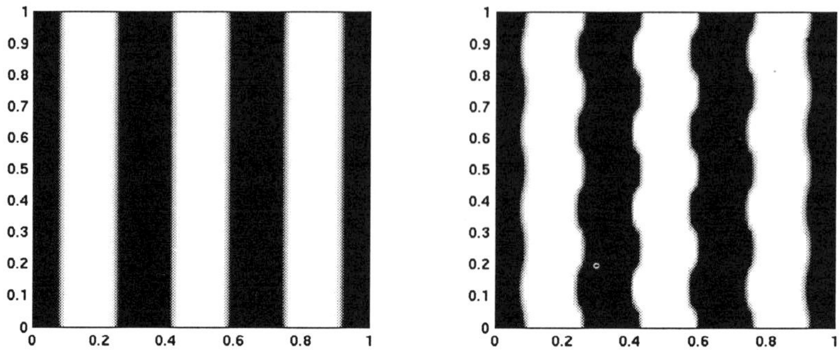

Figure 2. A perfect lamellar solution and a wriggled lamellar solution. In the dark regions the solutions are close to 1 and in the light regions the solutions are close to 0.

estimates C is always a positive constant independent of ϵ. Its value may vary from line to line. The L^2 inner product is denoted by $\langle \cdot, \cdot \rangle$ and the L^p norm by $\| \cdot \|_p$.

To simplify the formulation of our results, we take $D = (0,1) \times (0,1)$ to be a 2-D square instead of a 3-D box. Generalization to 3-D is trivial.

References on the mathematical aspects of the block copolymer theory include, in addition to the ones cited already, Ohnishi *et al* [13], Choksi [3], Choksi and Ren [4], Fife and Hilhorst [6], Henry [7], Ren and Wei [16,17,21], on diblock copolymers, and Ren and Wei [19, 20] on triblock copolymers.

2. One-dimensional Local Minimizers

First, we consider (1.3) when $D = (0,1)$. In Ren and Wei [15] a family of lamellar solutions is found. When $D = (0,1)$, for each positive integer K there exists a 1-dimensional local minimizer of I if ϵ is sufficiently small. The findings there are summarized in the following theorem.

Theorem 2.1. *[Ren and Wei [15]] In 1-D for each positive integer K the functional*

$$I_1(u) := \int_0^1 \{ \frac{\epsilon^2}{2} (\frac{du}{dx})^2 + \frac{\epsilon \gamma}{2} |(-\frac{d^2}{dx^2})^{-1/2} (u-a)|^2 + W(u) \} \, dx,$$

in $\{u \in W^{1,2}(0,1) : \overline{u} = a\}$, has a local minimizer u near u^0, under the L^2 norm, when ϵ is sufficiently small. It satisfies the Euler-Lagrange equation

$$-\epsilon^2 u'' + f(u) - \overline{f(u)} + \epsilon \gamma G_0[u-a] = 0, \;\; u'(0) = u'(1) = 0,$$

and has the properties

$$\lim_{\epsilon\to 0}\|u-u^0\|_2=0,\ \ \textit{and}\ \lim_{\epsilon\to 0}\epsilon^{-1}I_1(u)=\tau K+\frac{\gamma}{2}\int_0^1|(-\frac{d^2}{dx^2})^{-1/2}(u^0-a)|^2\,dx.$$

Let H be the solution of

$$-H''+f(H)=0 \text{ in } \mathbf{R},\ H(-\infty)=0,\ H(\infty)=1,\ H(0)=1/2. \tag{2.1}$$

The constant τ in the theorem is defined by

$$\tau:=\int_{\mathbf{R}}(H'(t))^2\,dt. \tag{2.2}$$

τ is often called the surface tension in the literature. u^0 is a step function of K jump discontinuity points, defined to be $u^0(x)=1$ on $(0,x_1^0)$, 0 on (x_1^0,x_2^0), 1 on (x_2^0,x_3^0), 0 on (x_3^0,x_4^0), 1 on (x_4^0,x_5^0), ... with (recall $b=1-a$)

$$x_1^0=\frac{a}{K},\ x_2^0=\frac{1+b}{K},\ x_3^0=\frac{2+a}{K},\ x_4^0=\frac{3+b}{K},\ x_5^0=\frac{4+a}{K},\ldots. \tag{2.3}$$

G_0 is the solution operator of $-v''=g$, $v'(0)=v'(1)=\overline{v}=0$, i.e. $v=G_0[g]=(-\frac{d^2}{dx^2})^{-1}g$.

There is another K-interface 1-D local minimizer whose limiting value as $\epsilon\to 0$ is 0 instead of 1 on the first interval $(0,b/K)$. It is just $1-\tilde{u}$ where $\tilde{u}$ is a solution constructed in Theorem 2.1, but with $\overline{\tilde{u}}=b$ instead. $1-\tilde{u}$ has the same properties as u does, so we focus on u. u is found periodic in the following sense.

Theorem 2.2. *[Ren and Wei [18]] Let u be a 1-D local minimizer constructed in Theorem 2.1. When ϵ is small, for every $x\in(0,1/K)$,*

$$u(x)=u(\tfrac{2}{K}-x)=u(x+\tfrac{2}{K})=u(\tfrac{4}{K}-x)=u(x+\tfrac{4}{K})=\ldots$$
$$=\begin{cases}u(1-x) & \textit{if } K \textit{ is even}\\ u(x+\frac{K-1}{K}) & \textit{if } K \textit{ is odd.}\end{cases}$$

Moreover when ϵ is small, u is the unique local minimizer of I_1 in an L^2 neighborhood of u^0. If u on $((j-1)/K,j/K)$ for some $j=1,2,\ldots,K$ is scaled to a function on $(0,1)$, then it is exactly a one-layer local minimizer of I_1 with ϵ and γ replaced by $\tilde{\epsilon}=\epsilon K$ and $\tilde{\gamma}=\gamma/K^3$.

Let us denote this u of K interfaces by u_γ, to emphasize its dependence on γ.

Actually Theorem (2.2) is also true when $\gamma = \frac{1}{\epsilon}$. See [17]. We note that when $a = \frac{1}{2}, \gamma = \frac{1}{\epsilon}$, Muller [9] first proved the periodicity of global minimizers under the condition that W is symmetric. Further attempts in removing the condition $a = \frac{1}{2}$ can be found in [1], [13], where only partial results are obtained.

3. Stability of the Perfect Lamellar Solutions in 2D

We consider $D = (0,1) \times (0,1)$. The 1-D local minimizer u_γ of I_1 is now viewed as a function on D, through extension to the second dimension trivially, so $u_\gamma(x,y) = u_\gamma(x)$. It is a solution of (1.2) and $I_1(u_\gamma) = I(u_\gamma)$. In 2-D it has straight interfaces. We call it a perfect lamellar solution of (1.2).

The linearized operator of (1.2) at u_γ is

$$\begin{aligned} L_\gamma \varphi &:= -\epsilon^2 \Delta \varphi + f'(u_\gamma)\varphi - \overline{f'(u_\gamma)\varphi} + \epsilon\gamma(-\Delta)^{-1}\varphi, \\ &\varphi \in W^{2,2}(D),\ \partial_\nu \varphi = 0 \text{ on } \partial D,\ \overline{\varphi} = 0. \end{aligned} \tag{3.1}$$

This is an unbounded self-adjoint operator defined densely on $\{\phi \in L^2(D) : \overline{\phi} = 0\}$ whose spectrum consists of real eigenvalues only.

For an eigen pair (λ, φ) of L_γ separation of variables shows that $\varphi(x,y) = \phi_m(x)\cos(m\pi y)$ where m is a non-negative integer. Hence the eigenvalues λ are naturally classified by m. We therefore denote a λ that is associated with m by λ_m.

Let G_m be the solution operators of the differential equations

$$-X'' = \phi_0, \quad X'(0) = X'(1) = 0, \quad \overline{X} = 0, \text{ if } m = 0, \tag{3.2}$$

$$-X'' + m^2\pi^2 X = \phi_m, \quad X'(0) = X'(1) = 0, \text{ if } m \neq 0, \tag{3.3}$$

i.e. $G_m[\phi_m] = X$. We often identify the operators G_m with the Green functions of (3.2) and (3.3). Then we have following theorem which characterizes λ_m:

Theorem 3.1. *[Ren and Wei [18]]* [a] *The eigenvalues λ of L are classified into λ_m by m which is a non-negative integer. The following 3 statements hold when ϵ is sufficiently small.*

(1) There exists $M(K)$ depending on K but not ϵ so that when $|m| \geq M(K)$, $\lambda_m \geq C\epsilon^2$ for some $C > 0$ independent of ϵ.

[a] [18, Theorem 1.1] is formulated for a 3-D box. The similar conclusions hold true for the 2-D square D here.

(2) *When* $m = 0$, *there are* K *small positive* λ_0*'s. One of them is of order* ϵ *whose only eigenfunction is approximately* $\sum_j (h_j(x) - \overline{h_j})$. *The other* $K-1$ λ_0*'s are of order* ϵ^2. *Their only eigenfunctions are approximately* $\sum_j c_j^0 h_j(x)$ *for some vectors* c^0 *satisfying* $\sum_j c_j^0 = 0$. *The remaining* λ_0*'s are positive and bounded below by a positive constant independent of* ϵ.

(3) *When* $m \neq 0$ *and* $|m| < M(K)$, *there are* K λ_m*'s of order* ϵ^2, *which are not necessarily positive, whose only eigenfunctions are approximately* $\sum_j c_j^0 h_j(x)\cos(m\pi y)$. *The remaining* λ_m*'s are positive and bounded below by a positive constant independent of* ϵ. *Only when* K *is sufficiently large or* γ *is sufficiently small, all the eigenvalues of* L *are positive and* u *is stable.*

The eigenvalues λ_0 in Part 2 of Theorem 3.1 are just the 1-D eigenvalues of u_γ. That they are positive is consistent with the fact that u_γ is a local minimizer of I_1. Bifurcations occur at 0 eigenvalues, so we are more interested in the λ_m's of Part 3. In [18, Sections 6 and 7] we obtained asymptotic expansions of the K pairs (λ_m, ϕ_m) in Part 3. When $m \geq 1$,

$$\lambda_m = \epsilon^2(\frac{\gamma}{\tau}(\Lambda - \frac{ab}{K}) + m^2\pi^2) + o(\epsilon^2), \ \phi_m = \sum_j c_j h_j + \epsilon^2 \phi_m^\perp. \quad (3.4)$$

In (3.4) ϕ_m is decomposed to $\sum_j c_j h_j$ in the subspace spanned by h_j, $j = 1, 2, ..., K$, and $\epsilon^2\phi_m^\perp$ in the orthogonal complement of the subspace. Moreover $\|\phi_m^\perp\|_2 = O(|c|)$ [b]. As $\epsilon \to 0$, $c_j \to c_j^0$. Here (Λ, c^0) are the K eigenpairs of the K by K matrix $[G_m(x_j^0, x_k^0)]$. $[G_m(x_j^0, x_k^0)]$ is diagonalized in [18, Section 7]. When $K = 1$, it has, for each $m \geq 1$, one eigenvalue pair

$$\Lambda = \frac{1}{m\pi(\tanh(m\pi a) + \tanh(m\pi b))}, \ c^0 \propto 1. \quad (3.5)$$

When $K = 2$, there are two eigenpairs

$$\Lambda = \frac{1}{m\pi(\coth(m\pi a) + \cot(m\pi b) - \operatorname{csch}(m\pi a) + \operatorname{csch}(m\pi b))}, c^0 \propto (-1, 1),$$
$$\Lambda = \frac{1}{m\pi(\coth(m\pi a) + \cot(m\pi b) - \operatorname{csch}(m\pi a) - \operatorname{csch}(m\pi b))}, c^0 \propto (1, 1). \quad (3.6)$$

When $K \geq 3$, there are K eigenpairs

$$\Lambda = \frac{1}{d - q}, \ c^0. \quad (3.7)$$

[b] See [18, Formula (6.55)].

Here q is one of the K eigenvalues of the tridiagonal matrix

$$Q = \begin{bmatrix} \alpha \; \beta & & & \\ \beta \; 0 \; \alpha & & & \\ & \alpha \; 0 \; \beta & & \\ & & \beta \; 0 \; \alpha & \\ & & \cdots & \end{bmatrix} \tag{3.8}$$

where

$$\alpha = m\pi \operatorname{csch} \frac{2m\pi a}{K}, \; \beta = m\pi \operatorname{csch} \frac{2m\pi b}{K}, \; d = m\pi (\coth \frac{2m\pi a}{K} + \coth \frac{2m\pi b}{K}),$$

and c^0 is a corresponding eigenvector of Q. We remark that such kind of matrices also appeared in the study of K-peaked solutions [8] and [11].

4. Existence of Wriggled Lamellar Solutions

We use bifurcation analysis to construct wriggled lamellar solutions. We use γ as a bifurcation parameter. Let $\lambda(\gamma)$ be one of the K eigenvalues of order ϵ^2 found in Part 3 of Theorem 3.1, associated with a positive integer m. Generically this eigenvalue is simple. To have multiplicity there would be another $m' \neq m$ so that $\lambda(\gamma) = \lambda_{m'}$ for a $\lambda_{m'}$ associated with m'. Because of (3.4) the latter case happens rarely, so we assume that $\lambda(\gamma)$ is simple. It is continued smoothly to a curve of simple eigenvalues $\lambda(\gamma)$ of L_γ as γ varies. Let γ_{B} be a particular value of γ so that $\lambda(\gamma_{\mathrm{B}}) = 0$. The existence of such γ_{B} follows from (3.4). The sign of $\lambda(\gamma)$ is determined, to the leading order term, by $\frac{\gamma}{\tau}(\Lambda - \frac{ab}{K}) + m^2\pi^2$. This quantity is positive when γ is small and negative when γ is large. See [18, Section 7] for more details. Denote the eigenfunction associated with $\lambda(\gamma_{\mathrm{B}})$ by $\varphi_{\mathrm{B}}(x,y) = \phi_{\mathrm{B}}(x)\cos(m\pi y)$. We write $u_{\mathrm{B}} := u_{\gamma_{\mathrm{B}}}$ and $L_{\mathrm{B}} := L_{\gamma_{\mathrm{B}}}$ for simplicity. Let

$$X := \{w \in W^{2,2}(D) : \partial_\nu w = 0 \text{ on } \partial D, \, \overline{w} = 0\}, \;\; Y := \{z \in L^2(D) : \overline{z} = 0\}. \tag{4.1}$$

Here X is a dense subspace of Y. Y is an Hilbert space with the usual inner product $\langle \cdot, \cdot \rangle$ inherited from $L^2(D)$.

A nonlinear map $F : (0, \infty) \times X \to Y$ is defined by

$$F(\gamma, w) := -\epsilon^2 \Delta(u_\gamma + w) + f(u_\gamma + w) - \overline{f(u_\gamma + w)} + \epsilon\gamma(-\Delta)^{-1}(u_\gamma + w - a). \tag{4.2}$$

Obviously the "trivial branch" $(\gamma, 0)$ is a solution branch of $F(\gamma, w) = 0$. It corresponds to the K-interface, perfect lamellar solution u_γ of (1.2), parameterized by γ. We look for another solution branch, a bifurcating branch, $(\gamma(s), w(s))$ of F. It gives another solution $u_{\gamma(s)} + w(s)$ of (1.2).

Using the classical Crandall and Rabinowitz's bifurcation theorem [5], we obtain the following

Theorem 4.1. *At $\gamma = \gamma_B$ another solution branch $(\gamma(s), w(s))$ bifurcates from the "trivial branch" $(\gamma, 0)$. Here $w(s) = s\varphi_B + sg(s)$ where the parameter s is in a neighborhood of 0 with $\gamma(0) = \gamma_B$ and $w(0) = 0$. Moreover $g(s) \in X$ satisfies $g(s) \perp \varphi_B$ and $g(0) = 0$.*

Note that $u_{\gamma(s)} + w(s)$ is approximately $u_{\gamma(s)}(x) + s\phi_B(x)\cos(m\pi y)$ since $g(s)$ is a smaller term compared to $\varphi_B(x, y) = \phi_B(x)\cos(m\pi y)$. Plot 2 of Figure 2 is made based on this observation.

5. Stability of the Bifurcating Solutions

The eigenvalue $\lambda(\gamma)$ of the "trivial" branch u_γ corresponds to an eigenvalue $\lambda_*(s)$ of the bifurcating solution $u_{\gamma(s)} + w(s)$. The sign of $\lambda_*(s)$ may be determined from the shape of $\gamma(s)$. Thus we proceed to compute $\gamma'(0)$ and $\gamma''(0)$. However the overall stability of $u_{\gamma(s)} + w(s)$ is interesting only when $\lambda(\gamma)$ is the principal, i.e. the smallest, eigenvalue of L_γ. Otherwise, both u_γ and $u_{\gamma(s)} + w(s)$ are unstable.

Place $w(s) = s\varphi_B + sg(s)$ into $F(\gamma, w) = 0$ and divide by s:

$$\begin{aligned}-\epsilon^2\Delta\left(\tfrac{u_{\gamma(s)}}{s} + \varphi_B + g(s)\right) + \tfrac{f(u_{\gamma(s)}+w(s))}{s} + \\ \epsilon\gamma(s)(-\Delta)^{-1}\left(\tfrac{u_{\gamma(s)}-a}{s} + \varphi_B + g(s)\right) = \text{Const.}\end{aligned} \tag{5.1}$$

where Const. refers to the term coming from the average of f, which is independent of (x, y). Here we do not need its exact value. On the other hand divide the equation (1.2) of $u_{\gamma(s)}$ by s and subtract the result from (5.1):

$$\begin{aligned}-\epsilon^2\Delta(\varphi_B + g(s)) + \tfrac{f(u_{\gamma(s)}+w(s)) - f(u_{\gamma(s)})}{s} + \\ \epsilon\gamma(s)(-\Delta)^{-1}(\varphi_B + g(s)) = \text{Const.}\end{aligned} \tag{5.2}$$

Differentiate (5.2) with respect to s and set $s = 0$ afterwards:

$$L_B g'(0) + \gamma'(0)\{f''(u_B)\frac{du_\gamma}{d\gamma}|_{\gamma=\gamma_B}\varphi_B + \epsilon(-\Delta)^{-1}\varphi_B\} + \frac{1}{2}f''(u_B)\varphi_B^2 = \text{Const.}. \tag{5.3}$$

Then we multiply (5.3) by φ_B and integrate over D:

$$\gamma'(0)\int_D \{f''(u_B)\frac{du_\gamma}{d\gamma}|_{\gamma=\gamma_B}\varphi_B^2 + \epsilon\varphi_B(-\Delta)^{-1}\varphi_B\} = -\int_D \frac{1}{2}f''(u_B)\varphi_B^3. \tag{5.4}$$

Clearly the right side of (5.4) is 0 since $\varphi_{\rm B}(x,y) = \phi_{\rm B}(x)\cos(m\pi y)$ and integration with respect to y yields 0. A simple computation shows that

$$\gamma'(0) = 0. \tag{5.5}$$

(5.5) implies that the bifurcation digram has the shape of a pitchfork. There are two possibilities illustrated in Figure 3. To determine which of

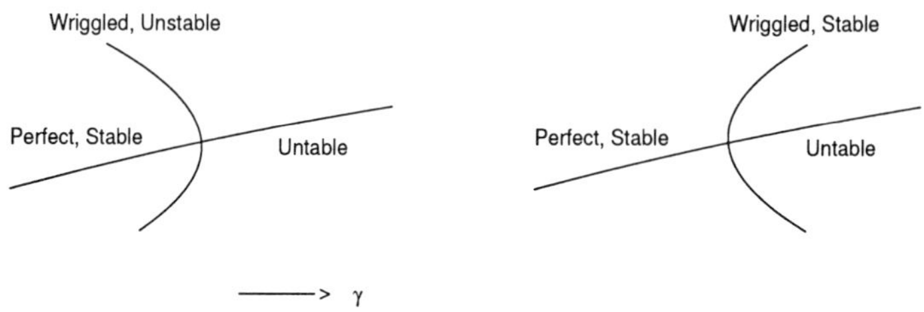

Figure 3. The two possible diagrams of wriggled lamellar solutions bifurcating out of perfect lamellar solutions. The bifurcating solutions are unstable in the first case where $\gamma''(0) < 0$, and stable in the second case where $\gamma''(0) > 0$.

the two cases occurs, we need to find $\gamma''(0)$ which can be calculated as follows:

$$\begin{aligned}&\gamma''(0)\int_D \{f''(u_{\rm B})\tfrac{du_\gamma}{d\gamma}|_{\gamma=\gamma_{\rm B}}\varphi_{\rm B}^2 + \epsilon\varphi_{\rm B}(-\Delta)^{-1}\varphi_{\rm B}\}\\ &= -\int_D\{2f''(u_{\rm B})\varphi_{\rm B}^2 g'(0) + \tfrac{1}{3}f'''(u_{\rm B})\varphi_{\rm B}^4\}.\end{aligned} \tag{5.6}$$

We need to compute the integrals on both sides of (5.6). The computations are formidable. We have to expand the quantity to the ϵ^5 order term, because all the lower order terms up to ϵ^4 vanish. Our main idea is to expand $u_{\rm B}, \phi_{\rm B}, 2g'(0)$ as $(...) + \epsilon^2(...)$ near each interface x_j. This is a very long computation. We don't know if there is a simpler proof.

To state our results, we now assume that at $\gamma = \gamma_{\rm B}$, the principle eigenvalue $\lambda(\gamma_{\rm B})$ of $u_{\rm B}$ is 0, and this eigenvalue is associated with a particular m. There are K eigenvalues of order ϵ^2 associated with this particular m. Here the 0 eigenvalue is the smallest. Hence Λ now is the smallest eigenvalue of $[G_m(x_j^0, x_k^0)]$. According to [18, Section 7]

$$\begin{aligned}\Lambda &= \frac{1}{m\pi(\tanh(m\pi a)+\tanh(m\pi b))}, \quad \text{if } K=1,\\ \Lambda &= \frac{1}{m\pi(\coth(m\pi a)+\cot(m\pi b)-\operatorname{csch}(m\pi a)+\operatorname{csch}(m\pi b))}, \quad \text{if } K=2,\\ \Lambda &= \frac{1}{d+\sqrt{\alpha^2+\beta^2+2\alpha\beta\cos\theta}}, \quad \theta = 2\pi/K, \quad \text{if } K\geq 3.\end{aligned} \tag{5.7}$$

Recall that α, β and d are defined after (3.8).

Define

$$S(a,K) := \sum_{j=1}^{K} \left(\frac{c_j^0}{|c^0|}\right)^4 \left[\frac{2+\cosh(2m\pi)}{8\sinh(2m\pi)} + \frac{\cosh(2m\pi(1-2x_j^0))}{8\sinh(2m\pi)} + \frac{\cosh(m\pi(1-2x_j^0))}{4\sinh(m\pi)} - \frac{5(m\pi)^3\tau}{8{\gamma_{\rm B}}^0} \right] \tag{5.8}$$

where ${\gamma_{\rm B}}^0/\tau$ can be determined and m is associated with the principal eigenvalue 0. Note that $S(a,K)$ depends on a and K only. It does not depend on τ. Since τ depends on the shape of W, $S(a,K)$ is independent of the exact shape of W. Then we obtain the following result:

$$\lim_{\epsilon\to 0} \epsilon^{-2}\gamma''(0) = \frac{2({\gamma_{\rm B}}^0)^2|c^0|^2}{m\pi\tau} S(a,K). \tag{5.9}$$

As a consequence, we have

Theorem 5.1. *When ϵ is sufficiently small, the bifurcating solution $u_{\gamma(s)} + w(s)$ of K wriggled interfaces is stable if $S(a,K) > 0$ and it is unstable if $S(a,K) < 0$.*

Let us use Theorem 5.1 to work out some examples. The quantity $S(a,K)$ may be accurately calculated. Tables 1 and 2 report our numerical calculations for the cases $a = 1/2$ and $7/8$. In each table the first column is the number of the interfaces in the perfect lamellar solution $u_{\rm B}$. The second column gives the value of m associated with the principal eigenvalue 0 of $u_{\rm B}$. Note that m does not increase as fast as K does. The third column has the value of ${\gamma_{\rm B}}^0/\tau$. We will explain the fourth in a moment. The fifth column has the value of $S(a,K)$. The last column indicates the stability of the bifurcating solution with K wriggled interfaces.

There an interesting relationship between the perfect lamellar solution $u_{\rm B}$ whose principal eigenvalue is 0, and the 1-D global minimizer. In [18, Section 8] it is shown that the 1-D global minimizer (the global minimizer of I_1 in Theorem 2.1, also a perfect lamellar solution on D after trivial extension), which is one of the 1-D local minimizers, has the number of interfaces K_{opt} which minimizes (among positive integers N) $\tau N + \gamma a^2 b^2/(6N^2)$. If we take $\gamma = \gamma_{\rm B}$ so that the K-interface, perfect lamellar solution $u_{\rm B}$ has 0 principal eigenvalue, we find the 1-D global minimizer corresponding to $\gamma_{\rm B}$. The number of interfaces K_{opt} of this 1-D global minimizer is reported in the fourth columns in Tables 1 and 2. For most a and K the 1-D global minimizer has one more interface than $u_{\rm B}$ does. In some other cases the

K	m	${\gamma_B}^0/\tau$	K_{opt}	$S(1/2, K)$	Stability
1	1	1.2906e+02	2	-2.0964e-03	Unstable
2	2	8.6349e+02	3	-2.1740e-02	Unstable
3	3	2.7193e+03	4	-3.3075e-02	Unstable
4	3	5.3823e+03	5	1.0764e-02	Stable
5	3	9.7086e+03	6	2.1578e-02	Stable
6	4	1.6165e+04	7	1.2129e-02	Stable
7	4	2.4091e+04	8	1.5804e-02	Stable
8	4	3.4492e+04	9	1.6739e-02	Stable
9	4	4.7728e+04	10	1.6541e-02	Stable
10	4	6.4156e+04	11	1.5885e-02	Stable

K	m	${\gamma_B}^0/\tau$	K_{opt}	$S(1/8, K)$	Stability
1	3	1.7317e+03	2	-1.5234e-01	Unstable
2	5	1.3418e+04	4	-8.9872e-03	Unstable
3	2	1.0218e+04	3	5.7102e-02	Stable
4	3	2.3798e+04	5	5.1520e-02	Stable
5	3	4.3553e+04	6	3.4174e-02	Stable
6	3	7.3607e+04	7	2.9597e-02	Stable
7	4	1.1373e+05	8	2.4505e-02	Stable
8	4	1.6489e+05	9	2.2058e-02	Stable
9	4	2.3061e+05	10	1.9979e-02	Stable
10	4	3.1284e+05	11	1.8189e-02	Stable

1-D global minimizer is exactly u_B. Because u_B is on the borderline of 2-D stability, the 1-D global minimizer sits near the 2-D stability borderline.

Acknowledgments. I thank Professor Chen Hua for inviting me to give a talk at this conference. The work reported here is a joint project with Prof. Xiaofeng Ren. This research is supported in part by a Direct Grant from CUHK and an Earmarked Grant of RGC of Hong Kong.

References

1. G. Alberti and S. Muller. A new approach to variational problems with multiple scales. *Comm. Pure Appl. Math.*, 54:761-825, 2001.

2. D. Andelman, F. Broçhard, and J.-F. Joanny. Phase transitions in Langmuir monolayers of polar molecules. *J. Chem. Phys.*, 86(6):3673–3681, 1987.
3. R. Choksi. Scaling laws in microphase separation of diblock copolymers. *J. Nonlinear Sci.*, 11:223–236, 2001.
4. R. Choksi and X. Ren. On the derivation of a density functional theory for microphase separation of diblock copolymers. *J. Statist. Phys.*, 113(1&2):151–176, 2003.
5. M.G. Crandall and P.H. Rabinowitz. Bifurcation, perturbation of simple eigenvalues, and linearized stability. *Arch. Rational Mech. Anal.*, 52(2):161–180, 1973.
6. P.C. Fife and D. Hilhorst. The Nishiura-Ohnishi free boundary problem in the 1D case. *SIAM J. Math. Anal.*, 33(3):589–606, 2001.
7. M. Henry. Singular limit of a fourth order problem arising in the micro-phase separation of diblock copolymers. *Adv. Differential Equations*, 6(9):1049–1114, 2001.
8. D. Iron, M. J. Ward, J. Wei. The stability of spike solutions to the one-dimensional Gierer-Meinhardt model. *Physica D* 150(1-2):25-62, 2001.
9. S. Muller. Singular perturbations as a selection criterion for periodic minimizing sequences. *Calc. Var. Partial Differential Equations* 1:169-204, 1993.
10. C.B. Muratov. Theory of domain patterns in systems with long-range interactions of Coulomb type. *Phys. Rev. E*, 66:066108, 2002.
11. Y. Nishiura. *Coexistence of infinitely many stable solutions to reaction-diffusion system in the singular limit, Dynamics Report: Expositions in Dynamical Systems, Edited by C.R.K.T. Jones and U. Kirchgraber*, volume 3. Springer-Verlag, New York, 1995.
12. Y. Nishiura and I. Ohnishi. Some mathematical aspects of the microphase separation in diblock copolymers. *Physica D*, 84:31–39, 1995.
13. I. Ohnishi, Y. Nishiura, M. Imai, and Y. Matsushita. Analytical solutions describing the phase separation driven by a free energy functional containing a long-range interaction term. *Chaos*, 9(2):329–341, 1999.
14. T. Ohta and K. Kawasaki. Equilibrium morphology of block copolymer melts. *Macromolecules*, 19(10):2621–2632, 1986.
15. X. Ren and J. Wei. On the multiplicity of solutions of two nonlocal variational problems. *SIAM J. Math. Anal.*, 31(4):909–924, 2000.
16. X. Ren and J. Wei. Concentrically layered energy equilibria of the di-block copolymer problem. *European J. Appl. Math.*, 13(5):479–496, 2002.
17. X. Ren and J. Wei. On energy minimizers of the di-block copolymer problem. *Interfaces Free Bound.*, 5:193–238, 2003.
18. X. Ren and J. Wei. On the spectra of 3-D lamellar solutions of the diblock copolymer problem. *SIAM J. Math. Anal.*, 35(1):1–32, 2003.
19. X. Ren and J. Wei. Triblock copolymer theory: Free energy, disordered phase and weak segregation. *Physica D*, 178:103–117, 2003.
20. X. Ren and J. Wei. Triblock copolymer theory: Ordered ABC lamellar phase. *J. Nonlinear Sci.*, 45(2):175–208, 2003.
21. X. Ren and J. Wei. Stability of spot and ring solutions of the diblock copolymer equation on a disc. submitted.

22. X. Ren and J. Wei. The soliton-stripe pattern in the Seul-Andelman membrane. *Phsica D*, in press, 2004.
23. X. Ren and J. Wei. Soliton-stripe patterns in charged Langmuir monolayers. *J. Nonlinear Sci.*, in press, 2004.
24. X. Ren and J. Wei. Molecular chirality and soliton-stripe pattern in liquid crystal films. *Nonlinearity* 17:617-632,2004.
25. J.V. Selinger, Z.-G. Wang, R.F. Bruinsma, and C.M. Knobler. Chiral symmetry breaking in Langmuir monolayers and smectic films. *Phys. Rev. Lett.*, 70(8):1139–1142, 1993.
26. M. Seul and D. Andelman. Domain shapes and patterns: The phenomenology of modulated phases. *Science*, 267:476–483, 1995.